S0-CAH-149

Chiral Catalyst Immobilization and Recycling

Edited by
D. E. De Vos, I. F. J. Vankelecom, P. A. Jacobs

Other Titles of Interest

B. Cornils / W. A. Herrmann
Applied Homogeneous Catalysis with Organometallic Compounds
2 Volumes
1996. XXXVI. 1246 pages with 1000 figures and 100 tables.
Hardcover. ISBN 3-527-29286-1
Softcover. ISBN 3-527-29594-1

B. Cornils / W. A. Herrmann / R. Schlögl / C.-H. Wong
Catalysis from A to Z
2000. XVIII. 640 pages with more than 300 figures and 14 tables.
Hardcover. ISBN 3-527-29855-X

R. I. Wijngaarden / A. Kronberg / K. R. Westerterp
Industrial Catalysis
1998. XVII. 268 pages with 85 figures and 26 tables.
Hardcover. ISBN 3-527-28581-4

M. Beller / C. Bolm
Transition Metals for Organic Synthesis
2 Volumes
1998. LVIII. 1062 pages with 733 figures and 75 tables

Chiral Catalyst Immobilization and Recycling

Edited by D. E. De Vos
I. F. J. Vankelecom
P. A. Jacobs

Weinheim · New York · Chichester
Brisbane · Singapore · Toronto

Prof. Dr. Dirk E. De Vos
Dr. Ivo F. J. Vankelecom
Prof. Dr. Pierre A. Jacobs
Centre for Surface Chemistry and Catalysis
Department of Interphase Chemistry
Faculty of Agricultural and Applied Biological Sciences
Katholieke Universiteit Leuven
Kardinaal Mercierlaan, 3001 Leuven
Belgium

Library of Congress Card No. applied for.

British Library Cataloging-in-Publication Data: A catalogue record for this book is available from the British Library.

Die Deutsche Bibliothek – CIP Cataloging-in-Publication Data
A catalogue record for this publication is available from Die Deutsche Bibliothek

ISBN 3-527-29952-1

Printed on acid-free paper.

Composition: K+V Fotosatz GmbH, D-64743 Beerfelden. Printing: betz-druck gmbh, D-64291 Darmstadt. Bookbinding: J. Schäffer GmbH & Co. KG, D-67269 Grünstadt.

Printed in the Federal Republic of Germany.

Preface

For the most part of human history, nature has monopolized chirality. Over the last decades, however, enantioselective catalysis has become the godly finger of mankind, its own instrument for synthesis of natural compounds, and for synthesis of new molecules with a substantially beneficial impact on health and environment. The field is growing at an incredible pace in academia, which continuously produces new ligands and catalysts. Additionally, large scale preparation of single enantiomers has now become an objective within reach for industry.

In such an era, books are doomed to get outdated like boulders drowning in a rising tide. However, the mere example of Ojima's book, which was published in 1993 by VCH, demonstrates that clear overviews by world experts can be an enlightening guide for scores of chemists around the world. For the present book as well, it was a prime ambition of the editors to gather leading scientists from all over the globe. We feel honored by the outstanding contributions that our authors have delivered, and we owe special thanks to all these scientists.

As enantioselective catalysis is being integrated in process schemes throughout the chemical industry, issues such as separation and reuse of expensive catalysts now come to the foreground. Thus, the publication of this book itself reflects a certain technical maturity but will hopefully also entice chemists and chemical engineers to contribute to this challenging subarea of technical chemistry.

In an introductory chapter, Blaser and his colleagues draw on their wide experience to give us a perspective on the challenges ahead, both for researchers in industry and in academia. The biggest challenge might well be, as they suggest, for people from universities and companies to look together for solutions. The next four chapters present general approaches to immobilization and recuperation of enantioselective catalysts. While Jacobs presents the major types and uses of inorganic supports in Chapter 2, Bergbreiter provides the organic polymer counterpart in Chapter 3. The focus is on the availability and preparation of the support materials, and on strategies to immobilize enantioselective catalysts. Liquid biphasic catalysis is addressed in Chapter 4. Hanson tackles specific issues, such as the ligand modifications that are required to confine a soluble catalyst to a single liquid phase. Chapter 5 is an exception in that it discusses enzyme catalysis. Based on his own experience in penicillin antibiotics synthesis, Rasor compares a spectrum of methods that can lead to economical reuse of enzymes, a topic amply illustrated with realistic figures.

The remaining chapters highlight specific reactions. Hydrogenations over modified metallic surfaces are discussed in Chapters 6 to 8. Wells and Wells give a comprehensive overview of the present understanding of alkaloid modified Pt and Pd, and carefully balance the sometimes conflicting viewpoints in literature. Baiker highlights the strategies, *e.g.* computational methods, that can lead to the successful rational design of new synthetic modifiers for Pt hydrogenation catalysts. The Japanese contribution of Tai and Sugimura shows in detail the evolution of the modified Ni catalysts, starting from a very moderate enantioselectivity in the early days, up to the excellent e.e.'s and profound understanding that have been achieved in recent years.

Catalysis with heterogenized metal complexes or ligands is the focus of Chapters 9 to 11. Bayston and Polywka introduce the important group of phosphine ligands for enantioselective hydrogenations and hydroformylations. The Salvadori group evaluates, based on its rich experience in the field, heterogenized epoxidation and dihydroxylation catalysts, for instance of the Jacobsen and Sharpless types. Finally, the group of Brunel discusses the variety of heterogenized, enantioselective catalysts that can be used to create new carbon-carbon bonds.

Ultimately, Chapter 12 deals with heterogeneous diastereoselective synthesis. An increasing number of recent publications shows that, in certain cases, this strategy can be an economically attractive alternative to enantioselective catalysis.

Finally, we thank Nico Wuestenberg for his skillful assistance in handling files of bewildering formats as well as the publishing editors of VCH for their fruitful collaboration. Two of the editors (DEDV and IFJV) are indebted to F.W.O. Vlaanderen for post-doctoral fellowships.

Dirk E. De Vos
Ivo F.J. Vankelecom
Pierre A. Jacobs

Leuven,
March 9, 2000

List of Contributors

S. Abramson
Laboratoire de Matériaux Catalytiques
et Catalyse en Chimie Organique
UMR-CNRS-5618
Ecole Nationale Supérieure de Chimie
8, rue de l'Ecole Normale
34296 – Montpellier cédex 05
France

Alfons Baiker
Laboratory of Technical Chemistry
Swiss Federal Institute of Technology
ETH-Zentrum
Universitätsstrasse 6
8092 Zürich
Switzerland

Daniel J. Bayston
Oxford Asymmetry International plc
151 Milton Park
Abingdon
Oxon, OX14 4SD
UK

N. Bellocq
Laboratoire de Matériaux Catalytiques
et Catalyse en Chimie Organique
UMR-CNRS-5618
Ecole Nationale Supérieure de Chimie
8, rue de l'Ecole Normale
34296 – Montpellier cédex 05
France

David E. Bergbreiter
Texas A&M University
Department of Chemistry
P.O. Box 30012
College Station, TX 77842-3012
USA

Hans-Ulrich Blaser
SOLVIAS AG
R 1055.6
4002 Basel
Switzerland

Daniel Brunel
Laboratoire de Matériaux Catalytiques
et Catalyse en Chimie Organique
UMR-CNRS-5618
Ecole Nationale Supérieure de Chimie
8, rue de l'Ecole Normale
34296 – Montpellier cédex 05
France

Florian G. Cocu
Chemical and Pharmaceutical Research
Institute
Vitan Avenue 112
74373 Bucharest
Romania

Mario De bruyn
Centre for Surface Chemistry
and Catalysis
Department of Interphase Chemistry
Katholieke Universiteit Leuven
Kardinaal Mercierlaan 92
3001 Leuven
Belgium

Dirk E. De Vos
Centre for Surface Chemistry
and Catalysis
Department of Interphase Chemistry
Katholieke Universiteit Leuven
Kardinaal Mercierlaan 92
3001 Leuven
Belgium

Brian E. Hanson
Department of Chemistry
Virginia Polytechnic Institute and
State University
Blacksburg, VA 24061-0212
USA

Pierre A. Jacobs
Centre for Surface Chemistry
and Catalysis
Department of Interphase Chemistry
Katholieke Universiteit Leuven
Kardinaal Mercierlaan 92
3001 Leuven
Belgium

Monique Laspéras
Laboratoire de Matériaux Catalytiques
et Catalyse en Chimie Organique
UMR-CNRS-5618
Ecole Nationale Supérieure de Chimie
8, rue de l'Ecole Normale
34296 – Montpellier cédex 05
France

A. Mandoli
Dipartimento Chimica
e Chimica Industriale
Università di Pisa
Via Risorgimento 35
56126 Pisa
Italy

Patrice Moreau
Laboratoire de Matériaux Catalytiques
et Catalyse en Chimie Organique
UMR-CNRS-5618
Ecole Nationale Supérieure de Chimie
8, rue de l'Ecole Normale
34296 – Montpellier cédex 05
France

Vasile I. Parvulescu
Department Chemical Technology and
Catalysis
University of Bucharest
Bdul Regina Elisabeta 4–12
Bucharest 70436
Romania

A. Petri
Dipartimento Chimica
e Chimica Industriale
Università di Pisa
Via Risorgimento 35
56126 Pisa
Italy

D. Pini
Dipartimento Chimica
e Chimica Industriale
Università di Pisa
Via Risorgimento 35
56126 Pisa
Italy

Mario E.C. Polywka
Oxford Asymmetry International plc
151 Milton Park
Abingdon
Oxon, OX14 4SD
UK

Benoît Pugin
SOLVIAS AG
R 1055.6
4002 Basel
Switzerland

Peter Rasor
Industrial Biochemicals Business, BB-PS
Roche Molecular Biochemicals
Roche Diagnostics GmbH
Nonnenwald 2
82372 Penzberg
Germany

P. Salvadori
Dipartimento di Chimica
e Chimica Industriale
Università di Pisa
Via Risorgimento 35
56126 Pisa
Italy

Martin Studer
SOLVIAS AG
R 1055.6
4002 Basel
Switzerland

Takashi Sugimura
Faculty of Science
Himeji Institute of Technology
Kanaji, Kamigori
Hyogo 678-12
Japan

Akira Tai
Faculty of Science
Himeji Institute of Technology
Kanaji, Kamigori
Hyogo 678-12
Japan

Ivo F.J. Vankelecom
Centre for Surface Chemistry
and Catalysis
Department of Interphase Chemistry
Katholieke Universiteit Leuven
Kardinaal Mercierlaan 92
3001 Leuven
Belgium

Peter B. Wells
Department of Chemistry
Cardiff University
Cardiff, CF10 3TB
UK

Richard P.K. Wells
Department of Chemistry
Cardiff University
Cardiff, CF10 3TB
UK

List of Abbreviations

AA	Acetoacetate
7-ACA	7-Amino-cephalosporanic acid
Acac	Acetylacetone
AC-SR	Acetic acid type silicone rubber
AD	Asymmetric dihydroxylation
AdA	1-Adamantanecarboxylic acid
7-ADCA	7-Aminodeacetoxycephalosporanic acid
AE	Asymmetric epoxidation
Aib	α-Aminoisobutyric acid
AIBN	α,α'-Azoisobutyronitrile
6-APA	6-Aminopenicillanic acid
AQN	Anthraquinone
B/n	Branched/normal
BDPP	2,4-Bis(diphenylphosphino)pentane
BDPP-DS	Disulfonated 2,4-bis(diphenylphosphino)pentane
BDPP-MS	Monosulfonated 2,4-bis(diphenylphosphino)pentane
BDPP-TrS	Trisulfonated 2,4-bis(diphenylphosphino)pentane
BDPP-TS	Tetrasulfonated 2,4-bis(diphenylphosphino)pentane
BINAP	2,2′-Bis(diphenylphosphanyl)-1,1′-binaphthyl
BINAPHOS	[2-Diphenylphosphino-1,1′-dinaphthalen-2′-yl][1,1′-dinaphthalene-2,2′-diyl]phosphite
BINAS	Sulfonated NAPHOS
BINOL	1,1′-Bi(2-naphthol)
BIPHLOPHOS	4,6,4′,6′-Tetrachloro-2,2′-bis-(diphenylphosphinomethyl)-1,1′-biphenyl
BISBI	2,2′-Bis-(diphenylphosphinomethyl)-1,1′-biphenyl
BPPM	1-*tert*-Butoxycarbonyl-4-diphenylphosphino-(2-diphenylphosphino-methyl)pyrrolidine
*i*BuA	Isobutyric acid
CAL-B	Lipase from *Candida antarctica*, type B
CD	Cinchonidine
CLB	4-Chlorobenzoate ester
CLEC	Cross-linked enzyme crystal

CN	Cinchonine
COD	*cis,cis*-1,5-cyclooctadiene
m-CPBA	*meta*-Chloroperoxybenzoic acid
CPG	Controlled pore glass
CRL	Lipase from *Candida rugosa*
CSD	Crystal size distribution
D_c	Crystal diameter
DABCO	1,4-Diazabicyclo[2.2.2]octane
D-AOD	D-Amino acid oxidase
DBU	1,8-Diazabicyclo[5.4.0]undec-7-ene
d.c.	Dielectric constant
d.e.	Diastereomer excess
DHCD	9-Deoxy-10,11-dihydrocinchonidine
DIOP	1,4-Bis(diphenylphosphino)-1,4-dideoxy-2,3-O-isopropylidene-threitol
DIPAMP	1,2-Bis[(o-methoxy)phenylphosphino]ethane
DMI	Dimethylitaconate
DNi	Nickel prepared by the thermal decomposition of nickel formate
DOPA	(3-(3,4-Dihydroxyphenyl)-alanine)
DPEN	1,2-Diphenylethylenediamine
DPP	Diphenylpyrazinopyridazine diether
DP-PHAL	Diphenylphthalazine diether
DVB	Divinylbenzene
E_A	Activation energy
e.d.a.	Enantiodifferentiating ability
EDCA	Ethyldicyclohexylamine
e.e.	Enantiomeric excess
EGDMA	Ethylene glycol dimethacrylate
EL	Ethyl lactate
EMR	Enzyme membrane reactor
E-region	Enantiodifferentiating region on the catalyst
EtAc	Ethyl acetate
EtPy	Ethyl pyruvate
GA	Glycolic acid
Gl-acylase	Glutaric acid acylase
h_8-BINAP	Octahydro-BINAP
HCD	10,11-Dihydrocinchonidine
HDMS	Hexadimethylsilazane
HEMA	Hydroxyethyl methacrylate
HNi	Nickel prepared by the hydrogenation of NiO
HP	4-Hydroxy-2-pentanone
HQ	10,11-Dihydroquinine
HQD	10,11-Dihydroquinidine
i	Intrinsic e.d.a.
IPB	Insoluble polymer-bound
Kg	Kieselguhr
LDH	Layered double hydroxide

MA	Malic acid
MAA	Methyl acetoacetate
MA-Mni	Malic acid modified Ni
McA	1-Methyl-1-cyclohexanecarboxylic acid
O-MeDHCD	*O*-Methyl derivatives
Me-DUPHOS	1,2-Bis(2,5-dimethylphospholano)benzene
MEPY	Methyl pyroglutamate
MePy	Methyl pyruvate
MHB	Methyl 3-hydroxybutanoate
MMA	Methyl methacrylate
MNi	Modified nickel
MP	Methyl pyruvate
MPC	Methyl piperazine-2-carboxylate
MPr	Methyl propionate
MTS	Micelle-templated silica
MVK	Methyl vinyl ketone
NAPHOS	2,2′-Bis-(diphenylphosphinomethyl)-1,1′-binaphthyl
NEA	1-(1-Naphthyl)ethylamine
NMO	*N*-Methylmorpholine-*N*-oxide
NORPHOS	2,3-Bis(diphenylphosphino)bicyclo[2.2.1]hept-5-ene
N-region	Non-enantiodifferentiating region on the catalyst
OX-SR	Oxime type SR
1P	Interaction through one hydrogen bond
2P	Interaction through two hydrogen bonds
PA	Pivalic acid
PAA	Phenylacetic acid
PCL	Lipase from *Pseudomonas cepacia*
PD	2,4-Pentanediols
PDHCD	2-Phenyl-9-deoxy-10,11-dihydrocinchonidine
PDMS	Polydimethylsiloxane
PEG	Polyethylene glycol
PGA	Penicillin G amidase, also called Penicillin acylase
PHAL	Phthalazine
PhIO	Iodosylbenzene
PHN	Phenantryl
PHPG	D-*p*-Hydroxyphenylglycine
PLE	Esterase from pig liver
PNE	(*R*)-2-(1-Pyrrolidinyl)-1-(1-naphthyl)ethanol
PPL	Lipase from porcine pancreas
c-Pr	Cyclopropyl group
PYCA	2-Pyrrolidone-5-carboxylate
PYR	Diphenylpyrimidine diether
QD	Quinidine
QN	Quinine
RNi	Raney nickel catalyst
RNiA	Acid-treated Raney nickel

RNiH	Raney-type leaching at high temperature
RNiL	Raney-type leaching at low temperature
RNiU	Ultrasound-irradiated RNi
SALEN	Bis(salicylidene)ethylenediamine
SAP	Supported aqueous phase
SAPC	Supported aqueous phase catalyst
SDS	Sodium dodecyl sulfate
SEM	Scanning electron microscopy
SLP	Supported liquid phase
SPC	Supported nickel catalyst
SR	Silicone rubber
TA	Tartaric acid
TADDOL	2,3-*O*-Isopropylidene-1,1,4,4-tetraphenyl-threitol
TAH_2	Free acid of TA
TAHNa	Monosodium salt of TA
$TANa_2$	Disodium salt of TA
THF	Tetrahydrofuran
Ti-PILC	Titanium-pillared montmorillonite
TMC	Transition metal complexes
TOF	Turnover frequency
TON	Turnover number
TPE	1-(9-Triptycenyl)-2-(1-pyrrolidinyl)ethanol
U	Units
xyl	Xylene

1 Enantioselective Heterogeneous Catalysis: Academic and Industrial Challenges

Hans-Ulrich Blaser, Benoît Pugin, and Martin Studer

1.1 Introduction

The trend towards the application of single enantiomers of chiral compounds is undoubtedly increasing. This is especially the case for pharmaceuticals but also for agrochemicals, flavors and fragrances [1, 2]. Among the various methods to selectively produce one single enantiomer of a chiral compound, enantioselective catalysis is arguably the most attractive method. With a minute quantity of a (usually expensive) chiral auxiliary, large amounts of the desired product can be produced. Homogeneous metal complexes with chiral ligands are currently the most widely used and versatile enantioselective catalysts. From an industrial point of view, however, catalysts that are not soluble in the same phase as the organic reactant have the inherent advantage of being easily separated and often having better handling properties. Such catalysts can be truly heterogeneous, i.e. insoluble, or they can be soluble in a second phase that is immiscible with the organic one [3–6]. Here, we will use the term 'heterogeneous catalyst' for both cases.

In this overview, we will first discuss the situation and requirements for the industrial application of a catalytic method and more specifically of heterogeneous catalysts. Thereafter, the present scope and limitations of different types of enantioselective heterogeneous catalysts are documented (with reference to the appropriate chapters in this book) and assessed from an application point of view. Based on this analysis, academic and industrial challenges are then defined.

1.2 The Industrial Process in General and the Specific Prerequisites for Chiral Catalysts

In order to understand the challenges facing the application of chiral catalysts in the fine chemicals industry, one not only has to understand the essential industrial requirements but also how process development is carried out and which criteria determine the suitability of a catalyst [1, 2].

1.2.1 Characteristics of the Manufacture of Enantiomerically Pure Products

The manufacture of chiral fine chemicals such as pharmaceuticals or agrochemicals can be characterized as follows (numbers given in parentheses reflect the experience of the authors):

- Multifunctional molecules produced via multistep syntheses (from 5 to over 10 steps for pharmaceuticals and 3 to 7 steps for agrochemicals) with short product lives (often less than 20 years).
- Relatively small-scale products (1–1000 t/y for pharmaceuticals, 500–10000 t/yr for agrochemicals), usually produced in multipurpose batch equipment.
- High purity requirements (usually >99% and <10 ppm metal residue in pharmaceuticals).
- High added values and therefore tolerance of higher process costs (especially for very effective, small-scale products).
- Short development time for the production process (< few months to 1–2 years), since marketing time affects the profitability of the product. In addition, development costs for a specific compound must be kept low, since process development often starts at an early phase when the chances of product success are low.
- At least in European companies, chemical development is carried out by all-round organic chemists, sometimes in collaboration with technology specialists.

1.2.2 Process development: Critical Factors for the Application of (Heterogeneous) Enantioselective Catalysts

The first decision to be made at the start of process development is the choice of a strategy that promises the best answer in the shortest time. This strategy will depend on a number of considerations: the goal of the development, the know-how of the investigators, the time frame, the available manpower and equipment etc. In process development, there is usually a hierarchy of goals (or criteria) to be met. It is useful to divide the development of a manufacturing process into different phases:

Phase 1: Outlining and assessing possible synthetic routes on paper.
Phase 2: Demonstrating the chemical feasibility of the key step, often the enantioselective catalytic reaction.
Phase 3: Optimizing the key catalytic reaction.
Phase 4: Optimizing the overall process.

In the final analysis, the choice whether a synthesis with an enantioselective catalytic step is chosen depends very often on the answers to two questions:

- Can the costs for the overall manufacturing process compete with alternative routes?
- Can the catalytic step be developed in the given time frame?

Presuming that enantioselective catalysis is the method of choice, the next question in the context of our treatise is whether to choose homogeneous or heterogeneous catalysis.

Table 1.1 gives a very condensed summary of the strong and weak points of the two classes of catalysts. This table is a somewhat subjective view of the authors and mirrors their personal experiences. Moreover, the importance of the various factors changes for any specific catalytic transformation and, in some cases, might well be just the opposite!

1.2.3 Important Criteria for Enantioselective Catalysts

As a consequence of the peculiarities of enantioselective catalysis described above, the following critical factors often determine the viability of an enantioselective process:

Enantioselectivity, expressed as enantiomeric excess (e.e., %). The enantioselectivity of a catalyst should be in the range of 99% for pharmaceuticals if further enrichment is not possible (this is relatively rare). E.e.'s >80% are acceptable for agrochemicals or if further enrichment is easy, e.g. via recrystallization or via separation of diastereomers later in the synthesis; in our experience, this is very often the case.

Catalyst productivity, given as turnover number (TON), determines catalyst costs. In our experience, TONs for (homogeneous) enantioselective hydrogenation reactions ought to be >1000 for small-scale, high-value products and >50 000 for large-scale or less expensive products. For C-C coupling reactions and probably also for some other reaction types with high added value or for very inexpensive catalysts, lower TONs

Table 1.1. Strong and weak points of homogeneous and heterogeneous catalysts.

	Homogeneous	Heterogeneous
Strong points	Defined on molecular level (close to organic chemistry)	separation, recovery, recycling
	Scope, variability (design?) (commercial) preparation	stability, handling
Weak points	Sensitivity (handling, stability)	characterization (understanding on molecular level)
	Activity, productivity (of many literature procedures)	availability, preparation (needs special know-how), reproducibility diffusion to and within catalyst

might be acceptable. Much lower limits might apply if catalyst reuse is possible without much loss in selectivity and activity.

Catalyst activity. For preparative applications, a useful number is the turnover frequency (TOF) at high conversion. Because this value determines the production capacity, TOFs (especially for high pressure reactions) ought to be >500 h^{-1} for small-scale and >10000 h^{-1} for large-scale products. For applications in standard equipment, lower TOFs might be acceptable.

Separation should be achieved by a simple operation such as distillation, filtration or phase separation, and at least 95% of the catalyst should be recovered. Methods like ultrafiltration or precipitation (e.g. for separating soluble polymer supports) usually require expensive equipment.

Stability. If the advantage of the heterogeneous catalyst is its recyclability, it has not only to show a stable catalytic performance, but it should also be mechanically stable and the active component must not leach (chemical stability).

Price of catalysts. The catalyst price will only be important at a later stage, when the cost of goods of the desired product is evaluated. For homogeneous catalysts, the chiral ligand often is the most expensive component (typical prices for the most important chiral phosphines are 100–500 $/g for laboratory quantities and 5000 to >20000 $/kg on a larger scale). For heterogeneous systems, the dominant cost elements depend on the type of catalyst.

Availability of the catalysts. If an enantioselective catalyst is not available at the right time and in the appropriate quantity, it will not be applied due to the time limitation of process development. At present, only very few homogeneous catalysts and ligands are commercially available in technical quantities, so that their large-scale synthesis must be part of the process development. The situation for heterogeneous catalyst systems is even more difficult, because their preparation and characterization require know-how that is usually not available in a standard development laboratory.

Which of these criteria will be critical for the development of a specific process will depend on the particular catalyst and transformation, the scale of the process, the technical experience and the production facilities of a company as well as the maturity of the catalytic process.

1.3 The General Challenges

In many areas described in the following chapters, much remains to be done for both academia and industry. A special challenge for both communities is the interdisciplinary nature of the field of heterogeneous enantioselective catalysis. It comprises the preparation of such widely different materials as polymers, inorganic supports, small metal particles, colloids, complex organic molecules and organometallic complexes. We have listed some crucial points which, in our view, are important for progress in heterogeneous enantioselective catalysis. Needless to say, a good dialog between industry and academia is probably the most important factor for the rate of progress, because the different approaches and goals are very often complimentary.

1.3.1 For Academia

Generally, the central task of academic researchers is to find new concepts, catalysts and reactions, to demonstrate proof of concept of a catalytic reaction and to investigate its mechanism.

Development of new concepts, new catalysts and processes. Most of the existing enantioselective catalytic systems lack general applicability. Although some new concepts (such as artificial catalytic antibodies) and technologies (e.g., fluorous biphasic reactions, or immobilization on an aqueous layer on a porous support) have been developed and applied recently, there is a need for new ideas which eventually will lead to new catalysts – hopefully with broader applicability.

Determination of synthetic scope and limitations. Well characterized catalysts with clear scope and limitations are much more likely to be applied by the synthetic chemist (both at the university and industry) who usually has little time and patience for trial and error. In the literature, many new systems are tested only on one or two model substrates under a very narrow set of reaction conditions. For an immobilized catalyst, a realistic comparison with the corresponding homogeneous systems is quite often lacking and, in addition, little information on catalyst activity or productivity is provided.

Characterization, mechanistic investigation, understanding and interpretation. Many of the heterogeneous systems are very difficult to characterize and are not well understood. Improved characterization should lead to better reproducibility, whereas understanding on a molecular level (if possible) can often help to improve existing concepts and develop new catalytic systems.

1.3.2 For Industry

Its main task is to apply the know-how created by basic research to practical problems. For the catalyst user, this means to adapt catalysts and processes to industrial conditions, and for catalyst manufacturers, to make available more well-defined catalysts on a commercial basis.

Development, up-scale and commercialization of industrially useful catalysts and processes. New systems, as described in the literature, are often unsuited for industrial application (exotic solvents, reactions conditions, too low productivity and activity, etc.). Since the industrial chemist knows the specific prerequisites of the process, it is his or her task to determine the technical scope and limitations and to adapt catalytic systems to the technical problems and conditions. In addition, investigation of technical aspects such as catalyst stability, recycling, metal leaching are often necessary.

Toolbox for fast development and commercial availability of catalysts. In many cases, development of technical processes with heterogeneous or immobilized chiral catalysts is very tedious. Automation of both the development of the best suited catalyst as well as the optimization of reaction conditions should improve that considerably. For this endeavor, the ready and easy availability of a large collection of (tunable) catalysts is necessary to get results in a timely manner. Involvement of the catalyst producers and commercial availability of versatile catalysts would certainly help their application.

1.4 Chiral Heterogeneous Catalysts: State of the Art and Future Challenges

In this section, the present scope as well as the specific problems and challenges are analyzed for the most important types of enantioselective heterogeneous catalytic systems. One can roughly distinguish between three types of enantioselective heterogeneous catalysts:

- Heterogeneous catalysts with demonstrated catalytic activities that are rendered chiral by modification with a chiral auxiliary,
- Homogeneous catalysts with demonstrated enantioselectivity and activity modified in such a way as to become heterogeneous (as defined in the introduction),
- Catalysts with no known precedent in these two categories.

1.4.1 Heterogeneous Catalysts Modified with a Chiral Auxiliary

1.4.1.1 Metallic Catalysts on Chiral Supports

Metals supported on chiral biopolymers and natural fibers were the first somewhat successful approach to produce enantioselective heterogeneous catalysts. For a review, see Blaser and Müller [3]. With the exception of Pd/silk fibroin where e.e.'s of up to 66% were reported for the hydrogenation of an oxazolinone derivative, the optical yields were very low. Later, it was found that the results observed with silk fibroin were not reproducible and this approach was practically abandoned.

Assessment and challenge. Clearly, fresh ideas would be needed to revive this class of enantioselective catalysts but at the moment, no leads exist for a promising revival.

1.4.1.2 Metallic Catalysts Modified with a Low Molecular Weight Chiral Auxiliary

This is undoubtedly the most successful approach to render an already active catalyst enantioselective, and several recent informative reviews on different aspects have been published [7–12]. The investigation of heterogeneous chiral hydrogenation catalysts started in the late fifties in Japan and has seen a renaissance in the last few years. Despite many efforts, only two classes of modified catalyst systems have been found to be of industrial interest at this time: Ni catalysts modified with tartaric acid and Pt and, to a lesser degree, Pd catalysts modified with cinchona alkaloids and analogs thereof. However, several laboratories are working to expand the scope of this interesting and potentially very versatile class of chiral catalysts.

Since these catalytic systems are covered in Chapters 6–8, we will not discuss but only list reactions and catalysts that have either sufficiently high enantioselectivities for synthetic applications or are of conceptual importance (see Table 1.1). One exception: in two very recent papers the highly selective hydrogenation of a variety of α-ketoacetals with cinchona modified Pt catalysts was described with enantioselectivities up to 97% [13, 14]. Since chiral α-hydroxyacetals are versatile intermediates for a variety of chiral building blocks (e.g., 1,2-diols, α-hydroxy acids, 1,2-amino alcohols), the new enantioselective transformation is also of synthetic significance.

Table 1.2. State of the art for the synthetic application of modified metallic catalysts.

Substrate	R/R'	Catalyst	Modifier	E.e. (%)	TOF (1/h) [a]	Ref.
CH_3COR	Alk	Ra-Ni	Tartrate/NaBr [b)]	70–85	$\ll$1	[10]
$PhCOCF_3$		Pt/Al_2O_3	Cinchona alkaloid	56	150	[15]
$RCOCOOR'$ [b)]	R/Alk,H	Pt/Al_2O_3	Cinchona alkaloid	85–98	Low->50000	[7, 8, 16]
$RCOCH(OR')_2$	R/Alk	Pt/Al_2O_3	Cinchona alkaloid	50–97	Low->20000	[13, 14]
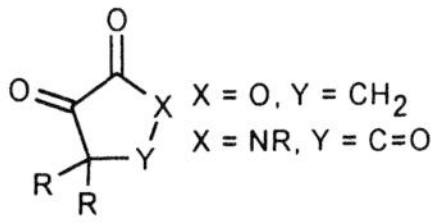 X = O, Y = CH_2; X = NR, Y = C=O	Alk	Pt/Al_2O_3	Cinchona alkaloid	92	50	[17, 18]
$RCOCH_2COOR'$ [c)]	Alk/Et	Ra-Ni	Tartrate/NaBr	83–98	<1	[19]
$CH_3COCH_2COCH_3$		Ra-Ni	Tartrate/NaBr	91(diol)	1	[20]
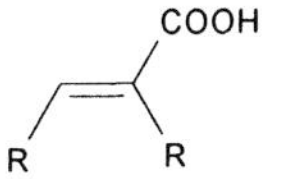 COOH, R, R	Alk, Aryl	Pd/TiO_2	Cinchona alkaloid	50–72	400	[8, 21]
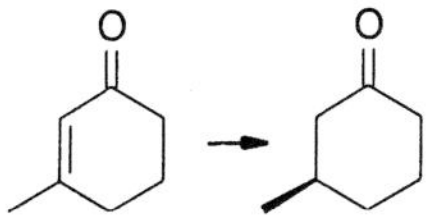		Pd black	Vinca alkaloid	53	–	[22]

a) TOFs for complete conversion, rough estimates. b) In presence of pivalic acid. c) Technical applications with R' = Et have been reported.

Assessment and challenges. Several transformations have already been developed for commercial applications or are mature to be used on a technical basis [3, 4]. There are some good and challenging ideas on the mode of action of the chiral catalysts, but by far no mechanism that explains all major effects or allows to design new catalysts. In the case of the Pt–cinchona system, both catalyst and some of the modifiers are available commercially or easy to prepare. Nevertheless, reproducibility is still an issue even here and especially for the Ni catalysts. Furthermore, the preparation procedures (soaking in dilute solutions, extractions etc.) and pretreatments (high temperature prereduction under hydrogen, sonication etc.) are often cumbersome. Besides these technical problems, the sensitivity for catalyst poisons and starting material quality is a major drawback. Last but not least, the scope of these systems is still very narrow, and only very few substrates give satisfactory activities and selectivities.

The challenge for academia is further progress in understanding mechanisms, identifying catalyst poisons and developing new catalytic systems. The challenge for catalyst producers is developing reproducible catalysts that do not need pretreatment and are less sensitive to poisoning, and for industrial process developers, optimizing existing systems with respect to technical applicability.

1.4.1.3 Metal Oxide Catalysts Modified with a Chiral Auxiliary having Low Molecular Weight

Titanium-pillared montmorillonite (Ti-PILC) modified with tartrates was described as a heterogeneous Sharpless epoxidation catalyst [23]. Unfortunately, the results could not be reproduced by other laboratories. Very recently, tantalum tartrate complexes grafted to silica were described with e.e.'s of up to 98% and promising activities for the epoxidation of allylic alcohols. Remarkably, the homogeneous Ta-complex was neither stable nor catalytically active [24]. Metal oxides modified with histamine showed modest efficiencies for the kinetic resolution of activated aminoacid esters ($k_R/k_S \approx 2$) [25]. Silica or alumina treated with diethyl aluminium chloride and menthol catalyzed the Diels-Alder reaction between cyclopentadiene and methacrolein with modest enantioselectivities of up to 31% [26]. Zeolite HY, modified with chiral sulfoxides, had remarkable selectivities for the kinetic resolution of 2-butanol by dehydration (k_S/k_R = 39). The enantioselectivity is due to the preferential acceleration of the dehydration of one enantiomer [27]. A NaY zeolite modified with norephedrine allowed the photocyclization of tropolone methyl ether with an e.e. of up to 50%, albeit not in a really catalytic fashion [28].

Assessment and challenges. Although solid acids and bases are increasingly applied for the catalytic synthesis of fine chemicals, chirally modified versions, though potentially interesting because of their variability, are definitely not ready for synthetic applications. In many cases, the preparation of the catalysts is not trivial and not always reproducible. There is very little known about their mode of action, and few new concepts are currently being discussed. Filling this gap is an important fundamental challenge for academic laboratories with a good background in metal oxide catalysis.

1.4.2 Immobilized and Functionalized Homogeneous Catalysts

Control of stereoselectivity is easier with homogeneous than with heterogeneous catalysts, and the majority of chiral catalysts are homogeneous. On the other hand, these soluble catalysts are more difficult to separate and to handle than the technically well established heterogeneous catalysts. One promising strategy to combine the best properties of the two catalyst types is the heterogenization or immobilization of active metal complexes on supports or carriers, which may be separated by filtration or precipitation [29–31]. Another strategy is the functionalization of ligands to allow separation from the product solution.

1.4.2.1 Immobilized Homogeneous Catalysts

Figure 1.1 summarizes the most important approaches to immobilize or heterogenize soluble catalysts that have been described in the literature. The following materials have been used as supports:

- Linear, non-cross-linked polymers and, more recently, also better defined dendrimers [32] are soluble in suitable solvents and give catalysts with high mobility and good mass transport properties. However, separation is not trivial (precipitation or ultrafiltration).
- Swellable, slightly cross-linked polymers such as polystyrene cross-linked with 0.5-3% 1,4-divinylbenzene, can easily be separated by filtration or sedimentation. To allow good mass transport, these polymers have to be used in solvents in which they swell.
- Highly cross-linked polymers (e.g. macroreticular polystyrenes or polyacrylates) and inorganic supports (metal oxides, e.g. silica gel) hardly swell and can be used in a large variety of different solvents without changes of texture or mass transport properties.

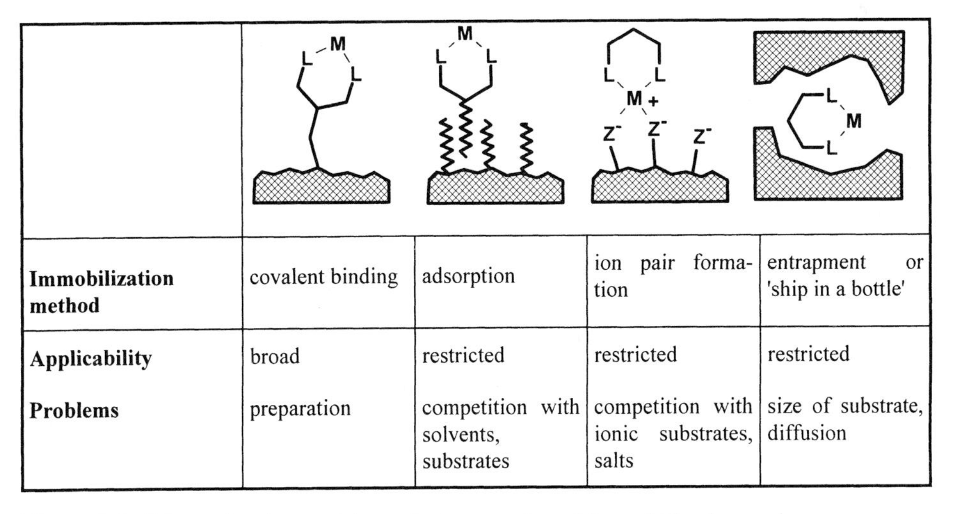

Immobilization method	covalent binding	adsorption	ion pair formation	entrapment or 'ship in a bottle'
Applicability	broad	restricted	restricted	restricted
Problems	preparation	competition with solvents, substrates	competition with ionic substrates, salts	size of substrate, diffusion

Figure 1.1. Schematic view and important properties of immobilized complexes.

- For immobilization via entrapment or intercalation, materials such as zeolites or clays with well defined pores and cavities have to be used to effect reliable confinement of the metal complex catalysts.

Many of the immobilized catalysts are discussed in greater detail in this book (see Chapters 2–4, 9–11). Therefore, we will only illustrate the state of the art with typical catalysts that are promising with respect to potential future applications and/or are of conceptual importance (the numbers refer to the entries in Table 1.3).

Heterogenization via covalently bound ligands

Covalent binding, by far the most frequently used strategy, can be effected either by copolymerization of functionalized ligands with a suitable monomer or by grafting functionalized ligands or metal complexes with reactive groups of a preformed support. The catalysts are much more complex than their homogeneous counterparts, and many additional parameters such as type of support, solvent, spacer length and flexibility, degree of surface coverage have to be optimized to obtain an acceptable catalytic performance [33].

The highest activities and productivities and very good e.e.'s have so far been obtained in hydrogenations with chiral Rh- and Ir-diphosphines that were immobilized on solid inorganic supports or on soluble polymers (entries 1–5, 10–12). These catalysts were built up using a modular system combining functionalized ligands with different supports and linkers [34–38]. Excellent enantioselectivities and practically unchanged activities have been obtained with enantioselective hydroformylation catalysts (entry 13) [39]. Polymer and silica gel supported Co-salen complexes exhibited excellent performance (high e.e., acceptable TOF, reuse) for the hydrolytic kinetic resolution of terminal epoxides, e.g., epichlorohydrin (entries 14, 15) [40]. Not yet very successful was the immobilization of Mn-salen complexes for the epoxidation of unfunctionalized olefins [41].

Heterogenization via adsorption and ion-pair formation

This approach relies on various adsorptive interactions between a carrier and a metal complex. Cationic Rh-diphosphine complexes were bound to anionic resins [42, 43]. The resulting hydrogenation catalysts could be recycled 20 times with almost constant activity and selectivity and with little leaching (entry 7). An innovative modular method was developed by Augustine who used heteropoly acids as anchoring agents to attach a large variety of metal complexes to different supports (entry 6) [44]. Compared with their homogeneous counterparts, immobilized Rh-catalysts exhibited approximately equal or in some cases better activities and enantioselectivities. In one case, a catalyst was recycled 15 times, giving a total TON of 600. This method is being further developed, among others, by Engelhard Corporation and may become commercialized in the near future.

A special case of adsorption is the "supported aqueous phase" (SAP), whereby a water-soluble catalyst dissolved in a very polar solvent is adsorbed on a hydrophilic support. With such a catalyst, the hydrogenation of a C=C double bond in the Naproxen synthesis with e.e.'s of up to 95.7% was described (entry 9) [45, 46].

Heterogenization via entrapment

Here, the size of the metal complex rather than a specific adsorptive interaction is important [5]. There are two different preparation strategies. Often called the 'ship in a bottle' approach, the first strategy is based on building up catalysts in well-defined cages of porous supports [47]. Mn-epoxidation catalysts with different salen ligands were assembled inside zeolites. In zeolite EMT [48] e.e.'s of up to 88% and in zeolite Y [49] e.e.'s of up to 58% were obtained with *cis-β*-methylstyrene. However, both entrapped catalysts were much less active than their homogeneous counterparts. The other approach is to build up an inorganic sol gel [50] or organic polymeric network around a preformed catalyst. A promising example was reported by the group of Jacobs [51] who entrapped Jacobsen's Mn-salen epoxidation catalyst and Noyori's Rubinap hydrogenation catalyst in polydimethylsiloxane. Leaching depended strongly on the size and the solubility of the metal complex and the swelling of the polymer [52]. The best e.e.'s were 50% for epoxidation of 1,3-cyclooctadiene and 92% for the hydrogenation of methyl acetoacetate in presence of toluene-p-sulfonic acid (entry 11) [53].

1.4.2.2 Alternative Methods Using Functionalized Ligands

Because immobilization on solid supports often involves unpredictable changes of the catalytic properties, there is a strong tendency to develop alternative methods for catalyst separation. Up to now, two approaches are showing some promise. The first is the use of functionalized soluble ligands separable by extraction [54], ultrafiltration [55], or by precipitation by changing the temperature [56] or the pH [57]. The second one uses functionalized catalysts to run the reactions in two non-miscible liquid phases [58]. The technical feasibility of all these methods not been demonstrated yet, but the use of only one solvent is an inherent advantage and seems more likely to succeed.

Assessment and challenges. There are now many examples demonstrating that efficient separation by filtration or sedimentation can be achieved. In a few cases the immobilized complexes have a similar activity and selectivity as their homogeneous counterparts, and sometimes reuse of the recovered catalysts is possible without loss of performance. In addition, immobilization can lead to improved catalytic performance by site isolation [59] or site cooperation effects [40].

All immobilization methods for homogeneous catalysts have their strong and weak points. Covalently bound catalysts probably have the broadest scope, since the link to the support is stable towards most solvents or additives such as salts or acids and bases, but this method requires the expensive functionalization of the chiral ligand. With regard to costs and time, adsorption or entrapment of metal complexes are clearly advantageous. However, the applicability of adsorbed catalysts will probably be limited by strong solvent and salt effects. Entrapped catalytic systems are still far from being practical useful. For every catalytic reaction, a new suitable support has to be found, thereby guaranteeing confinement of the catalyst but still allowing an efficient transport of the reactants. A promising alternative are catalysts used in the soluble state and separated after the reaction by precipitation or extraction. This method requires functionalization of the ligands but allows a better prediction of the catalytic properties which usually remain very similar to those of their unfunctionalized counterparts.

From an industrial point of view, immobilization of homogeneous catalysts is not yet a practical alternative and will only be taken into consideration, if the classical separation methods such as distillation or crystallization fail. The most important problems are the high complexity of heterogeneous systems leading to a poor predictability of the catalytic properties and the unavailability and/or the high price of functionalized and immobilized chiral ligands. For the moment, binap is the only commercially available, immobilized chiral ligand [60].

Table 1.3. Typical and/or potentially synthetically useful immobilized catalysts.

Entry	Reaction/ catalyst	Support	Method	E.e. 1)	TOF1)	Ref
Hydrogenation of C=C						
1	Rh-pyrphos	Silica gel	covalent	100	400	[34]
2		Polymer (Tentagel)	covalent	97	60	[35]
3	Various Rh-	Silica gel	covalent	94.5	1500	[37, 59, 62]
4	diphosphines	Soluble polymers	covalent	95	2000	[36]
5		Cross-linked polymers	covalent	97	200	[36]
6		Supported heteropolyacids	adsorbed	95	100	[43, 44]
7	Rh-phosphinite	Silica grafted with SO_3H	adsorbed	95	>1000	[41, 42]
8		and ion exchange resins				
9	Ru-sulfonated binap	Controlled pore glass	SAP	95.7	40	[46]
Hydrogenation of C=O						
10	Ru-binap	Cross-linked polystyrene	covalent	97	3	[50]
11		Polydimethylsiloxane	entrapped	92	42	[51–53]
Hydrogenation of C=N						
12	Ir-ferrocenyl-diphosphines	Silicagel	covalent	79	10000 2)	[38]
Hydroformylation						
13	Rh-phosphine-phosphinite	Cross-linked polystyrene	covalent	93	>200	[39]
Hydrolytic kinetic resolution of epoxides						
14	Co-salen	Cross-linked polystyrene	covalent	99	140	[40]
15		Silicagel	covalent	96	110	[40]
Dihydroxylation of alkenes						
18	Os-1,4-bis-(9-O-quinyl)-phthalazine	Silicagel	covalent	99	6	[63]
19		Polyacrylic esters	covalent	99	6	[64]

1) Best e.e. and TOF for complete conversion
2) Hydrogenation in presence of acid and iodide.

To improve the chances for the application of immobilized homogeneous catalysts many obstacles have to be addressed:

- Preparation of inexpensive chiral ligands with a suitable anchoring group.
- Development of well-defined, inert and mechanically stable supports with a highly specific surface area and chemical functions suitable for binding catalysts.
- Development of efficient and versatile immobilization methods.
- Better understanding of the interactions between supports and catalysts.

However, in the near future combinatorial solid phase synthesis and high-throughput screening of chiral ligands might give another impetus to the development of this still young field [61].

1.4.3 Catalysts with No Known Heterogeneous or Homogeneous Precedent

Whereas the aforementioned catalysts have either a non-chiral or a homogeneous counterpart, this is not the case for the catalysts described in this section. Generally, these are macromolecular systems where activation and stereocontrol are connected to supramolecular effects. Here, we would like to briefly discuss two cases: catalysis by polymers and gels, where two synthetically useful systems already exist, and artificial catalytic antibodies, where a convincing proof of concept is still lacking.

1.4.3.1 Insoluble Polypeptides and Gels

Synthetic polypeptide derivatives are highly selective catalysts for the asymmetric epoxidation of electron-deficient olefins with $NaOH/H_2O_2$ in a biphasic reaction system (Fig. 1.2) [65]. Significant parameters for the catalytic performance are the type of amino acid, the degree of polymerization, the substituent at the terminal amino group (R in Scheme 1.1), catalyst pretreatment and the organic solvent. In general, the catalysts are commercially available [66] or seem relatively easy to prepare even in larger quantities (>200 g) [67]. Generally, chemical yields are high but the activity of the catalysts is rather low (TOFs ca. 0.1 h^{-1}). Moreover, relatively long reaction times and/or high catalyst concentrations are necessary. An accelerating effect of ultrasound was recently reported [68].

R	R'	E.e.
(subst)-phenyl, 2-naphthyl 2'-styryl, cylopropyl, 2-furyl pyridyl	phenyl, 2-naphthyl, 2-furyl *t*-Bu, cyclopropyl	**>90 up to 99%**
Ph	*t*-BuO	**95%**

Scheme 1.1. Epoxidation of various activated C=C bonds using polypeptide catalysts [66].

Cyclic dipeptides, especially cyclo[(S)-phenylalanyl-(S)-histidyl], are efficient and selective catalysts for the hydrocyanation of aldehydes [68, 69] using HCN (see Scheme 1.2). The catalysts are only selective in a particular heterogeneous state, described as a "clear gel" [70, 71]. It seems that the method of precipitation is crucial [70, 72] and that reproducing literature results is not always easy [73]. The chiral dipeptide catalysts are not commercially available but can be synthesized by conventional methods, and their structure can be easily varied. Generally, 2% catalyst is needed, and best optical yields are obtained at room temperature or below. As summarized by North [68], the pattern and the nature of substitution strongly affect the enantioselectivity.

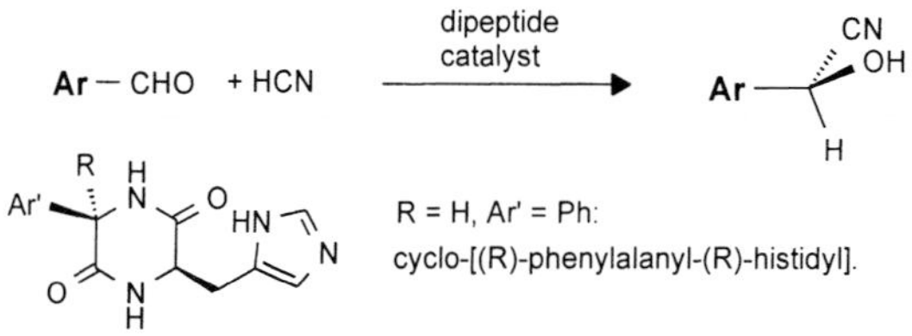

Ar—CHO + HCN —(dipeptide catalyst)→ Ar–C(CN)(OH)H

R = H, Ar' = Ph:
cyclo-[(R)-phenylalanyl-(R)-histidyl].

Ar	E.e.(%)
Ph, m-RO-Ph, m-tol, 2-naphthyl	91-97
p-MeO-Ph, o-MeO-Ph	78-84
heteroaromatic, aliphatic aldehydes	40-70

Scheme 1.2. Reaction scheme, structure and enantioselectivities of cyclo-[(R)-phenylalanyl-(R)-histidyl] and analogs [69, 72].

1.4.3.2 Artificial Catalytic Antibodies

The imprinting of organic [74] and inorganic materials [75] with transition state analog templates should, at least in principle, lead to what could be called artificial catalytic antibodies. This is a very exciting idea in principle, because it would allow the construction of catalysts tailored to a specific enantioselective transformation. While quite a few materials have already been reported with good discrimination ability, none has yet been described having both good chiral recognition and acceptable catalytic properties. Some examples with at least detectable effects are a zeolite β, partially enriched in polymorph 'A' [75], "chiral footprints" on silica surfaces [76], or several imprinted polymers [74].

Assessment and challenges. While the two polymeric catalyst systems have very good enantioselectivity, their activity is rather low. Nonetheless, both have been applied for synthesizing molecules with useful functional properties, albeit only on a small-scale. For both types of catalysts, the challenge is clearly for the academic laboratories to develop increased understanding and concepts and eventually new catalysts.

1.5 Conclusions

After a period of stagnation, the fascinating area of heterogeneous enantioselective catalysis has again become a fashionable topic. Many different research groups, both in industry and at universities, are now actively involved in this field. Three general strategies have been successful: i) The chiral modification of metallic and oxidic heterogeneous catalysts, ii) the immobilization of homogeneous catalysts and iii) the application of chiral polymers. The situation for the three areas of research is quite different. Modified catalysts and chiral polymers have already been applied for the synthesis of active compounds or have even been developed for technical applications. However, their scope is still very narrow, i.e., only few transformations are catalyzed with useful ees. Immobilized catalysts, on the other hand, have been applied only for model reactions. Relatively few of the highly selective ligands have been immobilized, but in principle, this is possible for many more. The potential scope of this approach is impressive because numerous catalytic transformations using homogeneous catalysts with high enantioselectivity are reported in the literature.

At the present time, only few of the catalysts described above are commercially available. Because some rather special know-how is required to prepare and characterize these heterogeneous materials, their application will probably be restricted for some time to the "specialists" among the synthetic organic chemists. We think that many of these heterogeneous chiral catalysts are feasible for technical applications in the pharmaceutical and agrochemical industry, where the advantages of heterogeneous catalysts with respect to separation and handling properties will be most significant.

From the authors' point of view, the most significant challenges for academia will be to develop and realize conceptually new catalytic systems and to improve the mechanistic understanding of existing ones. For catalyst producers in industry, the development of technically mature chiral heterogeneous catalysts to be used "straight from the shelf" should have highest priority, because otherwise, technical applications will be very unlikely. For the industrial catalyst specialist, the biggest challenge might well be to convince the chemists who develop the overall processes, that chiral heterogeneous catalysts are reliable and offer opportunities for more economical processes.

References

[1] R.A. Sheldon, *Chirotechnology*, Marcel Dekker, Inc., New York, 1993.
[2] A.N. Collins, G.N. Sheldrake, J. Crosby (eds), *Chirality in Industry* Vol. I and Vol. II, John Wiley, Chichester, 1992 and 1997.
[3] H.U. Blaser, M. Müller, *Stud. Surf. Sci. Catal.* 59 (1991) 73.
[4] A. Baiker, H.U. Blaser in G. Ertl, H. Knötzinger, J. Weitkamp (eds), *Handbook of Heterogeneous Catalysis*, p. 2422, 1997.
[5] T. Bein, *Curr. Opin. Solid State Mater. Sci.* 4 (1999) 85.
[6] A. Baiker, *Curr. Opin. Solid State Mater. Sci.* 3 (1998) 86.
[7] H.U. Blaser, H.P. Jalett, M. Müller, M. Studer, *Catalysis Today* 37 (1997) 441.

[8] A. Baiker, *J. Mol. Catal. A: Chemical* 115 (1997) 473.
[9] A. Pfaltz, T. Heinz, *Topics in Catalysis* 4 (1997) 229.
[10] T. Osawa, T. Harada, A. Tai, *Catal. Today* 37 (1997) 465.
[11] P.B. Wells, A.G. Wilkinson, *Topics in Catalysis* 5 (1998) 39.
[12] M. Schürch, T. Heinz, R. Aeschlimann, T. Mallat, A. Pfaltz, A. Baiker, *J. Catal.* 173 (1998) 197.
[13] M. Studer, S. Burkhardt, H.U. Blaser, *Chem. Commun.* (1999) 1727.
[14] B. Török, K. Felföldi, K. Balazsik, M. Bartok, *Chem. Commun.* (1999) 1725.
[15] T. Mallat, M. Bodmer, A. Baiker, *Catal. Lett.* 44 (1997) 95.
[16] X. Zuo, H. Liu, D. Guo, X. Yang, *Tetrahedron* 55 (1999) 7787.
[17] M. Schürch, N. Künzle, T. Mallat, A. Baiker, *J. Catal.* 176 (1998) 569.
[18] A. Szabo, N. Künzle, T. Mallat, A. Baiker, *Tetrahedron: Asymmetry* 10 (1999) 61.
[19] S. Nakagawa, T. Sugimura, A. Tai, *Chem. Lett.* (1997) 859.
[20] A. Tai, T. Kikukawa, T. Sugimura, Y. Inoue, T Osawa, S. Fujii, *Chem. Commun.* (1991) 795.
[21] Y. Nitta. K. Kobiro, *Chem. Lett.* (1996) 897.
[22] G. Farkas, K. Fodor, A. Tungler, T. Mathe, G. Toth, R.A. Sheldon, *J. Mol. Catal. A: Chemical* 138 (1999) 123.
[23] B.M. Choudary, V.L.K. Valli, A. Durga Prasad, *Chem. Commun.* (1990) 1186.
[24] D. Meunier, A. Piechaczyk, A. de Mallmann, J.-M. Basset, *Angew. Chem.* 111 (1999) 3738.
[25] T. Moriguchi, Y.G. Guo, S. Yamamoto, Y. Matsubara, M. Yoshihara, T. Maeshima, *Chem. Express* 7 (1992) 625.
[26] J.M. Fraile, J.I. Garcia, J.A. Mayoral, A.J. Royo, *Tetrahedron: Asymmetry* 7 (1996) 2263.
[27] R.P.K. Wells, P. Tynjälä, J.E. Bailie, D.J. Willock, G.W. Watson, F. King, C.H. Rochest, D. Bethell, P.C. Bulman Page, G.J. Hutchings, *Applied Catalysis A: General* 182 (1999) 75 and references cited therein.
[28] A. Joy, J.R, Scheffer, D.R. Corbin, V. Ramamurthy, *Chem. Commun.* (1998) 1379.
[29] F.R. Hartley in: F.R. Hartley (ed) *The Chemistry of Metal Carbon Bond*, Vol. 4, John Wiley & Sons Ltd, p. 1163 (1986).
[30] R. Selke, K. Häupke, H.W. Krause, *J. Mol. Catal.* 56 (1989) 315.
[31] S.J. Shuttleworth, S.M. Allin, P.K. Sharma, *Synthesis* (1997) 1217.
[32] C. Köllner, B. Pugin, A. Togni, *J. Am. Chem. Soc.* 120 (1998) 10274.
[33] P. Hodge, *Chem. Soc. Rev.* 26 (1997) 417.
[34] U. Nagel, E. Kinzel, *J. Chem. Soc. Chem. Commun.* (1986) 1098.
[35] U. Nagel, J. Leipold, *Chem. Ber.* 129 (1996) 815.
[36] B. Pugin, *EP 728768* (1996) assigned to Ciba-Geigy.
[37] B. Pugin, F. Spindler, M. Müller, *EP 496699* (1992) and *EP 496700* (1992) assigned to Ciba-Geigy AG.
[38] B. Pugin, *EP 729969* (1996), *WO 9632400* (1996), *WO 9702232* (1997), assigned to Novartis AG.
[39] K. Nozaki, Y. Itoi, F. Shibahara, *J. Am. Chem. Soc.* 120 (1998) 4051.
[40] D.A. Annis, E.N. Jacobsen, *J. Am. Chem. Soc.* 121 (1999) 4147.
[41] L. Canali, D.C. Sherrington, *Chem. Soc. Rev.* 28 (1999) 85.
[42] R. Selke, K. Häupke, H.W. Krause, *J. Mol. Catal.* 56 (1989) 315.
[43] R. Selke, M. Capka, *J. Mol. Catal.* 63 (1990) 319.
[44] R. Augustine, S. Tanielyan, S. Anderson, H. Yang, *Chem. Commun.* (1999) 1257.
[45] K.T. Wan, M.E. Davis, *J. Catal.* 148 (1994) 1.
[46] K.T. Wan, M.E. Davis, *J. Catal.* 152 (1995) 25.
[47] D.E. De Vos, F. Thibault-Starzyk, P.P. Knopps-Gerrits, R.F. Parton, P.A. Jacobs, *Macromol. Symp.* 80 (1994) 157.
[48] S.B. Ogunwumi, T. Bein, *Chem. Commun.* (1997) 901.
[49] M.J. Sabater, A. Corma, A. Domenech, V. Fornés, H. Garcia, *Chem. Commun.* (1997) 1285.
[50] J. Blum, D. Avnir, H. Schuhmann, *Chemtech* (1999) 32.
[51] I.F.J. Vankelecom, D. Tas, R.F. Parton, V. Van de Vyver, P.A. Jacobs, *Angew. Chem. Int. Ed. Engl.* 35 (1996) 1346.
[52] K.B.M. Janssen, I. Laquière, W. Dehaen, R.F. Parton, I.F.J. Vankelecom, P.A. Jacobs *Tetrahedron: Asymmetry*, 8 (1997) 3481.

[53] D. Tas, C. Thoelen, I. Vankelecom, P.A. Jacobs, *Chem. Commun.* (1997) 2323.
[54] F. Joó, Á. Kathó, *J. Mol. Catal. A: Chemical* 116 (1997) 3.
[55] M. Felder, G. Giffels, C. Wandrey, *Tetrahedron: Asymmetry*, 8 (1997) 1975.
[56] D.E. Bergbreiter, *Chemtech* (1987) 686.
[57] D.E. Bergbreiter, Y.S. Liu, *Tetrahedron Lett.* 38 (1997) 3703.
[58] See e.g. I. Toth, B.E. Hanson, M.E. Davis, *Tetrahedron: Asymmetry* 1 (1990) 913.
[59] B. Pugin, *J. Mol. Catal.* 107 (1996) 273.
[60] D.J. Bayston, J.L. Fraser, M.R. Ashton, A.D. Baxter, M.E.C. Polywka, E. Moses, *J. Org. Chem.* 63 (1998) 3137.
[61] B.M. Cole, K.D. Shimizu, C.A. Krueger, J.P.A. Harrity, M.L. Snapper, A.H. Hoveyda, *Angew. Chem, Int. Ed. Engl.* 35 (1996) 1668.
[62] B. Pugin, M. Müller, *Stud. Surf. Sci. Catal.* 78 (1993) 107.
[63] C.E. Song, J.W. Yang, H.J. Ha, *Tetrahedron: Asymmetry* 8 (1997) 841.
[64] C.E. Song, J.W. Yang, H.J. Ha, S. Lee, *Tetrahedron: Asymmetry* 7 (1996) 645.
[65] For a comprehensive review, see M. Aglietto, E. Chiellini, S. D'Antone, G. Ruggeri, R. Solaro, *Pure & Appl. Chem.* 60 (1988) 415.
[66] J.V. Allen, S. Bergeron, M.J. Griffiths, S. Mukerjee, S.M. Roberts, N.M. Williamson, L.E. Wu *Chem Soc, Perkin Trans I* (1998) 3171 and references cited therein.
[67] J.R. Flisak, K.J. Gombatz, M.H. Holmes, A.A. Jarmas, I. Lantos, W.L. Mendelson, V.J. Novack, J.J. Remich, L. Snyder, *J. Org. Chem.* 58 (1993) 6247.
[68] For a review, see M. North, *Synlett* (1993) 807.
[69] K. Tanaka, A. Mori, S. Inoue, *J. Org. Chem.* 55 (1990) 181.
[70] H. Danda, *Synlett* (1991) 263.
[71] C.R. Noe, A. Weigand, S. Pirker, *Monatshefte für Chemie* 127 (1996) 1081.
[72] H.J. Kim, W.R. Jackson, *Tetrahedron: Asymmetry* 5 (1994) 1541.
[73] R. Hulst, Q.B. Broxterman, J. Kamphuis, F. Formaggio, M. Crisma, C. Toniolo, R.M. Kellog, *Tetrahedron: Asymmetry* 8 (1997) 1987.
[74] For reviews, see G. Wulff, *Angew. Chem. Int. Ed. Engl.* 34 (1995) 1812; K.J. Shea, *Trends in Polymer Sci* 2 (1994) 166, and references cited therein.
[75] M.E. Davis, A. Katz, W.R. Ahmad, *Chem. Mater.* 8 (1996) 1820.
[76] K. Morihara, S. Kawasaki, M. Kofuji, T. Shimada, *Bull. Chem. Soc. Jpn.* 66 (1993) 906 and references cited therein.

2 Catalyst Immobilization on Inorganic Supports

Ivo F. J. Vankelecom and Pierre A. Jacobs

2.1 Introduction

By far the most important class of enantioselective catalysts heterogenized on inorganic supports consists of chirally modified metallic catalysts. Among them are the only two systems that have yet reached a certain commercial importance. One is a nickel catalyst modified with chiral organic acids – mostly tartaric acid and with bromide as co-modifier – for the hydrogenation of mainly β-ketoesters [1]. The other one is a precious metal system – mostly Pt – modified with cinchona alkaloids, like cinchonidine. They hydrogenate α- and α,β-diketones [2].

As these above mentioned catalytic systems are covered in detail in Chapters 6, 7 and 8 of this book, this chapter will focus only on the remaining systems. In most cases, this means immobilized chiral transition metal complexes (TMCs), but also some chirally modified catalysts – those not included in the categories mentioned above – will receive attention, as well as some very specific cases. Whereas this chapter mainly focusses on the immobilization aspect, the specific reactions with these systems will be discussed in more detail in the following chapters.

2.2 General Considerations

In order to be of practical use, immobilized enantioselective catalysts should meet the following general requirements:

- The catalyst preparation should be simple, efficient and as generally applicable as possible.
- The performance of the immobilized catalyst should be comparable to (or better) than that of the free catalyst.
- Separating the heterogeneous chiral catalyst from the reaction mixture after reaction should be possible via a simple filtration in which more than 95% of the catalyst should be recovered.

- Leaching of the active species from the heterogenized enantioselective catalyst should be minimal.
- Reuse should be possible without loss of activity.
- The supports carrying the chiral catalyst should be mechanically, thermally and chemically stable. They should be compatible with the solvent and – by preference – commercially available in a good and reproducible quality.
- From an environmental (increasing disposal costs) and economical (cost of raw materials and of downstream separation) viewpoint, selectivity of the catalyst might sometimes become more important than its activity and lifetime.

In the field of enantioselective heterogeneous catalysis, it is clear that only one reaction type, i.e. hydrogenation, has reached some degree of maturity as yet. However, this concerns almost exclusively the Ni and Pt systems covered elsewhere in this book. Most of the heterogeneous catalysts discussed in this chapter still need further improvement or should be better understood before becoming industrially relevant. Nevertheless, some clearly promising new systems have been proposed already. Table 2.1 gives an overview of the catalysts that will be discussed here, classified chronologically per reaction type in which they were applied. Again, hydrogenation obviously dominates, but the other reactions have clearly gained momentum in the last years. It is striking that in Table 2.1 several authors claim heterogeneous systems without reporting about or even performing regeneration experiments, undoubtedly an important aspect of a heterogeneous system. Although these publications are also considered in this chapter, the reader should bear in mind this important shortcoming.

Finally, it should also be considered that comparing homogeneous catalysts with their heterogenized analogs is sometimes difficult as both reactions cannot always be performed under the same conditions: e.g. a TMC might be insoluble in the reaction mixture in which the heterogeneous catalysts performs best. Another problem might occur when TMCs are attached covalently to a support via a tether. It is then sometimes difficult to find out about the exact influence of this tether on the intrinsic catalytic activity or enantioselectivity of the derivatized species, especially if the tether is

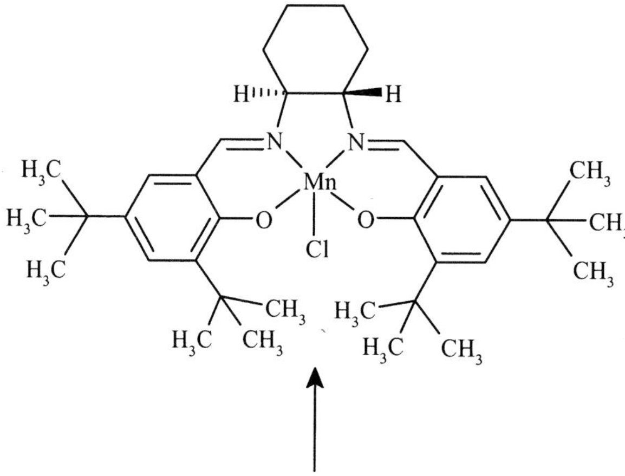

Figure 2.1. Prevented direction of olefin approach to the Mn-oxo bond in Jacobsen's catalyst.

attached to the complex close to the active site. This can be exemplified clearly with Jacobsen's catalyst (Fig. 2.1). Positions 3, 3′, 5 and 5′ of the complex are not only the positions of choice to derivatize the complex and subsequently link it to a support (cf. Chapter 10), but they are also crucial in inducing enantioselectivity in the reactions [3]. Indeed, the large substituents, like *tert.*-butyl groups, prevent the reacting olefins from approaching the active center via the direction (shown with an arrow in Fig. 2.1) that minimizes contact between the substrate and the chiral groups of the complex.

2.3 Supports

The inorganic supports to immobilize enantioselective catalysts are generally inert porous structures with highly specific surface area. Amorphous oxides, in particular silica and, to a lesser extent, alumina, zirconia or ZnO, are most routinely used. A broad range of such materials with a variation in pore size, pore size distribution and particle size are commercially available or can be synthesised easily [4].

Other commonly applied supports with a more defined structure are clay minerals, pillared clays and LDHs. Pillared clays contain stable metal oxide clusters which separate the layers that build the clay. A two-dimensional gallery is thus created with an opening that can be larger than 1 nm. LDHs, often denoted as hydrotalcite-like compounds, belong to the class of synthetic anionic clays. They show a positively charged layered structure with compensating anions between the sheets [5].

Other popular supports include zeolites. These crystalline materials, mostly aluminosilicates, have well-defined pores and channels in the micropore range (Fig. 2.2). The zeolites discussed in this chapter are zeolite β, zeolite Y and EMT. All have a three-dimensional pore system but with potentially different shape-selective properties. Zeolite β is a large pore high-silica zeolite with intersecting channels of 0.55 nm× 0.55 nm and

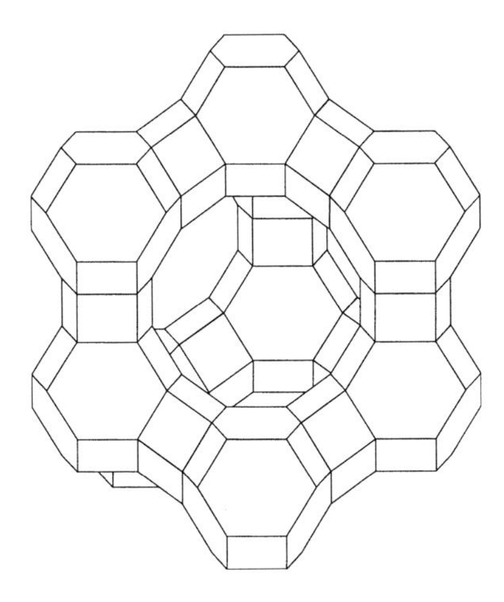

Figure 2.2. Schematic representation of a zeolite Y cage.

0.76 nm×0.64 nm [6]. The hypercages (1.3×1.3×1.4 nm) of EMT are accessible through three elliptical windows with free dimensions of 0.69×0.74 nm and two 0.74 nm circular apertures, while the hypocages miss these circular apertures. The pore structure of zeolite Y consists of almost spherical 1.3 nm cavities interconnected through smaller apertures of 0.74 nm [7]. Through steaming, zeolite Y can turn into a mesoporous structure, the so-called USY [8]. This changed porosity can be beneficial to prevent hindered mass transport and to enable reactivity towards more bulky molecules.

In contrast to the undefined mesoporosity of a USY zeolite, a rather recent type of regular mesoporous structures are the so-called M41S-materials: MCM-41 and MCM-48 possess a uni- and tri-dimensional pore system respectively with pore diameters varying between 1.5 nm and 100 nm [9].

2.4 Improved Activity of Heterogeneous Complexes

Immobilizing TMCs generates heterogeneous catalysts that are generally more complicated than the homogeneous catalysts. It is therefore not surprising that, in practice, the effect of immobilization is still very often unpredictable. Unfortunately, most heterogenized analogs of homogeneous catalysts are less active or lose part of their activity in every recycling cycle. However in some cases, the activities, chemo- or enantioselectivities of the heterogenized systems are superior to the ones of the homogeneous catalysts, as shown in Table 2.1.

Thomas et al. [10] state that an improved performance of a heterogeneous catalyst as compared with the homogeneous one, can sometimes be explained by the '*confinement concept*' (Fig. 2.3), in which the substrate's favorable interaction with both the pore wall and the chiral directing group is the key point. When compared to the

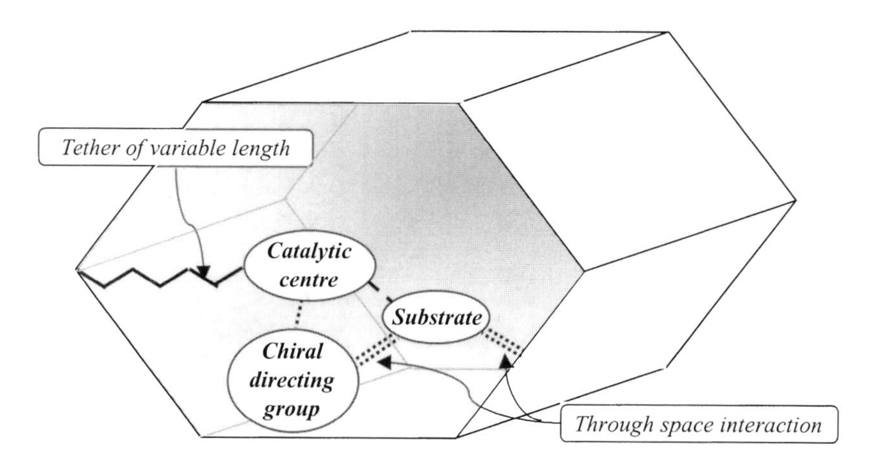

Figure 2.3. Schematic representation of the confinement concept (adapted from [10]).

Table 2.1. Overview of the reactions performed with enantioselective catalysts that were immobilized on inorganic supports.

	Substrate	Activity[a]	Ee (%)[a]	Re-use[b]	Type[c]	Remarks	Ref.
Hydrogenation							
1985	1-acetamidocinnamic acid/methylester 1-acetamidoacrylic acid Itaconic acid	+	+	–	Cov	Metal leaching	18
1986	1-acetamidocinnamic acid/methylester	–	≈	–	Cov	3rd use: decreased activity and ee due to catalyst oxidation	22
1989	1-acetamidocinnamic acid/methylester 1-acetamidoacrylic acid/methylester	+	–	+/–	Cov	No leaching (except in THF), increasing yields, and decreasing ee's upon reuse	17
1990	N-acetylamino-acrylic acid esters	≈	+	–	Ads	Minimized leaching if support preloaded with aniline	33
1991	N-acylphenylalanine derivatives	+	+	OK	Cov	USY better support than silica	19
1992	Ethyl (Z)-α-benzoylaminocinnamate	+	+	OK	Cov	USY better support than silica	20
1993	Methyl-1-acetamidocinnamate	=	=	OK	Cov	4–6% Rh lost in first reaction; <1% later	11
1994	Dehydro-naproxen	–	–	OK	SLP		55–58
1995	C=C, C=O, C=N double bonds and C≡C	=	=	OK	Cov		23–25
1996	Methyl (Z)-α-acetamido-cinnamate	+	+	OK	Trap	Only if surfactant added to homogeneous reaction: ee and activity equal to heterogeneous	48
	Methylacetamidecinnamate N-(2-methyl-6-ethylphen-1-yl)methoxymethylmethylketimin	=	=	+	Cov	Improvement on reuse	13
	Methyl acetoacetate	nr	nr	OK	Trap	No homogeneous reference reaction possible under heterogeneous reaction conditions	50
1997	Dimethylitaconate Geraniol	=	=	nm	Ads	No Ru leaching detected	40
	Methyl acetoacetate	–	–	OK	Trap	Co-heterogenization of additive (organic acid)	52

Table 2.1 (continued)

	Substrate	Activity[a]	Ee (%)[a]	Re-use[b]	Type[c]	Remarks	Ref.
1998	Methyl-2-acetamidoacrylate Dimethylitaconate	+	+	+	Ads	Improved performance on regeneration	43
	Methyl-2-acetamidoacrylate/cinnamate Dimethyl itaconate	+	+	+	Ads	Improved performance on regeneration	45
1999	Methyl-2-acetamidoacrylate Dimethyl itaconate	+	+	+	Ads	Improved performance on regeneration	44
	Geraniol Methylitaconate Methyl acetoacetate	+	+	nm	Ads	Small crystals of zeolite β in acid form are best supports + only for MAA	41
	Naphthylalanine precursor Itaconic acid Methylacetoacetate	+	+	–	Ads	Small crystals of zeolite β in acid form are best supports	42
Cyclopropanation							
1984	Styrene Indene	+	≈	OK	Cov	Styrene polymerization deactivates catalyst	16
1998	Stilbene + ethyldiazoacetate	–	–	OK	Ads		36
Addition of dialkylzinc to aldehyde							
1990	Benzaldehyde n-nonanal 2-naphtylaldehyde p-methyl-benzaldehyde	nm	nm	OK	Cov	Reuse after alkali treatment	26
1997	Benzaldehyde	nm	–	nm	Cov		27
Hydrolysis							
1992	Trans-stilbene oxide	nr	nr	nm	Sup	No homogeneous analogue available	66
	Amino acid esters	nm	nm	nm	Ads		34
Oxidation							
1993	Sulfides	–	=	–	Aux	Poisoning due to polymerization	64

Table 2.1 (continued)

	Substrate	Activity[a]	Ee (%)[a]	Re-use[b]	Type[c]	Remarks	Ref.
Dehydration							
1995	Butan-2-ol	nr	nr	nm	Aux		59–62
Norrish-Yang reaction							
1996	Ketones	nm	nm	nm	Aux	Max. ee 30%	63
Diels Alder							
1996	Cyclopentadiene + methacrolein or bromoacrolein	–		nm	Cov		29
Hydrogen transfer reduction							
1997	(o-methoxy)acetophenone + 2-propanol	–	+	nm	Cov		30
Dihydroxylation							
1997	Alkenes	nm	nm	nm	Cov	Leaching of Os	28
Aziridination							
1998	Alkenes	=	=	OK	Ads	Drying required before reusing catalyst	35
Epoxidation							
1996	Styrene	≈	≈	OK	Trap		50
	β-methylstyrene						
	1,3-cyclooctadiene						
1997	Styrene	–	=	nm	Caps		47
	β-methylstyrene						
	1,2-dihydronaphthalene	≈	≈	nm	Ads		38
	Trans-β-methylstyrene	–	–	–	Trap		54
	Alkenes	–	–	nm	Enc		46
1998	1,2-dihydronaphthalene	=	–	–	Ads	Gradual oxidation of salenligand	39
	Stilbene	+	–	–	Ads		37

Table 2.1 (continued)

	Substrate	Activity[a]	Ee (%)[a]	Re-use[b]	Type[c]	Remarks	Ref.
1999	1-phenyl-1-cyclohexene 1,3-cyclo-octadiene trans-β-methylstyrene 1,2-dihydronaphthalene	=	=	OK	Cov	Best results with PhIO	31
	Styrene derivatives and alkenes	+	+	–	Cov		32
Allylic amination							
1999	Cinnamyl acetate	+	+	nm	Cov		21
Allylation							
1998	Sodium dimethylmalonate + 2-butenyl-acetate	+	+	OK	Ads	Improved performance on regeneration	43, 45

[a] + = Heterogeneous reaction gives better results than the homogeneous one, even if this concerns sometimes only one substrate under specific conditions.
[b] OK = results are comparable to first run; + = better than in first run; – = worse than in first run.
[c] Type of heterogenization, following the classification in the text: cov = covalent attachment; ads = adsorption or ion pair formation; caps = encapsulation; trap = entrapment; SLP = supported liquid phase; aux = modification of an achiral heterogeneous catalyst with a chiral selector.
nr = not relevant.
nm = not mentioned.

situation in solution, this confinement of the substrate in small pores leads to a larger influence of the chiral directing group on the orientation of the substrate relative to the reactive catalytic center. Due to their very regular structure and tunable pore diameter, the MCM-type materials seem to be very interesting in this respect.

However, as will be shown later, this is a two-edged sword: in other cases, a tight caging of the immobilized catalyst molecules can cause lower enantioselectivities and activities. The lower activities can be due to reduced or even blocked mass transport in the small pores, imposing limits to the effective range of substrates that can be utilised [11]. The reduced enantioselectivity can be due to either a too strong physisorption of the complex on the wall of the support, or a restricted environment that prevents the chiral TMCs to take the spacial configuration that is required to induce chiral recognition. For Jacobsen's catalyst e. g., a crucial factor is its ability to assume a folded structure [12]. When attaching the complex too firmly to a solid, this possiblity is jeopardized, and the enantioselectivity decreases.

'*Site isolation*', i.e. attaching a catalyst to a support in such a way that the catalytic sites can no longer interact with each other, is another concept that might lead to better performing heterogeneous catalysts [13]. Its importance can be illustrated by work from Capka and co-workers [14, 15]. In this case, effective functional site isolation, and thus the formation of highly active species, was realized by matching three factors: a tight attachment of the ligand to the support, a low concentration of the surface ligands and an appropriate catalyst precursor.

Under homogeneous reaction conditions, the achiral Rh-phosphine complexes were found to be prone to formation of less active multinuclear complexes. Upon immobilization of such complexes on a silica particle, dimerization still occurred, leading to less active catalysts. The large motional freedom of the ligands on the silica surface allowed them to react as would a phosphine do, free in solution or attached to a polymer with low degree of cross-linking. However, the activity of the heterogeneous system could be increased drastically by decreasing this mobility of the ligand through reduction of the spacer length.

In general, the above mentioned low concentration of the ligands on the surface as a necessity to realize a well working catalyst, might sometimes be problematic. Indeed, the practical use of the catalyst can then be restricted due to the large amounts of support required to realize an envisaged conversion [13].

The aspect of an appropriate catalyst precursor can be well explained by distinguishing two different ways to functionalize (Fig. 2.4). When the whole complex was preformed in solution and subsequently reacted with the silica ('one-step functionalization'), formation of multinuclear Rh-species was observed, leading to the deactivation of the catalyst. A 'multi-step functionalization', in this case a two-step process in which a phosphinated silica was prepared first, and subsequently metallized with Rh in a next step, solved this problem. In this chapter, the one-step functionalizations are grouped in Fig. 2.5 and are represented by the molecule that was used to do the final immobilization. The other functionalizations are visualised as their final attachment to the inorganic support.

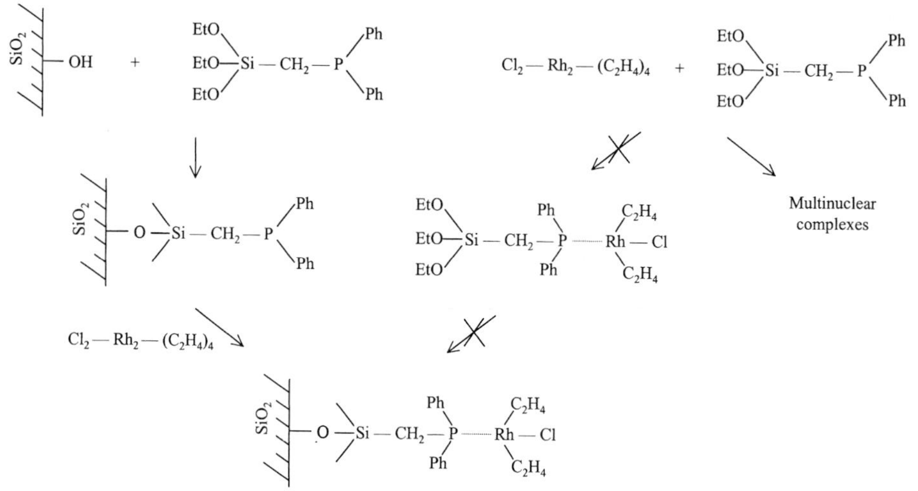

Figure 2.4. Attachment of a catalyst to a solid support, as exemplified by means of references 14 and 15: via multi-step functionalization (left side; in this case two-step) and the impossibility to use a one-step functionalization (right side).

2.5 Practical Examples

2.5.1 Covalent Attachment

Covalent attachment of the ligand to a solid support via a suitable bifunctional linker or tether is one of the most popular and versatile ways to heterogenize existing chiral TMCs. Both organic polymers (see Chapter 3) and inorganic solids are useful supports here, but the latter have the distinct advantage of being more robust materials.

The most important drawback of such heterogenized systems is the fact that the ligands have to be functionalized, which requires very often a large preparative effort, as both synthesis and subsequent purification can be tedious. An important advantage is that complexes attached via stable bonds experience no leaching from the support and form a stable catalyst, for as far as the metal ligation and all bonds are strong enough. In general, the point of attachment of the tether to the ligand should be as far as possible from the stereogenic center in order to disturb the chiral induction as little as possible.

An early example of a heterogeneous enantioselective catalyst that is more active than its homogeneous counterpart was given in 1984 by Matlin et al. [16] In a one step functionalization, the $SiCl_3$-derivative of a β-diketone, camphor more in particular, was reacted with the silanol groups of a silica (Fig. 2.5a). The Cu-form of this complex was used in the cyclopropanation of styrene. In a first run, the reaction proceeded 2.4 times faster than the corresponding homogeneous reaction, but due to the formation of a polystyrene coating around the silica, recycling failed. When this polymerization was absent, like in the cyclopropanation of indene, the authors claimed

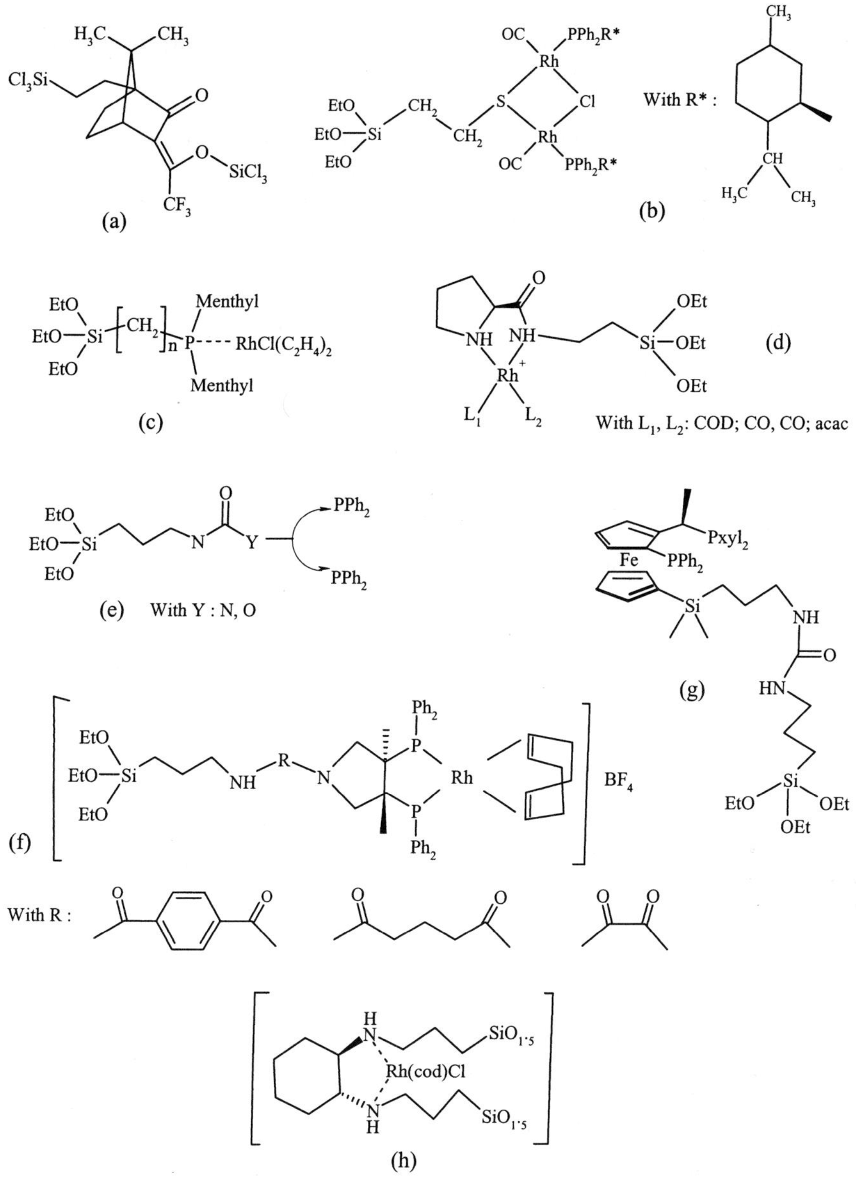

Figure 2.5. Catalysts or ligands covalently attached to the inorganic supports via a one-step functionalization.

continued activity (but no exact numbers were reported). The Ni-form proved to be even slightly more active than the Cu-form and was claimed to maintain its activity for at least 3 cycles.

Other heterogeneous catalysts with increased stability as compared to their homogeneous forms were reported by Eisen et al. [17] and Kinting et al. [18]. In the former (Fig. 2.5b), trimethoxysilyl-functionalized di-Rh-complexes were reacted with silica. Maximum activity was reached only after the first run. The heterogeneous catalysts proved to be thermally much more stable than the homogeneous analogs, and leaching was absent in all solvents tested. Kinting et al. [18] (Fig. 2.5c) observed a marked decrease in activity upon storing the original homogeneous complex. This was attributed to isomerization of the complexes or to the formation of multinuclear complexes. Comparable to Capka et al. [14, 15], a one-step functionalization was not feasible here due to this instability of the homogeneous complex. Instead, the silica was first phosphinated by reaction with (1-dimenthylphosphinomethyl)triethoxysilane and subsequently contacted with the rhodium-ethylene complex. The heterogenized catalyst thus formed kept its activity even after being stored for several weeks. However, catalyst leaching was encountered upon recycling. This could be minimized by using longer spacer molecules, but could never be avoided completely (a Rh-loss of 28% after 3 cycles was the best result reached). On the other hand, the use of these longer, more flexible spacer molecules entailed the formation of Rh-complexes bound bidentately to the surface.

Corma and co-workers anchored Rh-complexes (Fig. 2.5d) on silica and on USY [19, 20]. The induction period needed in the homogeneous hydrogenation of α-acylaminocinnamate derivatives was increased for the silica supported catalysts, but completely absent when using the USY-support. Either a concentration effect of the zeolite or the interaction of the substrate with the electrostatic fields in the zeolite were given as tentative explanations. The observed increase in e.e. was ascribed to the confined spaces of the zeolite pores.

A positive control effected by the surrounding support was also reported by Johnson et al. [21] where MCM-41 led to much better results than a silica support. An anchored ferrocene was coordinated to Pd (Fig. 2.6a) and showed an extremely high e.e. and conversion in the allylic amination of cinnamyl acetate, simultaneously changing drastically the regioselectivity of the reaction.

An example where confinement did not improve the catalytic performance was found in the immobilization of diphosphines on silica, reported by Pugin and Muller (Fig. 2.5e) [11]. In the large pore silicas (10–19 nm pore diameter), the unchanged activity of the immobilized complexes as compared with the homogeneous analogs, was attributed to the fact that the complexes carried a charge so that the formation of less active dimers, occurring when neutral complexes were densely loaded on a solid support, remained absent. For the silica with the smallest pores (<4 nm), however, a lower enantioselectivity was observed. This was ascribed to the tight caging of the immobilized catalyst molecules, while the reduced activity was ascribed to reduced or even blocked mass transport. Typically between 4 and 6% of the Rh was found in the reaction mixture after a first reaction, an amount that decreased to less than 1% in the regeneration experiments.

Nagel et al. [22] also immobilized a silylated Rh-phosphine on silica (Fig. 2.5f). The e.e. obtained with the heterogenized complexes increased as the link between the chiral site and the silica wall was longer (realized by means of the three different R-groups given in Fig. 2.5 f), finally reaching optical purities comparable with the

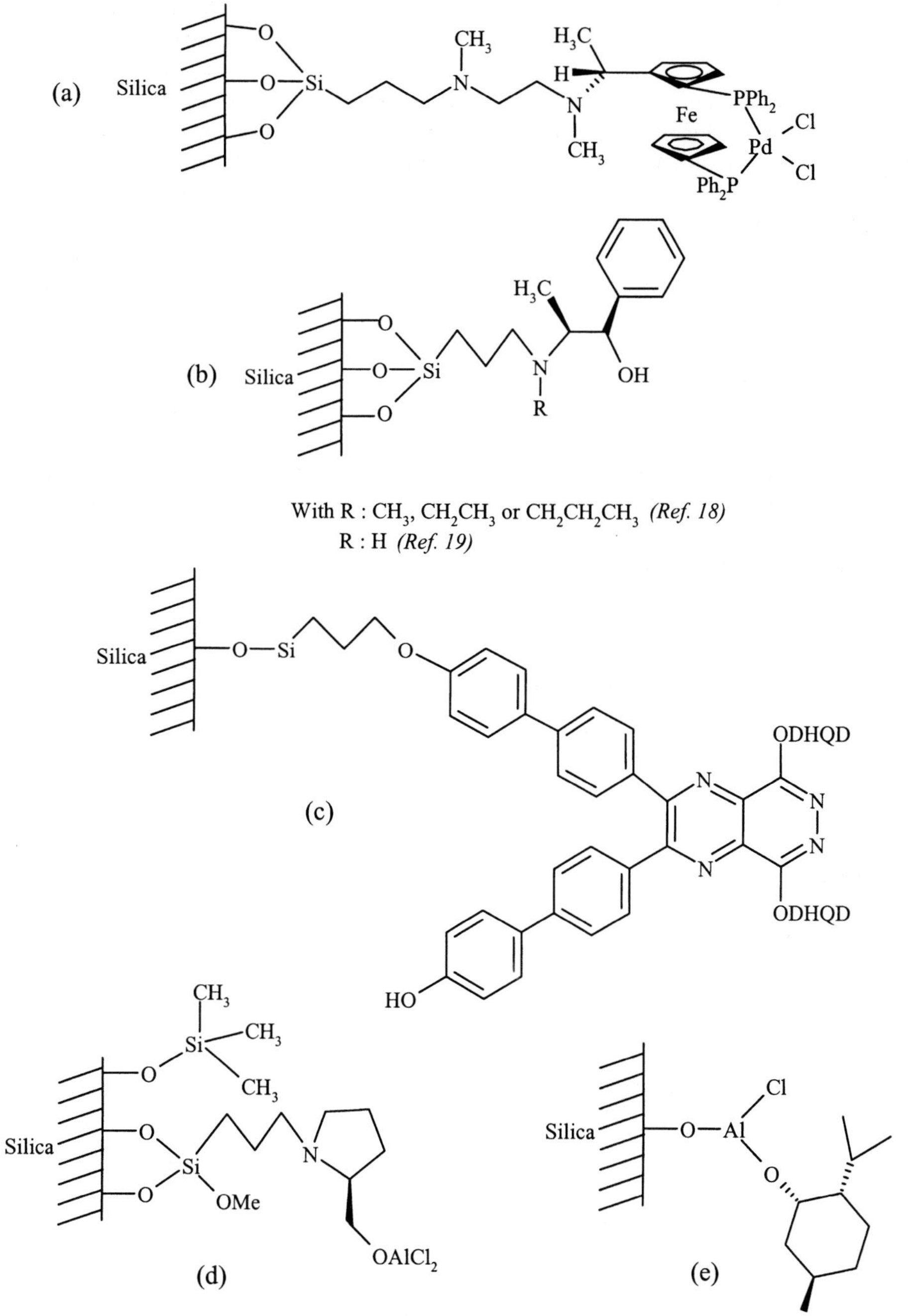

With R : CH_3, CH_2CH_3 or $CH_2CH_2CH_3$ *(Ref. 18)*
R : H *(Ref. 19)*

Figure 2.6. Catalysts or ligands covalently attached to the inorganic supports via a multi-step functionalization.

homogeneous reaction. However, the conversions were 4 times lower. The systems could be reused 3 times, but activity and e.e. decreased, possibly due to oxidation of the complex.

Ciba-Geigy (currently Solvias) patented some highly active ferrocenyldiphosphine complexes that were first functionalized with suitable silylating agents before attaching them to a variety of supports (Fig. 2.5 g) [23–25]. An example was given of a silica-bound ligand of which the Ir-form was reported to be a perfectly regenerable catalyst in the hydrogenation of imines.

Chiral N-alkylnorephedrines were bound to silylated silica and alumina in a two-step functionalization (Fig. 2.6 b) [26], but a regeneration (after filtration and treatment with alkali) was only mentioned for a silica coated with a layer of Cl-methylated polystyrene. Bellocq et al. [27] found that ephedrine (where the alkyl in N-alkylnorephedrine is a methyl group) covalently bound to mesoporous MCM-41 induced higher e.e.'s than when bound to silica.

Alkaloid ligands were attached to either chloropropyl, aminopropyl or diol functionalized silica using acylation or nucleophilic substitution (Fig. 2.6 c) [28]. Enantioselectivities and diol yields were excellent in the asymmetric dihydroxylation of alkenes. However, when using the amino-linked ligands, enantioselectivities dropped due to the formation of racemic diol groups. Ester hydrolysis under the basic reaction conditions prevented the recovery of those ligands which were linked to the silica via an ester bond. Consecutive reactions without loss of activity were only possible after addition of Os after each run.

Fraile et al. [29] studied the Diels-Alder reactions catalyzed by chiral Lewis acids immobilized on alumina and silica. Tyrosine or (S)-prolinol were used as chiral auxiliaries, the latter resulting in slightly better results due to its decreased flexibility (Fig. 2.6 d). When these chiral auxiliaries were grafted on the support, the catalytic activity with regard to the analogous homogeneous system increased, but the asymmetric induction decreased drastically. End capping of the unreacted surface silanol groups with HMDS could not improve the asymmetric induction, indicating that grafting of the chiral auxiliary on the solid support modified the structure of the chiral Lewis acid and that of its complex with the dienophile. A better strategy was found in using the metal, and not the chiral auxiliary – in this case (–)-menthol – to graft the Lewis acid to the solid (Fig. 2.6 e). This influenced less the conformation of the chiral auxiliary and hence the stereochemical reaction course.

A Rh-diamine complex (Fig. 2.6 i) immobilized at the surface of a silica gel was compared with a hybrid silsesquioxane solid in which the same chiral moieties were part of the hybrid network [30]. The latter was found to be both much more active and enantioselective in the hydrogen transfer reduction of acetophenone with propan-2-ol. Regarding e.e., the hybrid solid even performed better than the homogeneous analog, while showing a slightly reduced activity.

Salen complexes, among them Jacobsen's catalyst, have been immobilized on silica via three different approaches (Fig. 2.7 a). In a first one, a covalent link was established between the silica support carrying Si-H-groups and the vinylfunctionalized ligand. The choice of oxidant was crucial. NaOCl resulted in the formation of a considerable amount of chlorinated by-products, while m-CPBA resulted in only moderate enantioselectivities. Best results were obtained using PhIO. Under these conditions,

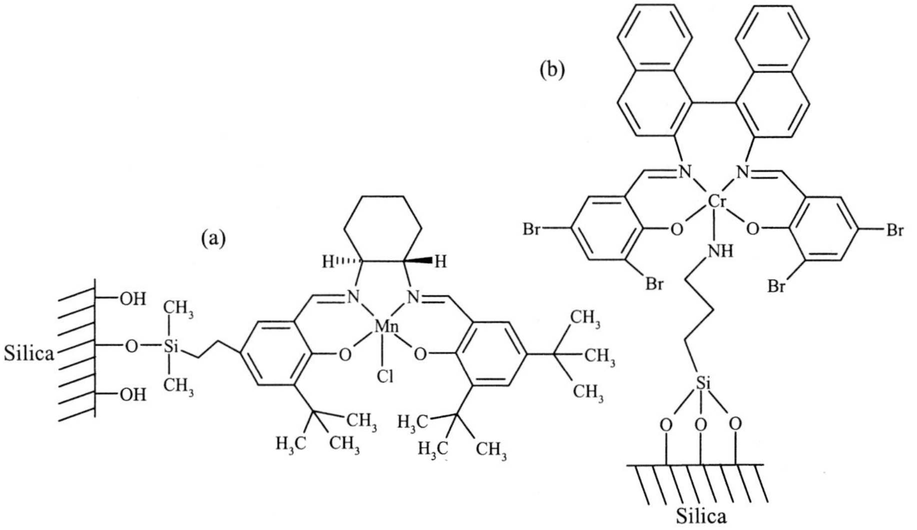

Figure 2.7. Different methods for covalent linking of salen complexes.

the heterogeneous catalyst showed increased chemo- and enantioselectivities. Activities – although slightly decreasing after regeneration – were comparable to those of the homogeneous catalyst [31].

Secondly, the same heterogeneous catalyst was prepared via a 1-step functionalization. The silica was reacted with a Si-Cl functionalized Mn-salen complex obtained after reacting the vinylderivatized ligand with dimethylchlorosilane. In a comparison with the first appoach, the resulting catalyst was found to be less efficient [31].

In the third approach (Fig. 2.7 b), an unmodified Cr-binaphthyl Schiff base was coordinated to an amine ligand that was previously attached covalently to the walls of an MCM-41 support [32]. After 4 regeneration steps, both activity and e.e. had decreased a lot, even though only 3% of Cr had leached into solution. Here again, the e.e. of the heterogenized system was higher than that of the homogeneous analog. An enhanced stability of the Cr-complex upon immobilization or a unique spatial environment constituted by both the chiral binaphtyl Schiff base ligand and the surface of the support, were given as tentative explanations.

2.5.2 Adsorption or Ion-Pair Formation

The easy preparation of these systems is a clear advantage, but their sensitivity to solvent effects is an important drawback.

Charges were created on silica via reaction with a SO_3^--terminated silylating agent (Fig. 2.8 a). The positively charged Rh(I)-complexes were then ion-exchanged and used in hydrogenations, where leaching could be avoided in toluene and water but not in methanol [33].

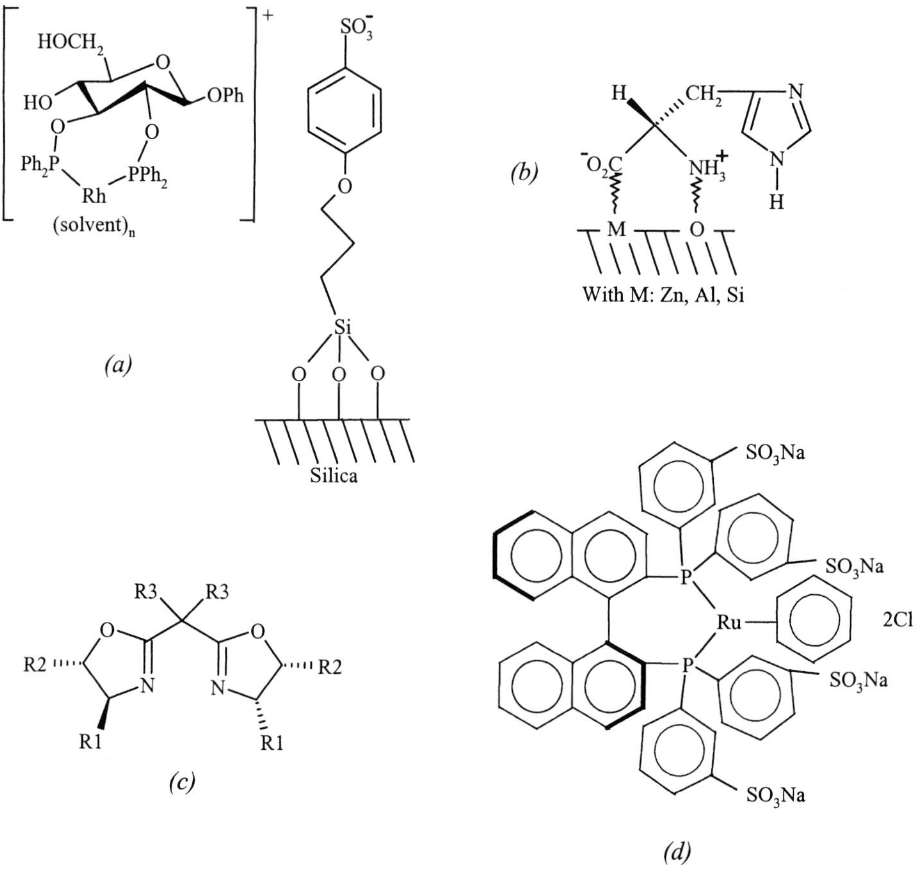

Figure 2.8. Enantioselective catalysts immobilized via adsorption or ion-pair formation.

Histidine was adsorbed on ZnO, Al_2O_3 and SiO_2 by simple dry grinding (Fig. 2.8b) [34]. E.e.'s of up to 20 and 37 respectively in alcoholysis and hydrolysis were reported, but lacking regeneration experiments, the stability of the adsorbed moieties on the surfaces cannot be evaluated.

Copper-exchanged zeolite Y was modified with chiral bis(oxazolines) (Fig. 2.8c) [35]. Remarkably, it was found that very low levels of the chiral modifier (less than one per supercage and in a Cu:modifier ratio of 2:1) sufficed already to have optimal yields and enantioselectivities. Excessive modifier even decreased yields, since it blocked the zeolite pores. It was further proven by using bulky substrates, such as *trans*-stilbene, or bulky ligands, like bis(oxazolines), that the reaction actually took place inside the zeolite pores. With continued use, some activity of these catalysts was lost unless the catalyst was dried before reuse. A *tert.-butyl* substituted chiral bis(oxazoline) apparently did not induce enantioselectivity. This was ascribed to the fact that acetonitrile binds more strongly to the Cu-ions than this particular bis(oxazoline), hence exemplifying the solvent sensitivity of such immobilized systems. Similar Cu-bis(oxazoline) complexes were also immobilized on different clays [36]. Only laponite, exchanged in the appropriate solvents and combined with those bis(oxazoline)complexes that give the strongest complex-support interactions, could be reused with comparable e.e.'s and activities.

In addition to the ways mentioned already above to heterogenize chiral salen complexes via a covalent link, Piaggio et al. [37] decribed an immobilization method for these complexes based on ion-exchange. Salen ligands were contacted with a Mn-exchanged Al-MCM-41, in such a way that 10% of the Mn was ligated. The turnover frequency of the heterogeneous catalyst was reported to be higher than the one of the homogeneous reaction, even though slightly less chemo- and enantioselective. The reaction filtrate proved inactive, as some salen leaching (but no Mn leaching) was observed. The activity as well as the enantioselectivity of the reused catalyst dropped significantly. The catalytic performance could only be fully restored after recalcining the used catalyst and adding new salen ligand.

Alternatively, tert.-butyl-salen complexes were immobilized by simple impregnation on Al-MCM-like structures [38]. Guest/host interactions, mainly between the aromatic rings of the complex and the internal surface silanol groups of the walls of the mesopores, were found to be strong enough to prevent leaching of the complex. The epoxidation of 1,2-dihydronaphthalene proceeded with exactly the same activity and enantioselectivity (but lower yields) as the homogeneous reference reaction. No complex leaching was observed, but regeneration experiments were not reported.

In a third approach, the *tert.*-butyl substituted salen complex was ion-exchanged with a Na-laponite clay and used in the same epoxidation [39]. Showing similar activity as the homogeneous complex, the e.e. was slightly reduced when using the clay supported complex. Since addition of pyridine as axial ligand did not change the e.e. of the heterogeneous reaction, it was speculated that the clay, itself, acted as an axial ligand. In contrast to the absence of leaching, regeneration experiments showed decreased activities and e.e. A gradual decomposition of the salen ligand – probably by oxidation – was postulated as the main reason, in addition to the formation of coke when applying certain reactants.

Layered double hydroxides (LDHs) were used to retain a sulfonated BINAP-complex – bearing 4 negative charges – via Coulombic interactions (Fig. 2.8d) [40]. Hydrogenation of dimethyl itaconate (DMI) proceeded with the same rate and e.e. as the homogeneous reaction. Highly specific interactions were found to be essential: similar ion exchange on organic resins resulted in inactive catalysts, alternative exchange methods failed and the use of a Zn,Al-NO_3 LDH strongly decreased e.e. in the hydrogenation of DMI. The importance of the surface properties of the support was confirmed when the unmodified Ru-BINAP complex was impregnated on zeolites [41, 42]. Only one type of zeolite β, with small particle size so that a monolayer coverage of the particles was created, resulted in a catalyst with good activity for the hydrogenation of MAA. Moreover, this heterogeneous catalyst was active already at much lower temperatures than those normally required for this type of hydrogenation. This remarkable activity of the catalyst was attributed to the presence of acidic sites on the outer surface of the zeolite. Via molecular graphics, a 'plug-socket'-type interaction between the BINAP complex and the zeolite pore mouths was proposed.

Surprising results were reported by Augustine et al. [43–45] These authors used heteropolyacids as a tethering agent for a whole set of support/catalyst combinations. The concept is claimed to be generally applicable as proved by the variety of systems prepared: montmorillonite, carbon, alumina and lanthana were used as supports, while P/W, Si/W and Si/Mo heteropolyacids were applied as tether and DIPAMP, BPPM, dppb, Wilkinson's catalyst, ProPHOS, Me-DUPHOS and BINAP were used as catalysts in chiral allylations and hydrogenations. The preparation of the heterogeneous catalysts simply involves mixing of the support with the heteropolyacid prior to adding a solution of the transition metal complex. None of the systems reported showed any leaching even after 15 cycles, despite the fact that the ligands were not modified. Moreover, both activity and enantioselectivity improved drastically in the first reuse and remained at these high levels from then onwards. Based on the observed color change, this was tentatively explained for the hydrogenation with Rh(DIPAMP)-(P/W)heteropolyacid-montmorillonite by some partial reduction of the tungsten in the heteropolyacid. Lanthana was the only support that did not show this improvement, which was ascribed to its basic nature.

2.5.3 Encapsulation

Encapsulating a TMC in the cages of a zeolite, for example, fully depends on the respective sizes of the complex and the cage in which it is held sterically to prevent leaching during reaction. Small and regular pores or cages are essential requirements for the support, leaving zeolites as excellent candidates for this 'ship in a bottle' approach. When applied for reactions with large molecules, encapsulation often leads to a strong resistance towards diffusion of substrates and products.

The encapsulation of chiral salen complexes in zeolites was described simultaneously by two groups, respectively using zeolite Y and EMT zeolite as a host.

Applying zeolite Y, only the smaller salen complex without *tert.*-butyl groups could be fitted into the cages [46]. The low reaction rates for the epoxidation of alkenes were explained in view of the restrictions imposed on the diffusion of substrate and products through the micropores of the solid. The slightly reduced e.e.'s were interpreted as a combination of a non-catalyzed, non-selective epoxidation reaction in the liquid phase and/or the existence of residual amounts of non-complexed Mn^{2+} acting as catalytic sites.

The larger cages of the EMT-structure, however, did allow the accommodation of a larger complex carrying methyl groups on positions 5 and 5′, and *tert.*-butyl groups on 3 and 3′ [47]. Here too, activities of the heterogeneous catalysts were lower, but the enantioselectivities were unchanged. Addition of pyridine N-oxide to the zeolites increased both activity and enantioselectivity significantly. The lower conversions, realized with the sterically more encumbered complex upon using a bulky oxidant like PhIO instead of NaOCl, suggested hindered access.

For all three systems, intrazeolite reaction and absence of leaching was proven, but no regeneration experiments were reported.

2.5.4 Entrapment

A catalytic Rh-complex was immobilized on silica and alumina via entrapment in a hydrophobic surface layer [48]. For this purpose, the silica and alumina supports were treated with surfactant which was bound either covalently (Fig. 2.9 a) or ionically. When comparing the activity and enantioselectivity of the immobilized catalyst with the homogeneous one, the homogeneous reactions gave lower activities and e.e.'s, unless surfactant was added. In that case, hardly any difference remained with the best heterogeneous systems. Recycling experiments kept the enantioselectivity constant, but a steady, though minor, decrease in activity occurred with subsequent regeneration cycles, despite the fact that hardly any leached Rh was found.

With an entirely inorganic backbone and only short organic side chains (Fig. 2.9 b), polydimethylsiloxane (PDMS) is situated on the borderline between organic and inorganic supports. The technique to embed TMCs in PDMS by simply mixing this TMC with the polymer and subsequent curing of the casted film [49], was first reported in 1996 for Ru-BINAP and Jacobsen's catalyst [50]. Later, the co-incorporation of an acid further improved the performance of the Ru-BINAP/PDMS membrane and simultaneously heterogenized the corrosive additive [51, 52]. Shaped as a membrane, these catalyst containing polymers were placed between two reagent phases, leaving solvents redundant and optimizing mass transport [53]. Leaching of the complexes often

(a) Silica–$(O)_3Si-CH_2-CH_2-CH_2-NH-C(=O)-NH-[CH_2]_p-C(=O)-O-[CH_2]_n-SO_3^- \; Na^+$

With n = 11 and p = 1 or 3

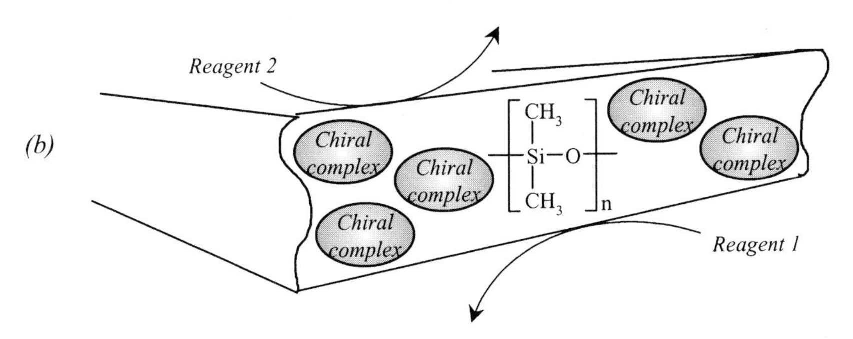

Figure 2.9. Enantioselective catalysts immobilized via entrapment in (a) a hydrophobic surface layer and (b) a membrane.

creates a problem [52, 54] which can be reduced by appropriately choosing solvents and reagents, thereby restricting swelling of the PDMS films. Another way to reduce leaching, leaving a broader choice in reaction conditions but demanding more preparative efforts, is to link the complexes covalently to the polymer chains [31] or to increase the size of the TMCs, e. g. by preparing dimers as reported for Jacobsen's catalyst [54].

2.5.5 Supported Liquid Phase (SLP)

A genuine hybrid of homogeneous and heterogeneous catalyst types is formed by an SLP where a solution of the catalyst is attached to the surface of a porous support and contacted with a solvent phase that contains the reagent. This triphasic system imposes serious solvent restrictions on reagents and catalyst, and mass transfer between the two liquid phases might be problematic due to the generally small contact area.

One of the best known heterogeneous enantioselective catalysts of this kind is the SAP (supported aqueous phase)-Ru-BINAP-4(SO_3Na) system (Fig. 2.10) developed by Wan and Davis [55–57]. Starting from a sulfonated Ru-BINAP with activities in

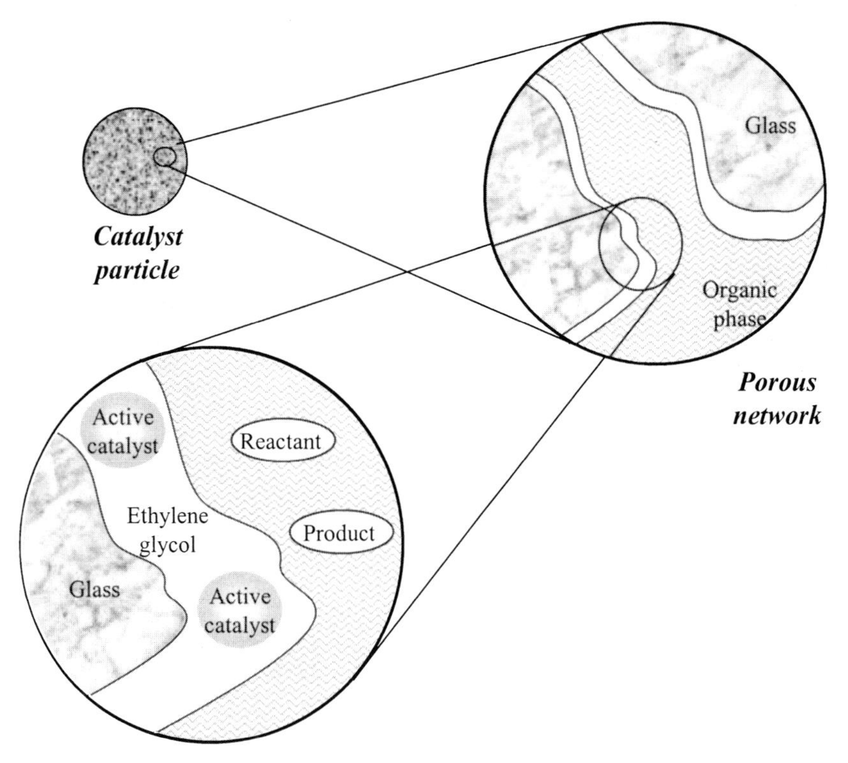

Figure 2.10. Supported liquid phase catalyst containing sulfonated Ru-BINAP, as developed by Wan and Davis (adapted from [55]).

methanol comparable to the non-sulfonated analog (TOF = 131 h^{-1}; 96% e.e.), biphasic hydrogenation of the naproxen precursor in water/ethyl acetate resulted in a very slow reaction due to the low solubility of the substrate in the water phase. The activity increased 50 times upon using a SAP-catalyst (TOF = 18.2 h^{-1}; 70% e.e.), thanks to the increased interfacial surface area. The e.e. was limited by the presence of water which caused the loss of the Cl-ligand of the complex through aquation of the Ru-Cl bond. On the other hand, a certain amount of water was needed in the SAP-system in order to give the complex some rotational mobility, crucial to reach activity. Ru-analysis and the absence of reactivity in the filtrates proved that the system was really heterogeneous. The SAP-catalyst was later improved by changing the water film on the glass beads for an ethylene glycol film (SLP): the TOF tripled compared to the earlier reported SAP and remained only 2-2.5 times below the TOF of the homogeneous analog [58]. An e.e. of 87.7%, equaling the one obtained under homogeneous conditions, was reached. The Ru-leaching, observed due to significant solubility of ethylene glycol in the organic phase, was eliminated by using a cyclohexane/chloroform mixture as organic phase. Enantioselectivities remained comparable to those of the homogeneous systems, and activity decreased only slightly compared to the earlier SLP.

2.5.6 Modification of an Achiral Heterogeneous Catalyst with a Chiral Auxiliary

The most known and successful applications in this area are the Ni-tartrate and the Pt-cinchonidine systems covered elsewhere in this book.

An interesting concept has been developed by Hutchings et al. [59] and applied in the first example of a heterogeneous enantioselective gas phase reaction. A chiral acidic zeolite was created by loading one molecule of *R*-1,3-dithiane-1-oxide (Fig. 2.11 a) per supercage of zeolite Y, either during, or after the zeolite synthesis [60]. In competing experiments, the dehydration of *S*-butan-2-ol was found to be up to 39 times faster than that of the *R*-enantiomer, much higher than what could be expected from the conversions of the separate enantiomers. Using computational simulation methods, this was attributed to a higher binding energy of the *S*-enantiomer on the catalytic site [61]. Later, a specific interaction was suggested between the dithiane oxide and both the extra-framework aluminium and the Brønsted acid site associated with the framework aluminium (Fig. 2.11 b) [62]. A similar effect was observed with *S*-2-phenyl-1,3-dithiane 1-oxide (*R* = phenyl in Fig. 2.11 b) as modifier.

Other chiral zeolites were formed by adsorbing ephedrine as a modifier on zeolites X and Y for the Norrish-Yang reaction [63]. Interestingly, sorbing (–)-ephedrine in NaY showed enantioselectivity in favor of the (+)-isomer, whereas the same chiral inductor favored the (–)-isomer in NaX. It was speculated that this effect was caused by the difference in supercage free volume between NaY and NaX and the number of type III cations. This interpretation was supported by results obtained with X-zeolites exchanged with different alkali metals.

A variety of Ti-catalysts, combined with various chiral auxiliaries and supported on alumina, silica, zirconia, clays and pillared clays, was used in the oxidation of sulfides

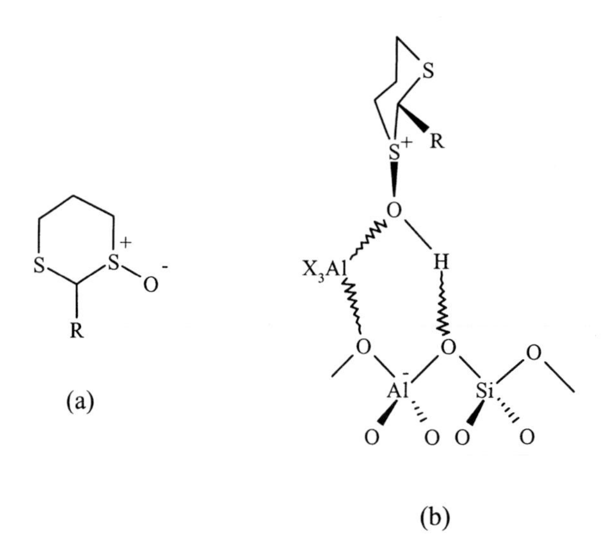

Figure 2.11. *R*-1,3-dithiane-1-oxide and coordination of the chiral auxiliary in zeolite Y (adapted from [62]).

[64]. The best results, approaching the ones obtained under homogeneous conditions, were realized with the pillared clays (based on montmorillonite). This was ascribed to the ease of complexation – similar to the homogeneous analog – of the titanium hydroxy oligomer with the reactant, oxidant and chiral auxiliary in the available interlayer space ($\approx$ 13 Å), and to the presence of Lewis acidity in the pillared clay. This hypothesis was confirmed by the inactivity observed when using larger auxiliaries.

2.5.7 Achiral Metal Catalysts on Chiral Supports

These systems are only mentioned for their historical meaning and for sake of completeness. Because most lack reproducibility and realize only low e.e.'s, their potential is very limited. Already in 1932, Pt and Pd were deposited on the enantiomers of quartz single crystals, and e.e.'s of 10% were reached in hydrogenations. Later, this was also done with natural fibers such as silk [65].

Some degree of success in catalysis was also accomplished by using the chiral polymorph of zeolite β [66]. This polymorph contains helical pores which are either left- or right-handed. By applying a chiral template in the zeolite synthesis, a sample could be prepared that was enriched in this chiral polymorph. Very low but still significant e.e.'s of 5% were reported for the hydrolysis of *trans*-stilbene oxide.

Acknowledgements

I.F.J.V. thanks the Fund of Scientific Research (F.W.O.) for a grant as a Post-Doctoral Researcher.

References

[1] Webb, G., Wells, P.B., *Catal. Today,* 12 (1992) 319.
[2] Wells, P.B., Wilkinson, A.G., *Topics in Catalysis*, 5 (1998) 39.
[3] Katsuki, T., *J. Mol. Catal. A: Chem.*, 113 (1996) 87.
[4] Brinker, C.J., Scherer, G.W., *Sol-gel Science, the Physics and Chemistry of Sol-gel Processing*, Academic Press, 1990.
[5] Cavani, F., Trifiro, F., Vaccari, A., *Catal. Today,* 11 (1991) 173.
[6] http://www.iza-sc.ethz.ch/IZA-SC/Atlas/data/BEA.html.
[7] Thomas, J.M., Ramdas, S., Millward, G.R., Klinowski, J., Audier, M., Conzalec-Calbet, J., Fyfe, C.A., *J. Solid State Chem.*, 45 (1982) 368.
[8] Corma, A., *Chem. Rev.*, 97 (1997) 2373.
[9] Zhao, X.S., Lu, G.Q., Millar, G.J., *Ind. Eng. Chem. Res.*, 35 (1996) 2075.
[10] Thomas, J.M., Maschmeyer, T., Johnson, B.F.G., Shepard, D.S., *J. Mol. Catal. A: Chem.* 141 (1999) 139.
[11] Pugin, B., Müller, M., in *Stud. Surf. Sci. Cat.* Part 78: Heterogeneous Catalysis and Fine Chemicals III, Ed. Guisnet, M., Barbier, J., Barrault, J., Bouchoule, C., Duprez, D., Pérot, G., Montassier, C., Elsevier (1993) 107.
[12] Ito, Y.N., Katsuki, T., *Tetrahedron Lett.*, 39 (1998) 4325.
[13] Pugin, B., *J. Mol. Catal. A: Chem.* 107 (1996) 273.
[14] Czakova, M., Capka, M., *J. Mol. Catal.*, 11(1981) 313.
[15] Michalska, Z.M., Capka, M., Stoch, J., *J. Mol. Catal.*, 11 (1981) 323.
[16] Matlin, S.A., Lough, W.J., Chan, L., Abram, D.M.H., Zhou, Z., *J. Chem. Soc., Chem. Commun.* (1994) 1038.
[17] Eisen, M., Blum, J., Schumann, H., Gorella, B., *J. Mol. Catal.*, 56 (1989) 329.
[18] Kinting, A., Krause, H., Capka, M., *J. Mol. Catal.*, 33 (1985) 215.
[19] Corma, A., Iglesias, M., del Pino, C., Sanchez, F., *J. Chem. Soc., Chem. Commun.* (1991) 1253.
[20] Corma, A., Iglesias, M., del Pino, C., Sanchez, F., *J. Organometallic Chem.*, 431 (1992) 233.
[21] Johnson, B.F.G., Raynor, S.A., Shephard, D.S., Maschmeyer, T., Thomas, J.M., Sankar, G. Bromley, S., Oldroyd, R., Gladden, L., Mantle, M.D. *Chem. Commun.* (1999) 1167.
[22] Nagel, U., Kinzel, E., *J. Chem. Soc., Chem. Commun.* (1986) 1098.
[23] Pugin, B., *WO 9632400 A1* (1996).
[24] Pugin, B., *WO 9702232 A1* (1997).
[25] Pugin, B., *EP 729969 A1* (1996).
[26] Soai, K., Watanabe, M., Yamamoto, A., *J. Org. Chem.*, 55 (1990) 4832.
[27] Bellocq, N., Brunel, D., Lasperas, M., Moreau, P., in *Stud. Surf. Sci. Cat.* Part 108: Heterogeneous Catalysis and Fine Chemicals IV, Ed. Blaser, H.U., Baiker, A., Prins, R., Elsevier (1997) 485.
[28] Bolm, C., Maischak, A., Gerlach, A., *Chem. Commun.* (1997) 2353.
[29] Fraile, J.M., Garcia, J.I., Mayoral, J.A., Royo, A.J., *Tetrahedron Asymm.* Vol. 7, No. 8 (1996) 2263.
[30] Adima, A., Moreau, J.J.E., Man, M.W.C., *J. Mater. Chem.*, 7, 12 (1997) 2331.
[31] Janssen, K., *PhD thesis*, K.U.Leuven, Belgium (1999).
[32] Zhou, X.-G., Yu, X.-Q., Huang, J.-S., Li, S.-G., Li, L.-S., Che, C.-M., *Chem. Commun.* (1999) 1789.
[33] Selke, R., Capka, M., *J. Mol. Catal.*, 63 (1990) 319.
[34] Moriguchi, T., Guo, Y.G., Yamamoto, S., Matsubara, Y., Yoshihara, M., Maeshima, T., *Chemistry Express*, Vol. 7, No. 8 (1992) 625.
[35] Langham, C., Piaggio, P., Bethell, D., Lee, D.F., McMorn, P., Bulman Page, P.C., Willock, D.J., Sly, C., Hancock, F.E., King, F., Hutchings, G.J., *Chem. Commun.* (1998) 1601.
[36] Fraile, J.M., Garcia, J.I., Mayoral, J.A., Tarnai, T., *Tetrahedron Asymm.*, 9 (1998) 3997.
[37] Piaggio, P., McMorn, P., Langham, C., Bethell, D., Bulman Page, P.C., Hancock, F.E., Hutchings, G.J., *New J. Chem.* (1998) 1167.

[38] Frunza, L., Kosslick, H., Landmesser, H., Höft, E., Fricke, R., *J. Mol. Catal. A: Chem.*, 123 (1997) 179.
[39] Fraille, J.M., Garcia, J.I., Massam, J., Mayoral, J.A., *J. Mol. Catal. A: Chem.* 136 (1998) 47.
[40] Tas, D., Jeanmart, D., Parton, R.F., Jacobs, P.A., in *Stud. Surf. Sci. Cat.* Part 108: Heterogeneous Catalysis and Fine Chemicals IV, Ed. Blaser, H.U., Baiker, A., Prins, R., Elsevier (1997) 493.
[41] Tas, D., *PhD thesis,* K.U.Leuven, Belgium (1997).
[42] Van Brussel, W., Renard, M., Tas, D., Parton, R., Jacobs, P.A., Rane, V.H., *US Patent 5997840* (1999).
[43] Tanielyan, S.K., Augustine, R.L., in *Chemical Industries*, Chem. Ind., Dekker, 75 (1998) 101.
[44] Augustine, R., Tanielyan, S., Anderson, S., Yang, H., *Chem. Commun.* (1999) 1257.
[45] Augustine, R.L., Tanielyan, S.K., Anderson, S., Li, H., *Proc. Chiratech '98*, Novembre 9-11, Barcelona, Spain (1998).
[46] Sabater, M.J., Corma, A., Domenech, A., Fornes, V., Garcia, H., *Chem. Commun.* (1997) 1285.
[47] Ogunwumi, S.B., Bein, T., *Chem. Commun.* (1997) 901.
[48] Flach, H.N., Grassert, I., Oehme, G., Capka, M., *J. Colloid. Polym. Sci.*, 274 (1996) 261.
[49] Vankelecom, I.F.J., Vercruysse, K., Neys, P., Tas, D., Janssen, K.B.M., Knops-Gerrits, P.-P., Jacobs, P.A., *Topics in Catal.*, 5 (1998) 125.
[50] Vankelecom, I.F.J., Tas, D., Parton, R.F., Van de Vyver, V., Jacobs, P.A., *Angew. Chem. Int. Ed. Engl.* 35, 12 (1996) 1346.
[51] Tas, D., Thoelen, C., Vankelecom, I.F.J., Jacobs, P.A., *Chem. Commun.* (1997) 2323.
[52] Vankelecom, I.F.J., Jacobs, P.A., *Catal. Today*, 1905 (1999) 1.
[53] Parton, R.F., Vankelecom, I.F.J., Casselman, M.J.A., Bezoukhanova, C.P., Uytterhoeven, J.B., Jacobs, P.A., *Nature*, 370 (1994) 18.
[54] Janssen, K.B.M., Laquière, I., Dehaen, W., Parton, R.F., Vankelecom, I.F.J., Jacobs, P.A., *Tetrahedron Asymm.*, 8, 20 (1997) 3481.
[55] Wan, K.T., Davis, M.E., *J. Catal.*, 148 (1994) 1.
[56] Wan, K.T., Davis, M.E., *US Patent 5736480* (1998).
[57] Wan, K.T., Davis, M.E., *US Patent 5827794* (1998).
[58] Wan, K.T., Davis, M.E., *J. Catal.*, 152 (1995) 25.
[59] Feast, S., Bethell, D., Bulman Page, P.C., King, F., Rochester, C.H., Siddiqui, M.R.H., Willock, D.J., Hutchings, G.J., *J. Chem. Soc., Chem. Commun.* (1995) 2409.
[60] Feast, S., Rafiq, M., Siddiqui, M.R.H., Wells, R.P.K., Willock, D.J., King, F., Rochester, C.H., Bethell, D., Bulman Page, P.C., Hutchings, G.J., *J. Catal.*, 167 (1997) 533.
[61] Feast, S., Bethell, D., Bulman Page, P.C., King, F., Rochester, C.H., Siddiqui, M.R.H., Willock, D.J., Hutchings, G.J., *J. Mol. Catal. A: Chem.*, 107 (1996) 291.
[62] Hutchings, G.J., *Chem. Commun.*, (1999) 301.
[63] Sundarababu, G., Leibovitch, M., Corbin, D.R., Scheffer, J.R., Ramamurthy, V., *Chem. Commun.* (1996) 2159.
[64] Choudhary, B.M., Shobha Rani, S., Narender, N., *Catal. Lett.*, 19, 4 (1993) 299.
[65] Refs. in Hutchings, G.J., Willock, D.J., *Topics in Catal.*, 5 (1998) 177.
[66] Davis, M.E., Lobo, R.F., *Chem. Mater.*, 4 (1992) 756.

3 Organic Polymers as a Catalyst Recovery Vehicle

David E. Bergbreiter

3.1 General Introduction

Homogeneous catalysis has several advantages over heterogeneous catalysis. The work in the other chapters of this book illustrates this by highlighting the virtues of asymmetric catalysis, chemistry most easily achieved with homogeneous catalysts, the activity and selectivity of which can be tuned by varying ligand structure. However, while the choice stereoselectivity and understanding of reaction mechanism resulting from homogeneous catalytic processes is notable, recovery of homogeneous catalysts or of their ligands continues to be both a problem and a sought after goal. Approaches in which soluble or insoluble polymers are used to address this problem are discussed and reviewed below.

The principles for catalyst immobilization and recovery on polymers are the same for enantioselective catalysts and non-enantioselective catalysts. Most work on catalyst immobilization beginning in the 1960s has focused on achiral catalysts. Since there is no substantial conceptual difference between using organic polymers to recover chiral and achiral catalysts, we begin with a general discussion of catalyst or ligand separation and recovery. The subsequent discussion cites examples where enantioselective catalysts were used, because such examples aptly illustrate the advantages of polymers as vehicles for homogeneous catalyst and ligand recovery. Polymer-supported catalysts including enantioselective catalysts on polymers have also been the subject of several other previous reviews, which more fully summarize the historical context of the many recent advances in asymmetric catalysis with recoverable, reusable polymeric catalysts [1–6].

There are several reasons to consider a scheme that uses polymers for catalyst and ligand recovery. One reason is cost. Chiral ligands are usually expensive. Transition metal catalysts often use expensive catalysts prepared from noble metals. A second reason is product purity. Residual catalysts or residual catalyst ligands, present even at low concentrations, compromise product purity. This is important, for example, in cases where the eventual product application involves a drug. In this instance, trace heavy metals and trace phosphine ligands are unacceptable contaminants. A third reason for designing new processes for catalyst and ligand recovery is the environmental problem associated with catalyst and ligand waste or associated with the processes necessary to separate and recover a catalyst or ligand from a product. Finally, organic polymer supports that are used for catalyst recovery can serve as handles for ligands leading to catalysts that dissolve in environmentally more desirable solvents like water or supercritical carbon dioxide. Most of these justifications for enantioselective catalyst and ligand recovery are examples of the trend towards "green chemistry" processes [7], a general theme in process design that is gaining favor both for societal and practical reasons.

In considering possible schemes for catalyst and ligand immobilization and recovery with organic polymers, several issues must be initially addressed. One must decide if it is more desirable to use a polymer that is insoluble throughout the reaction or if a polymer that is soluble at some point in the reaction scheme would be more useful. Because of the considerable differences between these approaches, they are considered separately below. One also has to decide whether solid/liquid separation or a liquid/liquid separation of catalysts and ligands from products is most desirable. In the latter case, a soluble polymeric catalyst must be used. In the former case, both soluble and insoluble polymer-bound catalysts have been used.

The most commonly used method in which polymers facilitate recovery of a catalyst and ligand from a reaction is based on a solid/liquid separation. Four approaches are typically applied (Fig. 3.1) [2, 8]. The most common method is to use a cross-

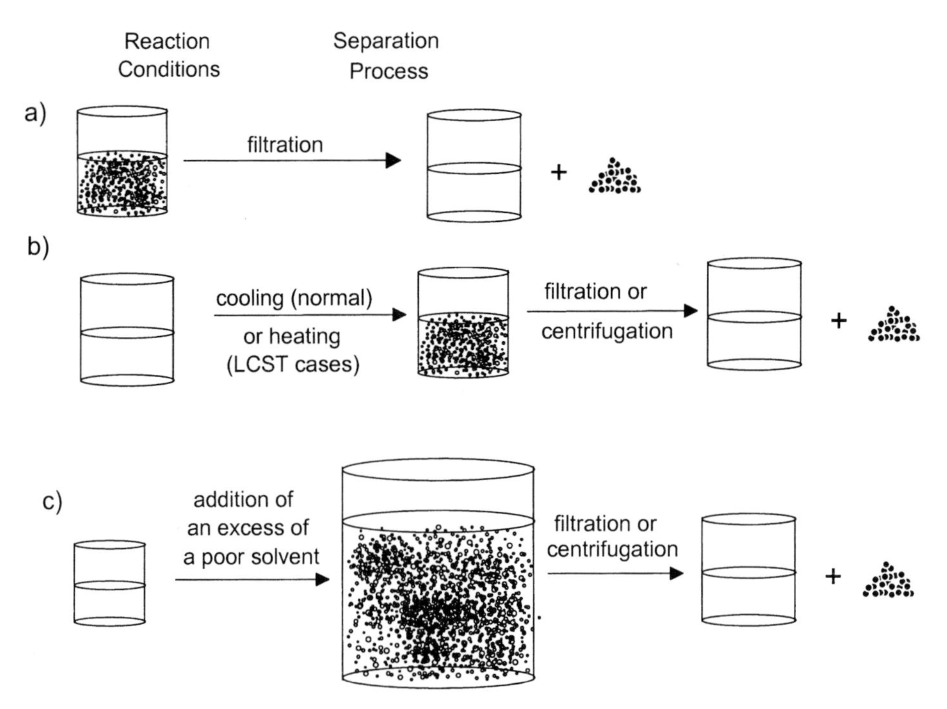

Figure 3.1. Separation schemes in which a soluble polymer is recovered and separated from a solution as a solid.

linked insoluble polymer as a support. Such polymers are always insoluble. Often a divinylbenzene cross-linked polystyrene support is used. In these cases, catalyst and ligand recovery is effected by filtration. A second approach uses polymers with temperature-dependent solubility [8]. Terminally functionalized polyethylene oligomers, having molecular weights in the 2000–3000 Da range, are an example of this approach [8]. In these cases, the oligomers are completely insoluble at room temperature in all solvents, because these chains are essentially polyethylene with one end modified to serve as a ligand. Such oligomers are derived from anionic polymerization and contain a CH_3-terminal group at one end and a ligand on the other end. The solubility of such oligomers is essentially independent of the single end group that serves as the ligand, provided that the oligomer contains about 150–200 carbons. In appropriate solvents like toluene or xylene, these oligomers like polyethylene dissolve when heated. When terminally functionalized oligomers whose M_n is <2500 Da are used, catalyst loadings sufficient to prepare 10^{-2}–10^{-3} M solutions are easily attained. The third separation scheme (Fig. 3.1 b) also uses a soluble polymer. However, in this instance, the soluble polymer precipitates upon heating. Centrifugation, decantation or filtration then leads to polymer (and polymer-bound catalyst) recovery [10]. The final example in Figure 3.1 c is the most general way in which soluble polymer-bound catalysts are recovered and is a common technique used in isolating polymers in laboratory synthesis. In this scheme, a solution of the polymer in a good solvent is added to an excess of the poor solvent to form a precipitate of the polymer that, in turn, is recovered by filtration [11].

Liquid/liquid separations as a strategy to recover a polymer-bound catalyst are less common [8]. Perhaps the most common of these techniques is membrane filtration shown in Figure 3.2 a which is typically applied in biochemistry and has been used in some examples with soluble homogeneous catalysts. A second technique is the use of a biphasic system. Aqueous biphasic systems are widely used but represent a relatively new technique. Fluorous biphasic catalysis, in particular, is currently receiving special attention (see Chapter 4) [12–14]. Polymeric ligands can be used in either biphasic system to insure that the catalyst and ligand remain in a specified phase. A third more recent technique that has not been employed yet with enantioselective catalysts is based on a biphasic system, the phase miscibility of which changes with heating and cooling. This so-called thermomorphic system has many of the advantages of the biphasic separations schematically shown in Fig. 3.2 b with the advantage that the actual catalytic chemistry occurs under homogeneous conditions. This latter technique has only been described to date with polymeric catalysts [15]. However, it should be possible to use other water-soluble catalysts (cf. Chapter 4) in this chemistry, too, as is suggested by the solubility behavior of nonpolymeric catalysts containing a sulfonated phosphine ligand [16].

The original approaches to use organic polymers as vehicles for enantioselective catalyst recovery originated from the success of solid-phase peptide syntheses and the commercial development of ion-exchange resins. These approaches and much of the subsequent work in the area primarily focused on three sorts of reactions – asymmetric hydrogenation of a prochiral alkene, asymmetric reduction of a ketone and 1,2-additions to carbonyl groups (most often with organozinc chemistry). Later work has included additional studies dealing with Lewis acid-catalyzed Diels-Alder reactions,

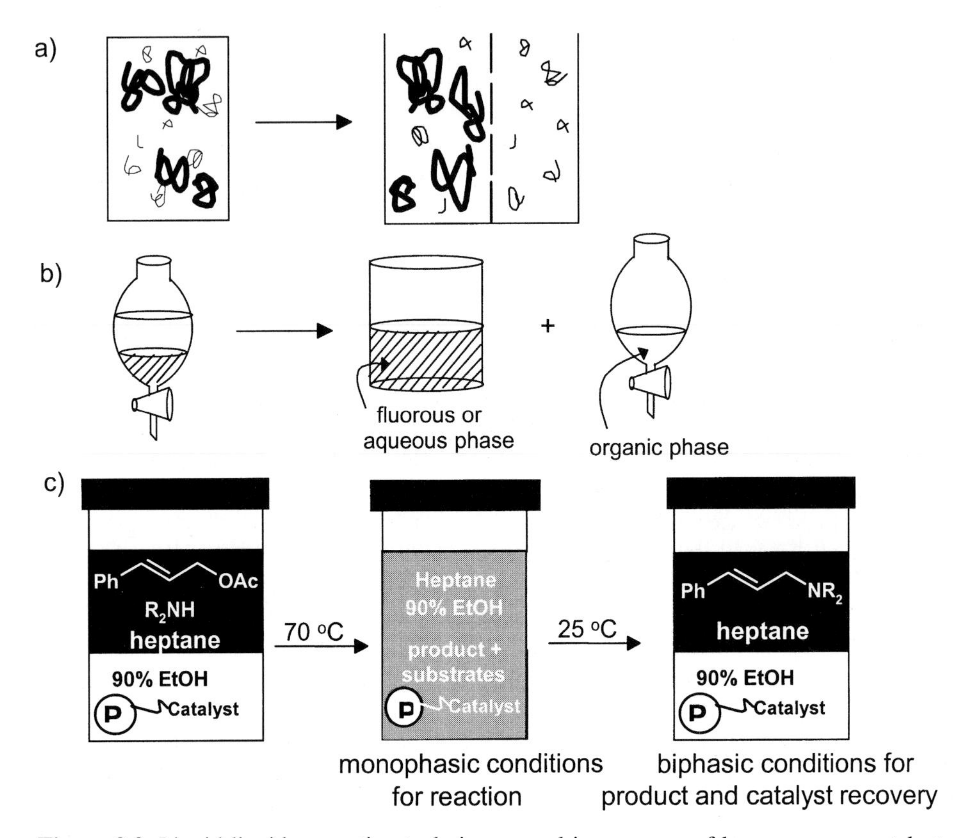

Figure 3.2. Liquid-liquid separation techniques used in recovery of homogeneous catalysts.

asymmetric enolate chemistry, asymmetric epoxidation and asymmetric dihydroxylation reactions. Other reactions like hydroformylation have also received attention. This chapter mainly focuses on describing the many ways in which organic polymers can be used to facilitate enantioselective catalyst recovery and reuse. Cases where the asymmetric reaction may not have proceeded with high enantioselectivity are included, if the strategy for catalyst immobilization and recovery is conceptually or historically significant. The discussion below emphasizes all the main points in catalyst recovery using polymers and the advantages that result from using polymers as a support in asymmetric catalysis. It is not, however, intended to be an encyclopedic coverage of every report of enantioselective catalysis using polymer-supported catalysts.

3.2 Alkene Hydrogenation

Asymmetric alkene reduction was one of the first asymmetric catalytic reactions to attract the broad attention of the synthetic community. Work by Kagan and the application of the asymmetric hydrogenation by Monsanto for *L*-DOPA synthesis established this chemistry as a practical goal and led groups to develop polymer supports that could simplify this chemistry by simplifying catalyst isolation, recovery and reuse [17].

While much of the early work on polymer-supported catalysts in the 1970s and 1980s simply extended Merrifield's chemistry by using polystyrene as a support, Stille recognized the importance of a polar protic environment for enantioselective DIOP-like diphosphine-ligated Rh(I) catalysts. Thus, his group prepared their own copolymers using a styrene monomer containing a chiral DIOP-like ligand and several vinyl monomers (Schemes 3.1 and 3.2) [18, 19]. In this work, the chiral DIOP-like ligand was prepared after polymerization, because substitution of the tosylate groups of the 2-*p*-styryl-4,5-bis(tosyloxymethyl)-1,3-dioxolane (**1**) with a diphenylphosphide anion led to unwanted anionic polymerization. In this case, the presence of ethylene glycol dimethacrylate in the hydroxyethyl methacrylate produces a cross-linked polymer that swells in polar solvents like alcohol but which is easily recovered by filtration. Optical yields in hydrogenation of 2-acetamidoacrylic acid or 2-acetamidocinnamic acid in a benzene-ethanol mixture at 25 °C were 52–60 and 86%, respectively, values that compared favorably with the enantioselectivity of an analogous homogeneous Rh(I)-DIOP catalyst (73 versus 81%, respectively). The polymeric catalyst was ca. 5-fold less reactive than its homogeneous analog and was sensitive to air (presumably because of air oxidation of the phosphine ligand). Another problem with this catalyst is that extended use could lead to polymer degradation through transesterification or hydrolysis (this process would, however, be slow at 25 °C given that the polymer is a methacrylate derivative). To avoid this problem and to further study the effects of polymer structure on asymmetric hydrogenation, Stille's group prepared other polar polymers containing a chiral alcohol group (Scheme 3.2).

In this case, hydrogenation of a prochiral alkene, like 2-acetamidoacrylic acid, in alcohol solvent gave the same *R*-alanine derivative as is obtained with the homogeneous DIOP catalyst (76–74% e.e. in 4 successive runs with the *S*-configuration polymeric alcohol; 83% e.e. of the *R*-configuration with the *R*-polymeric alcohol). However, the configuration of the polymer's pendant alcohol group did not have much effect on stereoselectivity in reductions in alcohol solvent. The presence of the polymeric –OH group did have a notable effect in THF. In THF, the homogeneous catalyst is ineffective (7% e.e.) but the polymeric catalyst still had modest selectivity (40% e.e., 28% conversion (*R*-pendant alcohol); 24% e.e., 22% conversion (*S*-pendant alcohol)). This was an early illustration of the constructive effect of polymer structure on the activity of a polymer-bound catalyst.

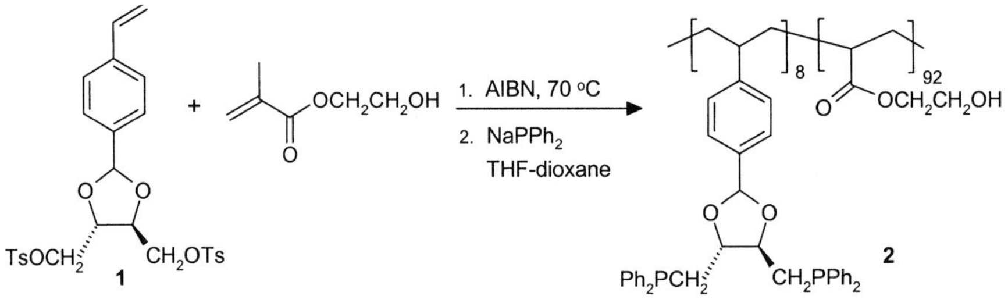
1. AIBN, 70 °C
2. NaPPh2
THF-dioxane
OCH2CH2OH
TsOCH2
CH2OTs
1
8
92
Ph2PCH2
CH2PPh2
2

(1)

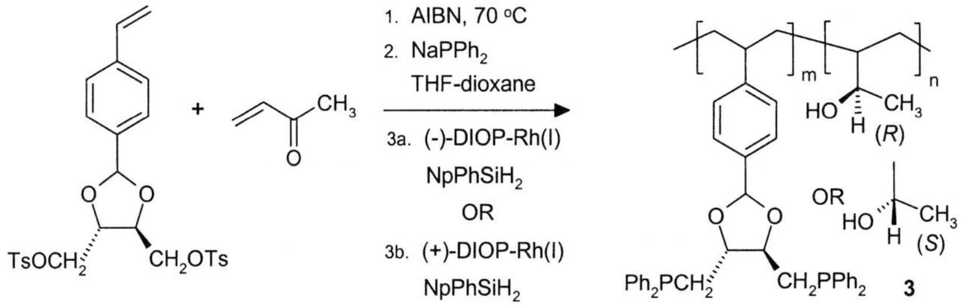
1. AIBN, 70 °C
2. NaPPh2
THF-dioxane
3a. (-)-DIOP-Rh(I)
NpPhSiH2
OR
3b. (+)-DIOP-Rh(I)
NpPhSiH2
CH3
TsOCH2
CH2OTs
m
n
(R)
(S)
OR
Ph2PCH2
CH2PPh2
3

(2)

Later work by Stille used phosphinopyrrolidine monomers like **4** copolymerized with hydroxyethyl methacrylate or 2,3-dihydroxybutyl methacrylate and a cross-linking agent, ethylene dimethacrylate to form the chiral polymeric ligand **5** (Scheme 3.3). Similar chiral phosphine-containing copolymers (**7**), derived from a carboxystyrene-alkyl methacrylate monomer (**6**) (Scheme 3.4), had earlier been used to prepare a Rh(I) hydrogenation catalyst which was used to hydrogenate oxapantolactone in a 75% yield [20].

In Stille's case, the cross-linked acrylate-acrylamide copolymers **5** were used to make a Rh(I) catalyst for hydrogenation of *N*-acyl dehydroaminoacids to yield chiral aminoacid derivatives [21]. As in Stille's earlier studies, the effect of the chiral hydroxyl group (in polymers derived from 2,3-dihydroxybutane) was studied. Again, significant positive effects were seen, especially in aprotic solvents like THF. However, these catalysts generally did not give practical e.e. values. Catalyst and ligand recovery was effected by simple filtration, and catalysts were reused with no loss in selectivity.

In further work, Stille also described the use of similar catalysts **8** in hydrogenation of cinnamic acid or amidoacrylic acid derivatives to yield chiral carboxylic acids with e.e. values ranging from 19–77% [21]. In this case, polymers were prepared such that they contained a glycerol derivative. Recycling the catalyst in these cases was accomplished by simple filtration of the insoluble cross-linked polymer. The catalysts derived from a homochiral (*R*)-2,3-dihydroxypropyl methacrylate gave slightly higher optical yields (6–8% e.e.) in comparison to polymers derived from the racemic- or (*S*)-2,3-dihydroxypropyl methacrylate. Again, this suggests a slight but measurable interaction of pendant alcohol groups on the polymer with the enantioselective catalyst.

More recent approaches to asymmetric alkene hydrogenation using polymer supported phosphine-ligated catalysts have focused on new catalyst synthesis strategies or on the use of ligands known to be more generally effective in solution. For example,

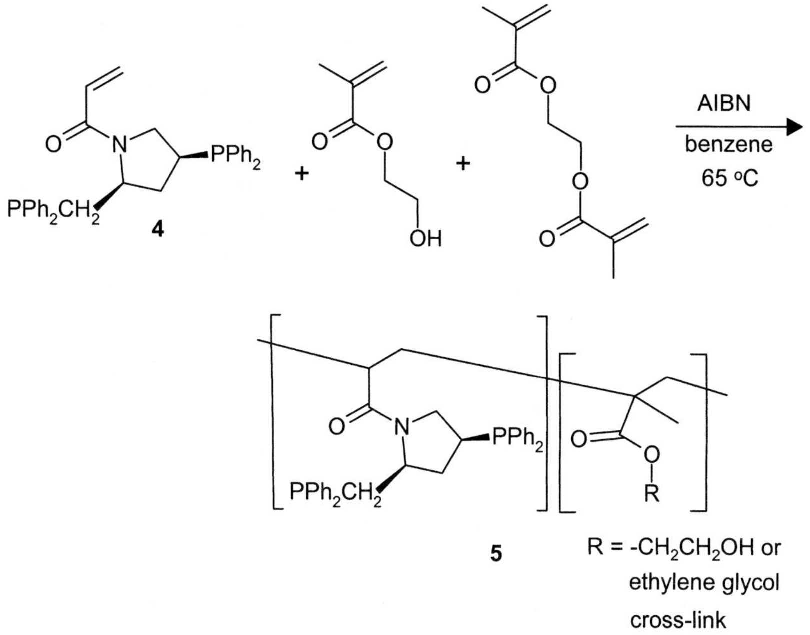

AIBN
benzene
65 °C

4 (3)

5 R = $-CH_2CH_2OH$ or ethylene glycol cross-link

AIBN

R = $-CH_3$ or $C(CH_3)_3$

6 **7** R = $-CH_3$ or $C(CH_3)_3$ 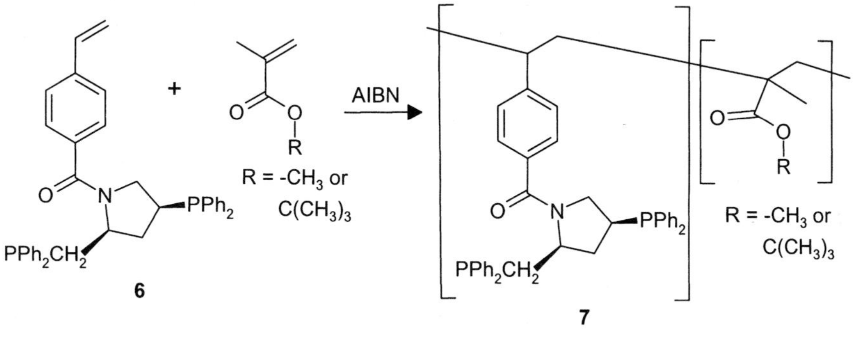 (4)

Bayston et al. prepared a BINAP-based diphosphine containing a functional group in the 6-position of a naphthalene ring (**9**) [22]. This diphosphine was then attached to an aminomethyl derivative of polystyrene (1% DVB cross-linking) via this carboxylic acid group (**10**) (Scheme 3.5). The resulting diphosphine, when attached to the polymer, was converted into a ruthenium catalyst, and this catalyst was effectively used in asymmetric alkene hydrogenation with 56% e.e. in reducing itaconic acid and 64% e.e. in reducing 2-acetamidoacrylic acid. More noteworthy in this case is the e.e. obtained upon reducing compounds like ketoesters. Reduction of methyl propionylacetate with this polymer-bound BINAP catalyst proceeded with 97% e.e. at 70 °C. Recycling the catalyst again involved a solid-liquid separation. Filtration under argon, washing with THF and reuse proved to be simple and effective. Reuse did require longer reaction times, however. In this instance, the authors analyzed the filtrate for ruthenium using ICP (inductively coupled plasma atomic emission spectroscopy).

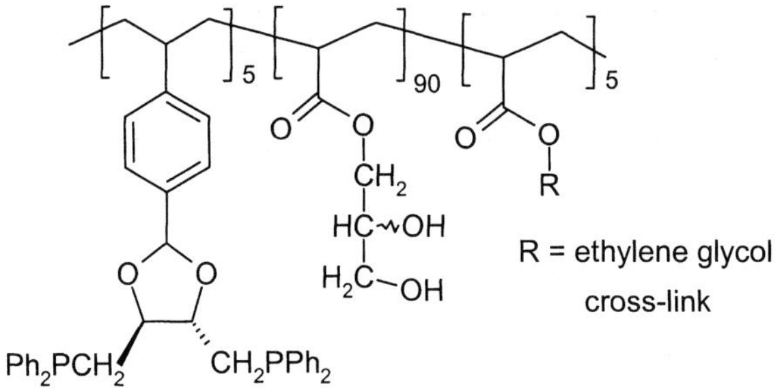

R = ethylene glycol cross-link

8

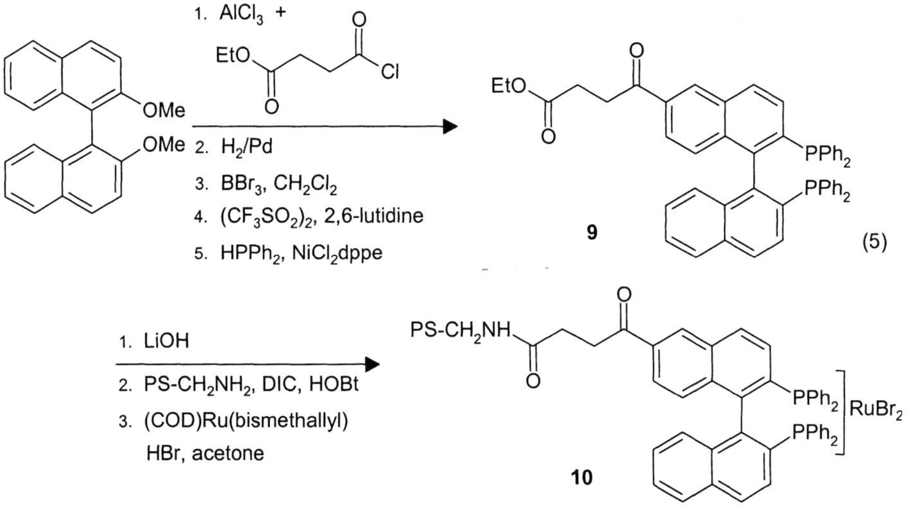

1. $AlCl_3$ + EtO(CO)CH_2CH_2COCl
2. H_2/Pd
3. BBr_3, CH_2Cl_2
4. $(CF_3SO_2)_2$, 2,6-lutidine
5. $HPPh_2$, $NiCl_2dppe$

9 (5)

1. LiOH
2. PS-CH_2NH_2, DIC, HOBt
3. (COD)Ru(bismethallyl) HBr, acetone

10

These results showed that <1% of the charged ruthenium catalyst was present in solution.

An undesirable but necessary feature of both the older work and the more recent work with enantioselective polymer-bound hydrogenation catalysts is the need to prepare ligands that have attachment points for coupling to a polymer. This is a general feature of chemistry using immobilized catalysts. In cases like the BINAP example illustrated in Scheme 3.5, this can involve a lot of steps and time, effort that is presumably compensated for by the facility with which the catalyst and ligand are reused. Such chemistry also has the disadvantage that it can stereochemically, electronically or sterically complicate an existing catalyst. This potential is illustrated by the chemistry above where the substituent groups for polymer immobilization desymmetrize the BINAP in a formal sense. However, in cases like these, the effect of the polymer and the substituent for tethering the ligand to the polymer is often of minimal consequence, if the tethering group is electronically or physically distant from the catalyst or stereogenic center. Often it is possible to test this before polymer attachment. For example, in the case of **9**, the soluble "pseudo C_2" BINAP ligand before polymer attachment was as effective as BINAP in a comparison test.

A more novel approach to polymeric phosphine ligands that can be used in asymmetric catalysis was described by Gilbertson's group at Washington University [23]. The aim of this work was the combinatorial synthesis of chiral phosphine ligands – an area where polymer supports have unique advantages and where they will be used to an even greater extent in the future. Using solid-phase peptide synthesis, a polystyrene-supported peptide-based bis(phosphine sulfide) Ac-Ala-Aib-Ala-Pps(S)-Ala-Ala-Cps(S)-Ala-Ala-Aib-Ala-polymer ligand was prepared (Pps = 3-(diphenylphosphino)-alanine, Cps = 3-(dicyclohexylphosphino)alanine; Aib = amino-isobutyric acid). The polymer was DVB-cross-linked polystyrene containing phenylacetamidomethyl groups to support the peptide. The conversion of phosphine sulfides to phosphines was accomplished by reaction of the polymeric phosphine sulfide with methyl triflate. Reaction with tris(dimethylamino)phosphine yielded a mixture of polymeric diphosphine, some unreacted polymer-bound bis(phosphonium) salt and (methylthio)tris(dimethylamino)phosphonium salt. This procedure avoids excessive handling of free phosphines and accidental oxidation of the phosphines by air to phosphine oxides and can be followed by ^{31}P NMR spectroscopy. Such accidental oxidation is a common problem in synthesis of immobilized phosphine ligands and can be a limitation in catalyst recycling. In this instance, the major interest was in a potential combinatorial route to polymer-bound phosphine ligands (Scheme 3.6). Nonetheless, the catalytic activity was tested. While 100% reduction of methyl 2-*N*-acetamidoacrylate was seen, the enantioselectivity achieved in reduction to *N*-acetyl alanine methyl ester was poor (4–9% e.e.).

Another approach to immobilize and recover phosphine ligands on polymers described recently used dendrimer-bound catalysts [24]. In this chemistry, 5-(*tert*-butyl-

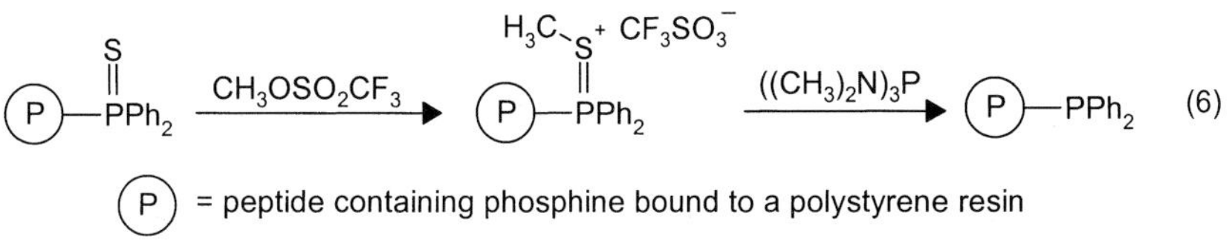

$$\text{(P)}\!-\!P(=S)Ph_2 \xrightarrow{CH_3OSO_2CF_3} \text{(P)}\!-\!P(=S^{+}CH_3)Ph_2\ CF_3SO_3^{-} \xrightarrow{((CH_3)_2N)_3P} \text{(P)}\!-\!PPh_2 \qquad (6)$$

(P) = peptide containing phosphine bound to a polystyrene resin

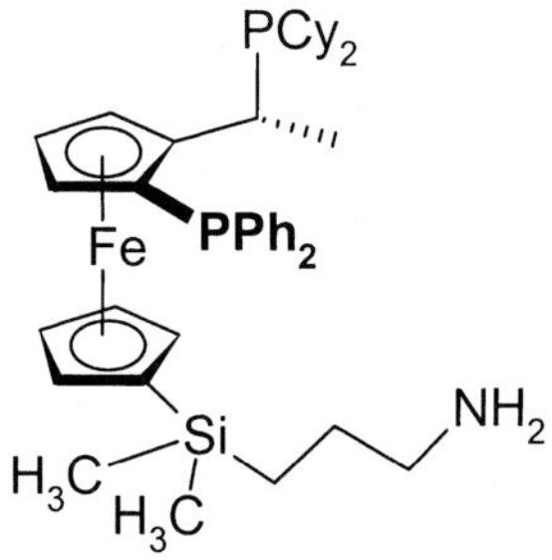

PCy$_2$ PPh$_2$ Fe Si NH$_2$ H$_3$C H$_3$C

11

dimethylsilyl)-isophthaloyl dichloride was allowed to react with the amine-containing ferrocenyl diphosphine **11**. Deprotection of the silyl groups yielded a phenol that, in turn, was allowed to react with adamantane-1,3,5,7-tetracarboxylic acid tetrachloride to yield a dendritic molecule containing 8 diphosphines **11** on its periphery. The enantioselectivity of a cationic rhodium catalyst (prepared from the dendritic octa(diphosphine) and $[Rh(COD)_2]BF_4$) in hydrogenation of dimethyl itaconate in methanol (1 mol% Rh catalyst, 1 bar) was 98%. While complete retention by a commercial nanofiltration membrane of these catalysts was reported, the activities of the recycled catalysts are not mentioned in this preliminary report. Similar ferrocenyl phosphines are known to be very effective catalyst ligands both in homogeneous catalysis and when immobilized on other insoluble supports. However, in the case of a dendrimer-bound catalyst, the immobilized chiral ferrocenyl diphosphine ligands will be soluble. Such catalysts can be recovered by membrane filtration because of the size of the dendrimer.

Immobilization of chiral pyrrolidine-containing diphosphines on soluble polymers is also possible. In unpublished work from our laboratory, Koshti has shown that poly(*N*-isopropylacrylamide)-*c*-(*N*-acryloxysuccinimide) (20:1) can be used to bind a chiral diphosphine like **12** to form a water soluble polymer **13** [25]. Coordination of Rh(I) to this polymer and hydrogenation of *N*-benzoyl dehydrophenylalanine yielded the chiral *N*-benzoyl phenylalanine product in 90% e.e. (Scheme 3.7). Catalyst recovery in this case was effected by membrane filtration. A problem in this case was the accidental oxidation of the polymeric phosphine due to the use of large volumes of solvent in the dialysis. This problem can presumably be minimized with appropriate experimental care. Ligand decomposition is, however, a general problem associated with catalyst recovery and reuse [26]. This issue was noted early in the development of polymer-bound catalysts. However, because it is relatively uncommon to find studies that attempt to recycle catalysts more than three times, the extent of this problem may not yet be fully realized by the community using polymers for catalyst recovery.

An earlier report of enantioselective hydrogenation in water used rhodium phosphine complexes bound to polyacrylic acid [27]. This example of a water-soluble polymeric catalyst used the (2*S*,4*S*)-4-diphenylphosphino-2-diphenylphosphinomethyl pyrrolidine ligand which was bound to polyacrylic acid via an amide bond. In this case, the remaining carboxylic acid groups of the polyacrylic acid impart high water solubility to the catalyst. This water solubility facilitates isolation of the polymer-

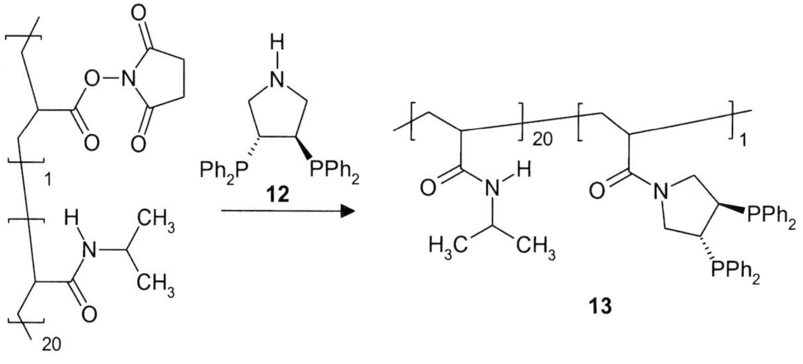

(7)

bound catalyst since the catalyst resides exclusively in the aqueous phase in an ethyl acetate-water mixture (Fig. 3.2b). The enantioselectivity of the catalyst was affected by catalyst loading on the polymer – the highest enantioselectivity was obtained at low loadings.

Naturally occurring polymers and other sorts of immobilization chemistry have received attention too for recovery/reuse of chiral polymer-bound catalysts for asymmetric hydrogenation. An early report described the use of a phosphinated cellulose as a support for both Pd(II) and Rh(I). The enantioselectivities seen using the Rh(I) catalyst bound to 6-*O*-triphenylmethylcellulose phosphinite in hydrogenation of 2-phenyl-1-butene to *R*-2-phenylbutane were modest (30–63%) but are nonetheless impressive given the simplicity and ready availability of the chiral polymer needed to prepare the phosphinite ligand [28].

Some very successful strategies using polymers to immobilize asymmetric hydrogenation catalysts have employed ion-pair immobilization on an ion-exchange resin instead of covalent coupling of the ligand to a polymeric support. One very promising example of this approach is the immobilization of phosphinite complexes of cationic rhodium catalysts **14** and **15** on a sulfonated ion-exchange resin [29]. In these cases, the asymmetric hydrogenation of the methyl ester of *α*-acetamidocinnamic acid to form the *R* phenylalanine derivative was quite successful (**14**, e.e. values from 76–85%; **15**, e.e. values for the product that were typically 95%). Ester-containing substrates were used to minimize metal leaching due to the carboxylic acid of the substrate; however, even then rhodium leaching still occurred to the extent of 0.6–3% per cycle.

A more successful immobilization/recovery method using ion-exchange resins was subsequently described by Hanson and Davis [30]. In this case, a cationic rhodium catalyst containing *p*-dimethylaminophenylphosphine ligands was used (**16**). This complex with an ion-exchange resin like Nafion-H (a sulfonated fluoropolymer) apparently immobilized the catalyst more effectively, because the catalyst-ligand complex was polycationic. However, the e.e. values for reduction of methyl *α*-acetamidocinnamate were slightly lower in this case: 73, 72, 73, 67, 64 and 63% in six successive runs with the same catalyst (conversions were typically 97–98%). This polyvalent approach to chiral catalyst immobilization seems a very attractive idea and should be further pursued given the broad acceptance of sequestering methodology using ion-exchange materials in high throughput organic synthesis [31].

The final example of a polymeric asymmetric hydrogenation catalyst discussed here was recently described by Chan [32]. In this report, the catalyst ligand was prepared as part of the polymer main chain by condensation of a chiral 5,5′-diamino-BINAP ligand with terephthaloyl chloride and 2*S*,4*S*-pentanediol. Polymeric ligands were prepared both with the (S)-BINAP derivative (shown below) or with the (R)-BINAP derivative (Scheme 3.8).

The molecular weight of the polymeric ligand was 10540 (M_w based on GPC with polystyrene standards). These non-cross-linked polymers were soluble in toluene, THF and methylene chloride, and the polymer was quantitatively recovered by precipitation in methanol. Ruthenium complexed by this polymeric ligand was studied in asymmetric hydrogenation of a 2-arylacrylic acid (Scheme 3.9). Conversions and enantioselectivities in this hydrogenation leading to (S)-Naproxyn were followed

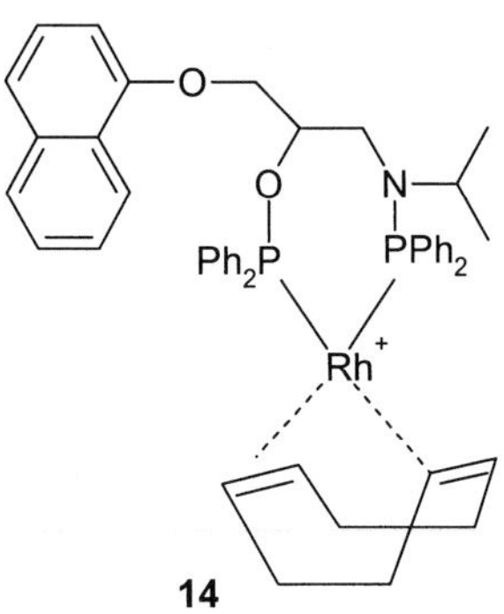

14 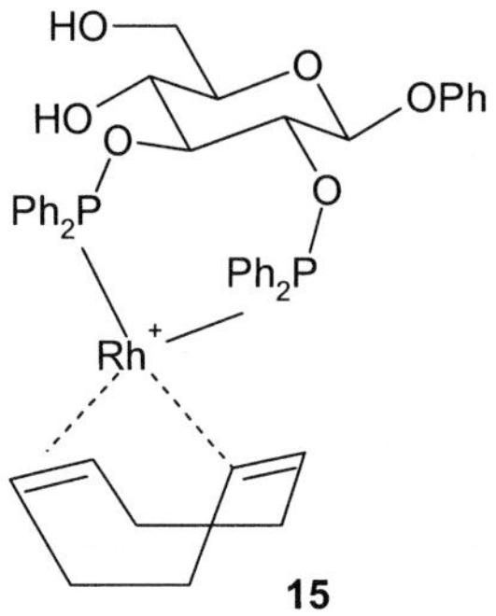 **15**

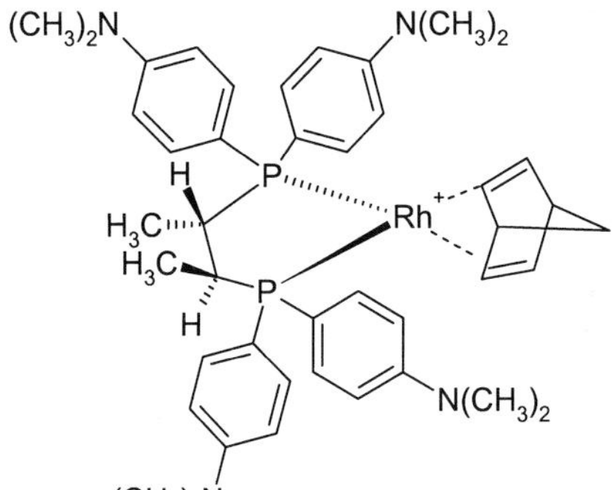

16

through 10 cycles. Conversions were typically in the 90–100% range with e.e. values ranging between 88 and 94%.

In a typical experiment, a 6.3×10^{-6} N solution of the ruthenium catalyst was dissolved in a 3:2 (v/v) toluene methanol mixture. Hydrogenation of 0.21 mmol of dehydronaproxyn was completed in 5 h under these conditions, at which point the catalyst was precipitated with 20 ml of cold methanol. Filtration recovered the precipitated catalyst. Analysis of the filtrate (atomic absorption spectroscopy) showed 16 ppb of Ru was present – a value that represents a >99.9% catalyst recovery. This polymeric system also exhibited higher reactivity than the corresponding monomeric catalyst.

3.3 Carbonyl and Imine Reduction

The strategies in which polymers were used to recover asymmetric hydrogenation catalysts illustrate the principal techniques used for recovery of enantioselective homogeneous catalysts and ligands with organic polymers. These same themes have also been applied in other types of catalytic reduction chemistry. A reaction similar to alkene reduction is asymmetric reduction of carbonyl compounds like ketones, aldehydes and imines. This reaction can be carried out in several ways by using asymmetric catalysts. Hydride transfer mediated by transition metal catalysts is one route for asymmetric reduction that has been explored with polymer-supported catalysts. The earliest examples of polymer-supported catalysts being used in reductions of ketones date back to the seventies. A second general route is asymmetric reduction using a main group metal hydride and a chiral ligand. This has been a more recent theme in asymmetric carbonyl reduction using recoverable, reusable polymer supports.

Reduction of carbonyl compounds with polymer-immobilized transition metal complexes often employs ketoesters as substrates. One example using a ruthenium phosphine-complex (**10**) is described above [22]. A second example is the reduction of pyruvate esters by anionic platinum or ruthenium metal carbonyl clusters attached ioni-

Ph_2P PPh_2 H_2N NH_2 **17** + Cl O O Cl + OH OH; pyridine, $ClCH_2CH_2Cl$ → (8)

CH_3 CH_3 O O O HN Ph_2P PPh_2 NH O O; 97; 3 **18**

CH_3O HO_2C → $[Ru(cymene)Cl_2]_2$, ligand **18**, H_2, (69 kg/cm^2), toluene-methanol → CH_3O CH_3 HO_2C (9)

H N $SiO_{1.5}$ Rd(cod)Cl N $SiO_{1.5}$ H

19

cally to DVB cross-linked polystyrene which contains as chiral modifier quaternary ammonium groups derived from alkaloids like cinchonidine [33]. Asymmetric induction varied in these reactions, with high e.e. values (>75% e.e.) only being obtained for methyl pyruvate. In these cases, the metal carbonyl is first activated by heating (loss of CO was reported). Regeneration of the starting, immobilized metal carbonyls by exposing used catalyst to CO (500 psi, 25 °C, 24 h) was confirmed by IR and UV-visible spectroscopy. This chemistry is similar to the use of heterogeneous platinum catalysts modified with cinchona alkaloids as chiral modifiers [34]. Higher enantioselectivities in the reduction of methyl pyruvate (97.6% e.e.) were seen when similar hydrogenations were carried out with polyvinylpyrrolidone-stabilized 14 Å-diameter Pt clusters modified with cinchonidine [35].

Hydride transfer (as opposed to reduction with H_2) is more commonly used as a route to ketone reduction. Several interesting approaches have been used in which polymers recover enantioselective catalysts. Hybrid organic-inorganic materials in the form of sol-gels prepared from chiral triethoxysilyl derivatives of (*R,R*)-1,2-diaminocyclohexane alone or with added tetraethoxysilane have been used to obtain up to 80% e.e. in the reduction of acetophenone by using transfer hydrogenation with isopropanol as the penultimate reducing agent [36]. The catalyst in this case is prepared from the starting triethoxysilane by applying $[ClRh(cod)]_2$, and the resulting material is then polymerized by adding water to obtain a porous solid described as **19** (BET surface areas of up to 280 m^2g^{-1} were reported for these materials).

Templating chemistry is a general strategy most often used in designing resins for separation and affinity chromatography [37]. The general idea is to copolymerize a substrate or substrate analog in a rigid but porous polymer matrix. The substrate or substrate analog is then chemically removed in such a way that a recognition site remains which specifically binds fresh substrate. This approach to molecular recognition with synthetic polymers is of interest in catalysis, since use of a transition-state analog for the substrate above generates what has been described as a 'plastic' antibody. The example in Scheme 3.10 below illustrates how this chemistry has been used to prepare a molecularly imprinted polymer (**20**) for asymmetric reduction of acetophenone [38]. A chiral diamine, a Rh(I) complex and the sodium salt of (*S*)-1-phenylethanol were polymerized with a mixture of di- and triisocyanates to form a cross-linked urea polymer. The resulting imprinted polymer reduced acetophenone to (*R*)-1-phenylethanol within 2 days with 70% e.e., a result that was comparable to the conversion of acetophenone in homogeneous solution (7 days, 67% e.e. with the same sort of catalyst). Since the authors reported that rhodium did not leach to the solution, reuse of the catalyst (though not explicitly described) should involve nothing more than simple filtration.

Another approach to carbonyl group reduction employs a main group hydride as the reducing agent. In this case, the polymer is used to support a chiral, chelating Lewis base that provides the asymmetric environment. This technique has been especially effective in solution-phase chemistry and works well with polymer supports, too.

One of the first examples of polymer-bound ligands being used in an asymmetric hydride reduction, often as stoichiometric auxiliaries, is found in work by Frechet [39]. In this work, a cross-linked chloromethyl-containing polystyrene resin was modi-

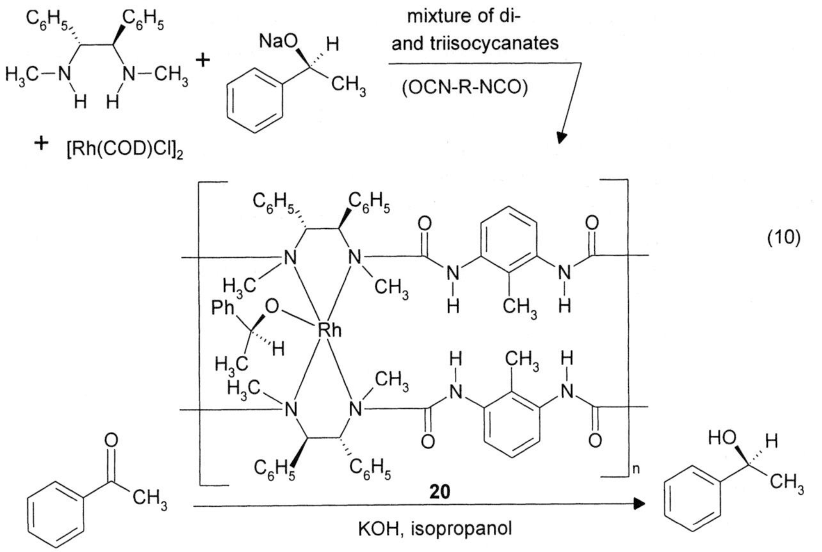

(10)

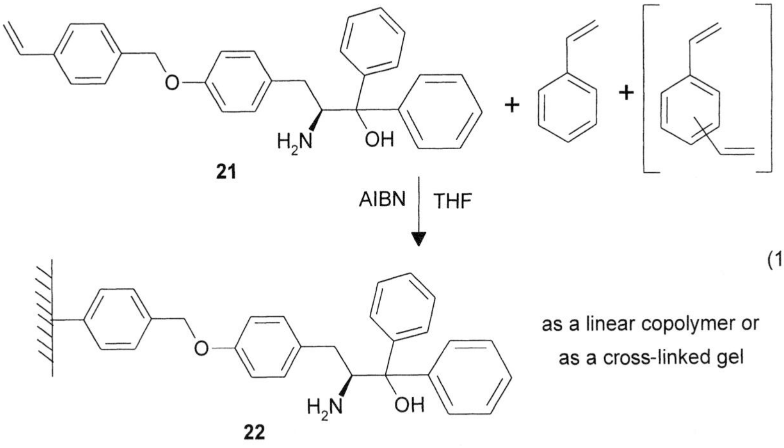

(11)

fied with a chiral aminoalcohol or a polyol. A $LiAlH_4$ derivative was then added to the resin to form a polymer-supported hydride reagent, whereby the chiral auxiliary group is covalently coupled to the resin. Addition of a prochiral ketone led to reduction and formation of a chiral alcohol. However, in this early work the e.e. values for the alcohol product using the recyclable polymer-bound reagent were inferior to those obtained with soluble chiral auxiliaries. A more successful early version of this chemistry is found in work by Hirao [40]. In this work, a chiral linear copolymer containing 33% of a chiral aminoalcohol was prepared by copolymerization of the chiral monomer **21** with styrene. Suspension polymerization with divinylbenzene as a cross-linking agent formed an insoluble, THF-swellable polymer **22** containing 10% of **21** (Scheme 3.11). Addition of borane to this polymer produced a chiral reducing agent that reduced the *O*-methyl oxime of acetophenone to (*S*)-1-phenylethylamine with

99% e.e. (85–95% isolated chemical yield). Both the e.e. and yield were unchanged upon recycling the recovered (filtration) resin. This paper also compared the DVB cross-linked copolymer prepared by using **21** with a linear copolymer prepared by using **21** and styrene. The linear copolymer also reduced acetophenone *O*-methyloxime to (*S*)-1-phenylethylamine in similar yield with >90% e.e. The linear copolymer was recovered by solvent precipitation and reused without significant loss of activity.

More recent methods using chiral auxiliaries on polymers in asymmetric hydride reductions have chosen approaches that parallel the earlier work of Frechet and Hirao but with BH_3 as a reducing agent and a chiral oxazaborolidine as a catalyst. Several examples have been described. Three examples which use different strategies for catalyst recovery are found in the work by Itsuno, Hodge and Wandrey [41–43].

Itsuno's work with the polystyrene-bound aminoalcohol **21** was one of the first examples where a polymer-bound aminoalcohol was used as a catalyst for an enantioselective borane reduction [41]. In this case, the catalytic properties of **22** were luckily discovered when the polymeric reagent **22** was carefully recovered by filtration under nitrogen. Under these conditions, addition of more borane in THF and valerophenone produced more (*R*)-1-phenylpentanol.

Attachment of chiral oxazaborolidines to a polymer via the boron atom has also been studied by Caze and Hodge [42]. In this case, a polymeric boronic acid (**23**) was formed from the copolymer of *p*-bromostyrene and styrene (2% DVB cross-linking) by BuLi metalation followed by trimethyl borate treatment and hydrolysis. This polymer (**23**) was allowed to react with (1*R*,2*S*)-(–)-norephedrine to form the catalyst **24** (Scheme 3.12). This polymeric catalyst contained 1.2 mmol/g of the oxazaborolidine and 0.7 mmol/g of unreacted boronic acid. This polymeric catalyst had to be used at fairly high levels (10–30 mol%) in the reduction of acetophenone to ensure that all the reduction occurred via the immobilized chiral oxazaborolidine. Under these conditions, good chemical yields (88%) were obtained with e.e. values of 70% throughout three cycles. Recovery of this cross-linked polymer was achieved by filtration.

The third example of an oxazaborolidine catalyst **25** was prepared using a soluble methyl hydrosiloxane-dimethylsiloxane copolymer containing 15% of the chiral pyrrolidine monomer [43]. This polymer was prepared by hydrosilylation of the allyl ether of a *trans*-4-hydroxyproline derived aminoalcohol (Scheme 3.13). Addition of BH_3 in THF to this polymer produced a soluble polymeric oxazaborolidine catalyst that was then used to reduce acetophenone, propiophenone or α-chloroacetophenone to the corresponding (*R*), (*R*) or (*S*) alcohols, respectively in high e.e. (97, 89 and 94%) with excellent chemical yield. In this case, the polymeric catalyst was recovered by nanofiltration. These studies showed that these kinds of soluble polymeric catalysts could be used in a continuously operated membrane reactor with good enantiomeric excess and high space-time yields. This particular system has also been compared and contrasted to a similar enzymatic reaction using dehydrogenase in a membrane reactor.

An unusual approach to asymmetric reduction using polymeric catalysts was reported in 1993 by Hodge [44]. In this chemistry, a prochiral ketone was absorbed onto a chiral polymer support and then reduced with an aqueous solution of potassium borohydride. Microcrystalline cellulose triacetate was used as the polymer support. Stereoselectivities were quite low (ca. 8–9% e.e. in reductions of 2-substituted naphthoquinones) but were improved somewhat using tetraalkylammonium salts as

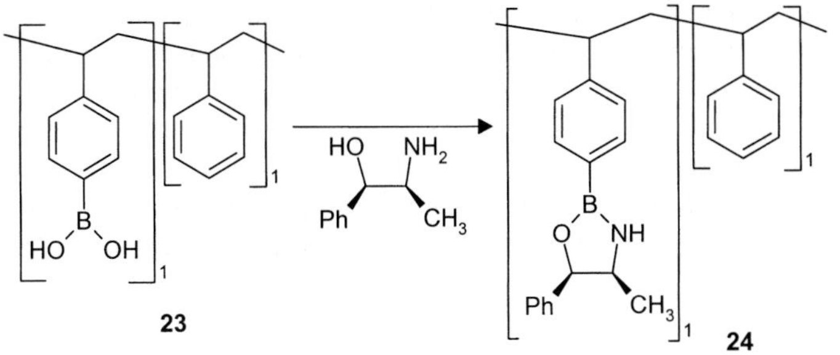

(12)

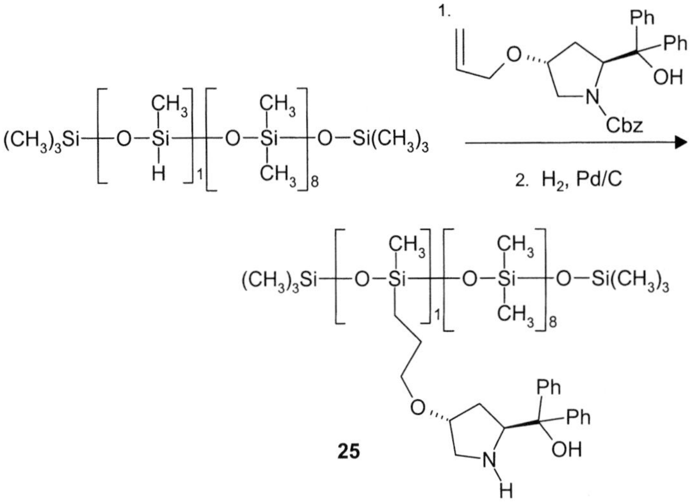

(13)

phase transfer catalysts. In this case, the polymer simply serves as an adsorbent, so that recovery and reuse merely requires filtration. While this approach is simple, the stereoselectivities pale in comparison to the hydride chemistry discussed above, and significant improvements, perhaps using template technology, will be necessary if this methodology is to compete with other asymmetric reduction chemistry. An advantage of this chemistry would be that it should be possible to design the polymeric adsorbate so that recovery/reuse should be especially easy.

Any approach to asymmetric catalysis has to consider other approaches to catalyst recovery. Biphasic approaches, including aqueous or fluorous biphasic hydrogenation, have to be considered in this regard and are discussed elsewhere in this volume (Chapter 4). Immobilization of chiral catalysts on inorganic supports must also be considered as an alternative approach (Chapter 2). Equally important is the question of whether catalyst recovery will always be necessary. Two recent examples of catalysts that reduce carbonyl groups or derivatives of carbonyl compounds illustrate this. In the first of these examples, the Novartis group has developed an enantioselective route to the herbicide (*S*)-metolachlor. In developing this process, this group considered enamide hydrogenation and imine reduction, settling on the latter approach. However, while they reportedly considered immobilization and catalyst/ligand recovery, the eventual process had turnover numbers of 1 000 000 and turnover frequencies of >200 000/h [45]. Likewise, Noyori has demonstrated asymmetric reductions of ketones using ruthenium catalysts and chiral diamine promoters where the TONs and TOFs are equally high [46]. When such catalyst activities are easily obtained, catalyst and ligand recovery may become economically and environmentally less important since the necessary catalyst and ligand concentrations could be extremely low. In such cases, catalyst recovery with polymers may only be necessary when product purification is simplified.

3.4 Carbon-Carbon Bond Formation

Reactions involving the synthesis of asymmetric carbon-carbon bonds have less commonly used polymer supports to recover transition metal catalysts. An early example of this chemistry is the asymmetric Grignard cross-coupling by means of polymer-supported optically active β-dimethylaminoalkylphosphine nickel catalysts. In this case, a DVB cross-linked polystyrene polymer support was used to prepare polystyrene-supported ligands **26** [47]. These chiral ligands were then used for the Ni-catalyzed asymmetric cross-coupling of *sec*-alkyl Grignard reagents with vinyl bromide to give optically active alkenes. Using 1-phenylethylmagnesium chloride as the Grignard reagent, (*S*)-3-phenyl-1-butene was obtained in 49% e.e. Catalyst recycling after a quick filtration in air reportedly led to a recycled catalyst that was equally effective. Treatment of the catalyst with acid rendered it inactive.

A second example of the synthesis of carbon-carbon bonds carried out with asymmetric catalysts on polymers is hydroformylation. This, too, was the focus of early work by Stille's group [48]. A very successful example of this chemistry was reported recently by Nozaki's group. In their work, a 50% divinylbenzene-cross-linked polystyrene catalyst was prepared as shown in Scheme 3.14 [49].

The resulting polymer containing a mixture of phosphite and phosphine ligands was then suspended in benzene in the presence of $Rh(acac)(CO)_2$ to form the polymer-bound version of a (*R,S*)-BINAPHOS Rh(I) catalyst. Polymerization of the preformed Rh(I) complex of the ligand also successfully produced a hydroformylation catalyst with comparable reactivity and enantioselectivity. Turnovers of 2000 over a 12 h period led to a mixture of branched and normal product from styrene with the branched product (*R*)-2-phenylpropanal predominating (84% first run, 79% in a repeated run). The enantioselectivities were high (89%) with only a slight decrease (86% e.e.) being seen with a recycled catalyst. Higher enantioselectivities were seen when preformed catalyst was polymerized than when the highly cross-linked ligand was first prepared and then treated with Rh(I). Catalyst recycling was accomplished by decanting the product solution away from the insoluble catalyst by forced siphon. Elemental analysis for catalyst residues in the product solution was not explicitly reported. Instead, the absence of yellow coloration in solution due to the catalyst was

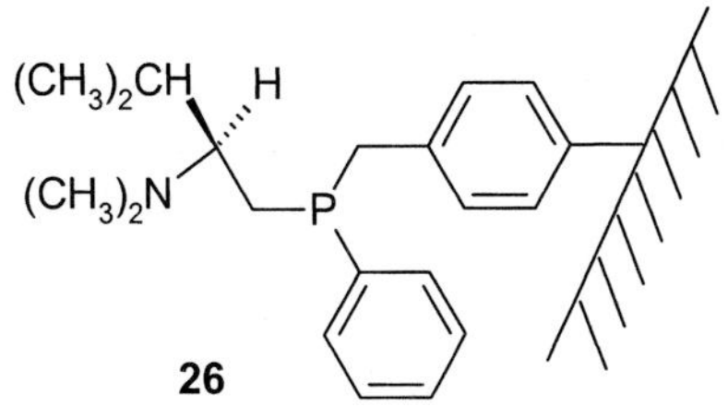

(CH3)2CH H
(CH3)2N
P
26

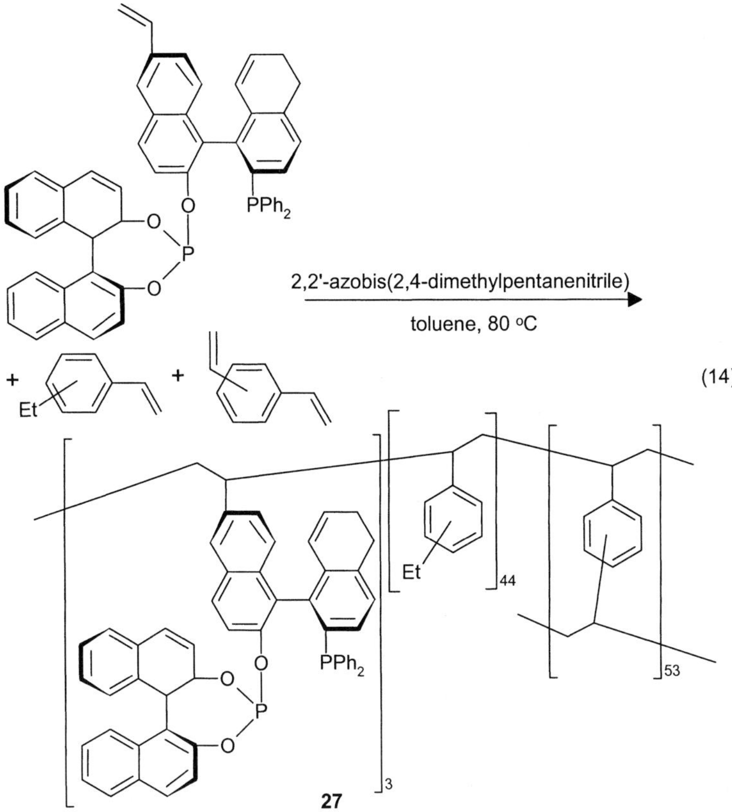

PPh2
O
O
P
O
2,2'-azobis(2,4-dimethylpentanenitrile)
toluene, 80 °C
+ Et +
(14)
Et
44
PPh2
O
O
P
O
27
3
53

cited as evidence for complete catalyst recovery. Vinyl acetate was similarly hydroformylated with this same catalyst using substrate/catalyst ratios of 500/1 to yield (*S*)-2-acetoxypropanal in about 80% yield with enantioselectivities in the 90% range.

A third example of asymmetric synthesis of carbon-carbon bonds using a chiral linear polyamide employed the polymer **28**, 5 mol% Pd(dba)$_2$ and the sodium salt of dimethyl malonate to form a chiral malonate derivative in modest yield (38%) but with good e.e. (80%) within 2 days [50]. A similar polyurea **29** was more effective in terms of yield of the malonate product (72% yield) but with lower e.e. (38% e.e.) [51]. Recovery of the palladium catalyst prepared by using these polymers was effected by filtration. However, the reported formation of palladium black in similar systems may limit the utility of these catalysts in repeated reactions.

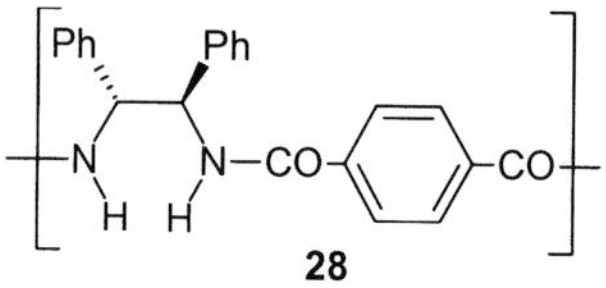

28

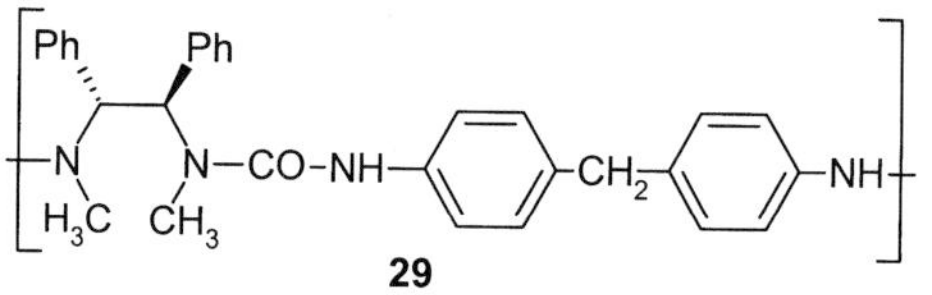

29

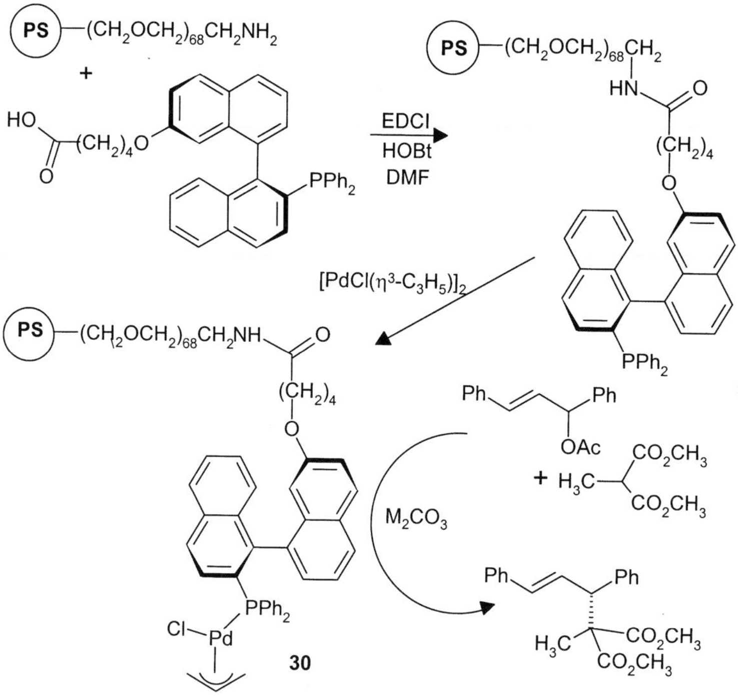

(15)

30

Better success in asymmetric allylic substitution has been reported recently using the amphiphilic resin-supported 2-diphenylphosphino-1,1′-binaphthyl Pd(0) catalyst **30** [52]. This chemistry occurs in aqueous media, the polystyrene resin having been rendered hydrophilic by the addition of poly(ethylene glycol) groups. While the results obtained varied somewhat with the size of the tether and with the nature of the inorganic base used (Scheme 3.15), good enantioselectivity (75–84%) and modest yields (50–75%) were typical. Recycling of the catalyst would presumably be accomplished by filtration. Examples of recycling were not provided in this initial report. Recycling in this case may be problematic because of the known sensitivity of Pd(0) phosphine complexes to adventitious oxygen.

A fourth example of carbon-carbon bond synthesis using an asymmetric catalyst on a polymer support was reported as part of collaborative work by the Bergbreiter and Doyle groups [53]. This work used a terminally functionalized polyethylene oligomer derived from anionic polymerization of ethylene to bind 2-pyrrolidone-5(*S*)-carboxylic acid (Scheme 3.16). This ligand, in turn, was used to prepare a dirhodium(II) 2-pyrrolidone-5(*S*)-carboxylate (PE-Rh_2(5(*S*)-PYCA)$_4$) (**31**) by ligand exchange with the enantiomerically pure 2-pyrrolidone-5-carboxylate catalyst. The resulting catalyst was then applied in asymmetric C-H insertion reactions and in intramolecular cyclopropanations. The isolated yield through 7 runs was essentially constant in Scheme 3.17. The initially high enantioselectivity (98% e.e.) gradually decreased to 61% by run 7, a value that was comparable to the value of 77% e.e. obtained with the low-molecular-weight analog $Rh_2(5(S)\text{-MEPY})_4$ in refluxing benzene in the intramolecular insertion shown in scheme 3.17. A novel feature of this chemistry is that the enantioselectivity was enhanced by the addition of some of the low-molecular-weight ligand. Prior work by the Bergbreiter group had shown that polyethylene ligand precipitation entropically favored metal or catalyst recovery [54]. Thus, the presence of 2.7 mol% low molecular weight ligand did not lead to rhodium leaching but could usefully enhance enantioselectivity.

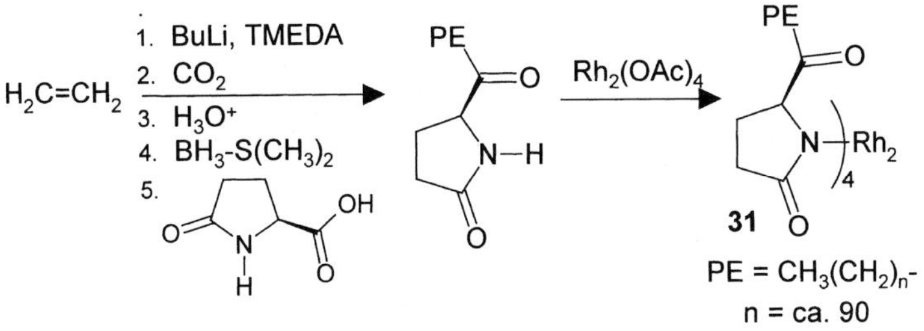

$H_2C{=}CH_2$ — 1. BuLi, TMEDA; 2. CO_2; 3. H_3O^+; 4. $BH_3\text{-}S(CH_3)_2$; 5. (pyroglutamic acid) → PE-amide N-H → $Rh_2(OAc)_4$ → **31** (16)

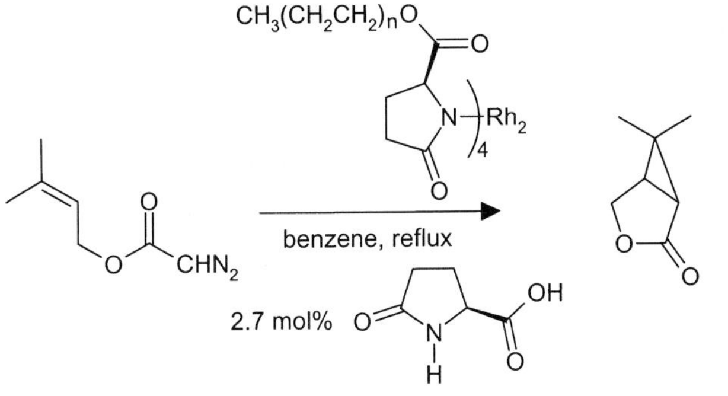

PE = $CH_3(CH_2)_n$-
n = ca. 90

$CH_3(CH_2CH_2)_nO$ … N–Rh$_2$ (4); benzene, reflux; 2.7 mol% (pyroglutamic acid); CHN_2 (17)

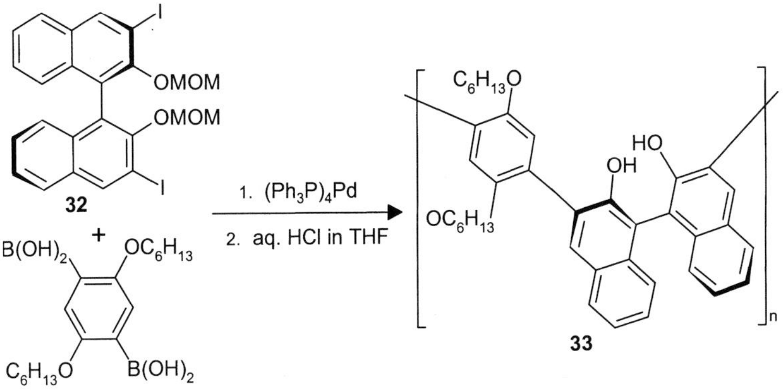

32 + $B(OH)_2$, OC_6H_{13}, $C_6H_{13}O$, $B(OH)_2$ — 1. $(Ph_3P)_4Pd$; 2. aq. HCl in THF → **33** (18)

3.5 Carbonyl Alkylation

Polymer-bound chiral ligands have also been used as catalysts for the 1,2-addition of dialkyl and diarylzinc reagents to carbonyl compounds. In a series of publications on chiral polymeric catalysts for this chemistry [55], Pu's group first prepared the chiral polymer **33** (Scheme 3.18) which, at 5 mol%, promoted the addition of diethylzinc to *p*-substituted benzaldehydes and to various aliphatic aldehydes with e.e. values in the 73–93% range [56]. However, subsequent comparison of the polymeric binaphthol with a low molecular weight analog **34** suggested that the alkoxy groups on the phenylene spacer serve as dual ligands for both adjacent binaphthyl units. Using chemistry like that shown in Scheme 3.18, a second polymer **36** was prepared from monomers **32** and **35**. As predicted, this polymer was even more effective than **33** with most e.e. values in the high 90% range for addition of diethylzinc to aliphatic and aromatic aldehydes [57].

In these cases, the polymers are rigid rod polymers and the stereochemical environment of the catalysts is constrained to resemble that of the low-molecular-weight ligand analog, thereby avoiding some of the ambiguities that may result in other flexible polymers where there are stereochemically irregular sites on the polymer backbone. Catalyst recovery in these cases is effected by precipitation of the catalyst with methanol.

The application of polymer **36** as a catalyst for 1,2-additions of organozinc reagents to carbonyl compounds has been extended recently to include the addition of diphenylzinc to 1-propanal to form (*S*)-1-phenylpropanol in 82% yield with 85% e.e [58]. Recycling the 20 mol% polymeric catalyst in this case was accomplished with methanol precipitation as in the earlier studies with these chiral polymeric catalysts.

Activity in the late eighties and early nineties from several groups highlighted the utility of chiral aminoalcohols bound to insoluble polymers (**37–40**) as catalysts for

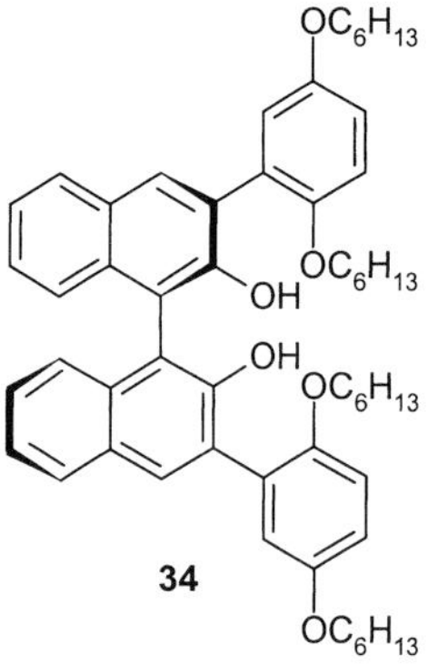

OC_6H_{13} OC_6H_{13} OH OH OC_6H_{13} OC_6H_{13} **34**

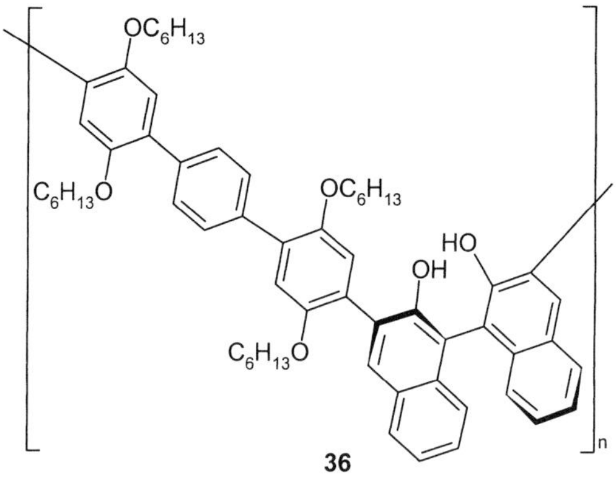

OC_6H_{13} $C_6H_{13}O$ OC_6H_{13} HO OH $C_6H_{13}O$ n **36**

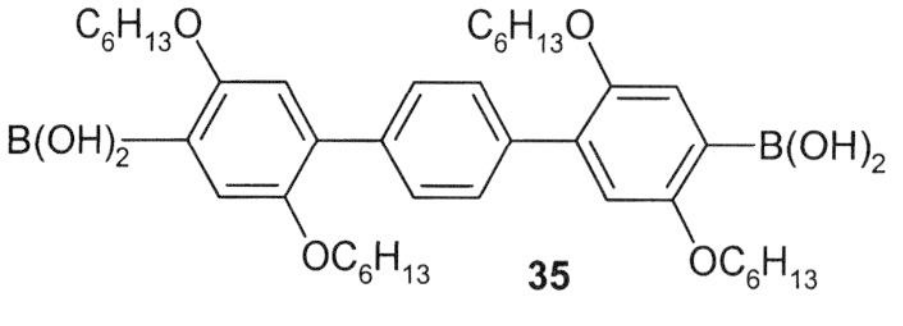

$C_6H_{13}O$ $C_6H_{13}O$ $B(OH)_2$ $B(OH)_2$ OC_6H_{13} **35** OC_6H_{13}

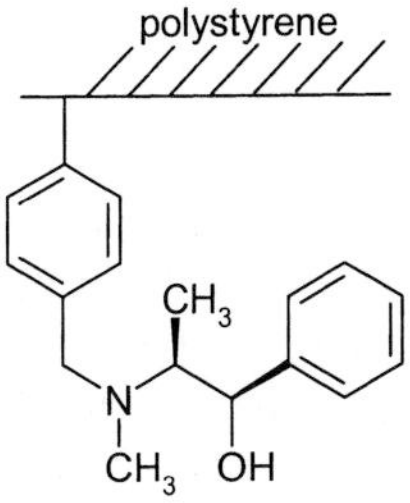

37 (83% yield, 89% e.e.)

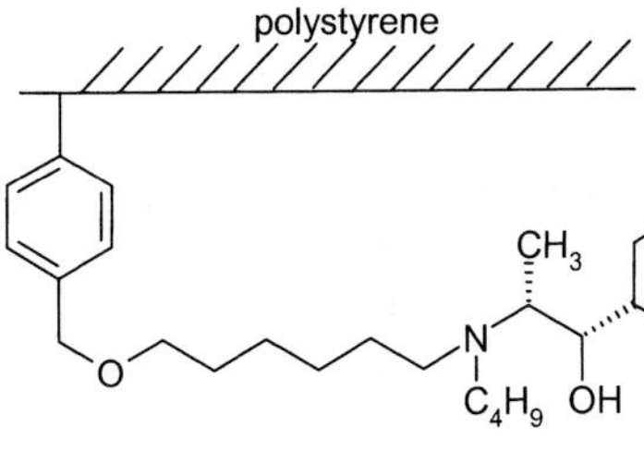

38 (82% yield, 91% e.e.)

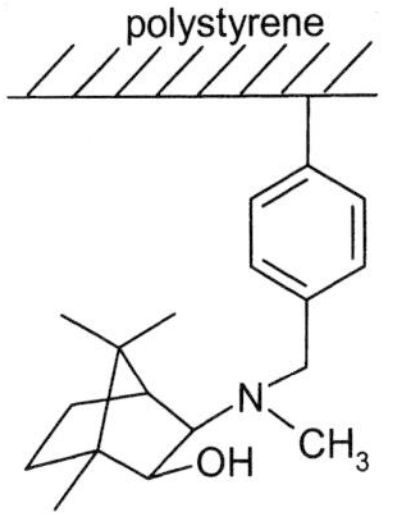

39 (91% yield, 92% e.e.)

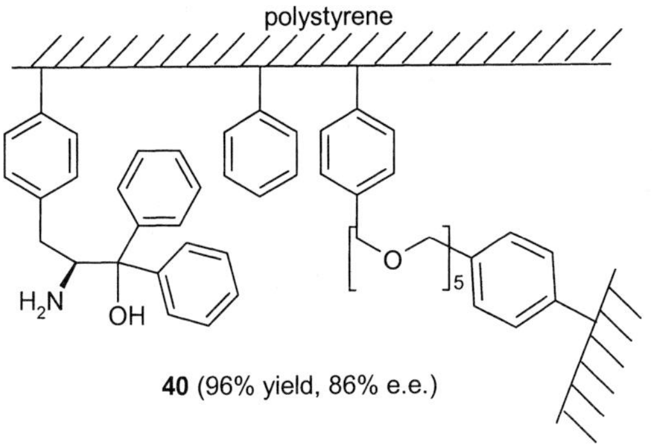

40 (96% yield, 86% e.e.)

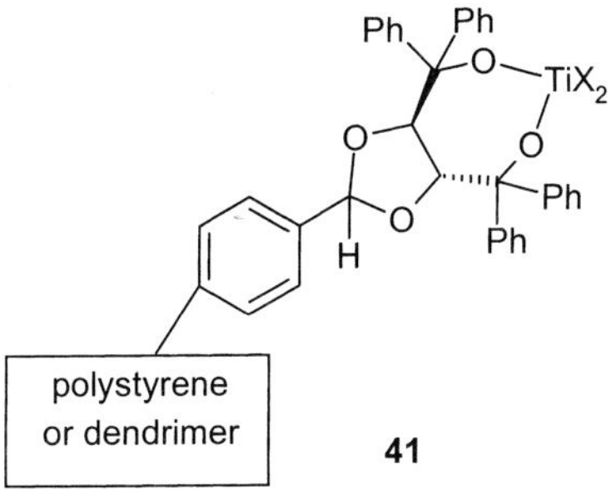

41

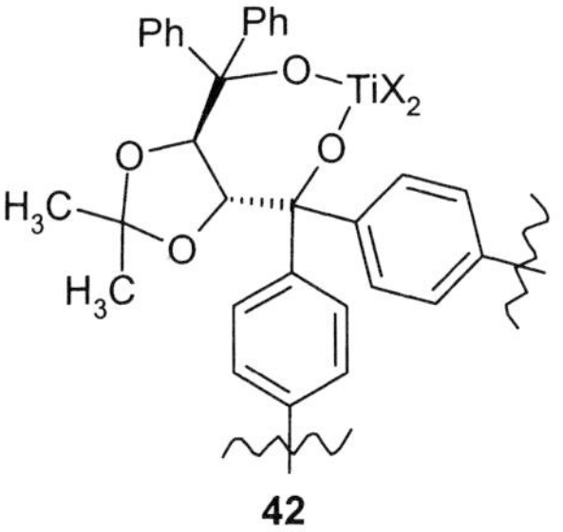

42

ethylation of aryl and aliphatic aldehydes [59–62]. Examples of the polymers that were used in these studies are shown below with yield and e.e. values listed for reaction of diethylzinc with benzaldehyde (a comparable substrate for all examples but not necessarily the best substrate in each study). In all of these cases, the supporting polymer was polystyrene. Catalyst recovery was generally effected by filtration. Little or no catalyst deactivation was reported in cases where catalyst recovery was explicitly studied. For example, using **40** as a ligand and benzaldehyde as a substrate, five cycles of ethylation of this aldehyde with diethylzinc proceeded with a 79% average chemical yield and an 86% average e.e. In one example, **38**, the usefulness of a tether in increasing the e.e. in ethylation of aliphatic aldehydes was noted. The polymeric catalyst **40** is also of interest in that it was prepared as a highly cross-linked resin with ethylene oxide cross-links. These cross-linking groups facilitated swelling in solvents and led to a significant improvement in asymmetric induction relative to a simpler resin lacking these ethylene oxide cross-links.

Mechanistic questions associated with asymmetric alkylations of aldehydes by dialkylzinc reagents have been discussed [63], but the issue of non-linearity in asymmetric induction previously noted for similar chemistry in solution has not been well studied with these polymeric catalysts for ketone alkylation. Such issues may have profound effects on the usefulness of these and other polymeric catalysts, and more recent work has noted these effects in cases where other polymer-bound enantioselective catalysts are used (*vide infra*).

Polymer- and dendrimer-bound titanium TADDOLates like **41** ($X=OCH(CH_3)_2$) have also been described. These catalysts are useful in asymmetric 1,2-addition of diethylzinc or of *in situ* formed alkyltitanium alkoxides to aldehydes [64]. In this chemistry, 20 mol% of the polymeric or dendrimer-bound TADDOLate was used as a Lewis acid. Polymers used were derived from chloromethylated polystyrene or were prepared by suspension polymerization of a styrene derivative of TADDOL in the presence of styrene and divinylbenzene. Typical catalyst loadings were 0.33–0.37 mmol/g although one example (**42**) was described with a very high loading of 1.22 mmol of TADDOL/g of polymer (the TADDOL monomer alone would have a 1.94 mmol/g loading). Interestingly, the heavily loaded polymer seemed to be almost

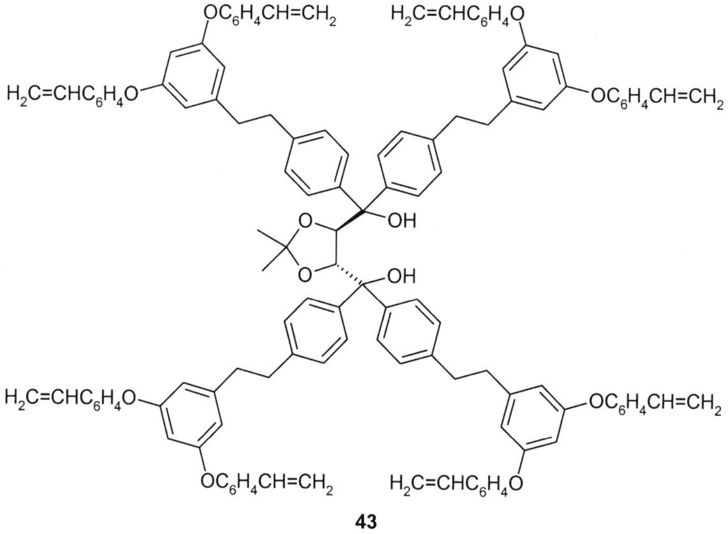

$OC_6H_4CH{=}CH_2$ $H_2C{=}CHC_6H_4O$ $H_2C{=}CHC_6H_4O$ $OC_6H_4CH{=}CH_2$ OH OH O O $H_2C{=}CHC_6H_4O$ $OC_6H_4CH{=}CH_2$ $OC_6H_4CH{=}CH_2$ $H_2C{=}CHC_6H_4O$

43

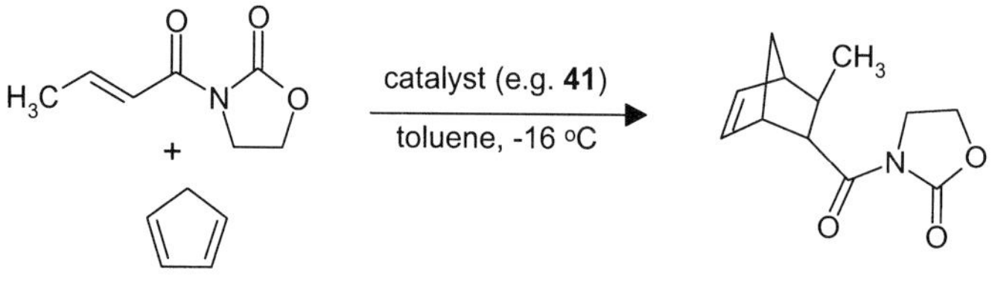

H_3C O O N O + catalyst (e.g. **41**) toluene, -16 °C CH_3 N O O O (19)

as effective in promoting the reaction of diethylzinc and benzaldehyde as the less loaded polymers. The dendrimers were derived from 3,5-dihydroxybenzyl alcohol and *p*-chloromethylbenzaldehyde. However, an excess (1.2 equiv) of $Ti(OCHCH_3)_4$ was required. Yields and enantioselectivities were very good with typical numbers for addition of diethylzinc to benzaldehyde being in the 94–99% range. Recycling of the polymeric catalyst up to four times with 2 mmol-scale reactions was also demonstrated.

More recent work by Seebach's group used has hyperbranched polystyrene-supported TADDOLates **43** for diethylzinc addition to benzaldehyde that are more effective than the catalysts **41** and **42** [65]. These polymers, at very modest loadings of 0.1 mmol/g of Ti, gave 1-phenylpropanol in consistently high (>96%) e.e. with rates that were not distinguishably different from those of a low-molecular-weight catalyst. In toluene, these polymers were swollen to the extent that they filled the entire reaction volume under the conditions applied. In reactions, the product was isolated by washing the polymer with excess toluene and removing the toluene solvent. An attractive aspect of this work is the demonstrated recoverability and recyclability of the catalysts throughout 20 cycles of catalyst use, washing of the insoluble polymer and reuse.

3.6 Diels-Alder Reactions

Titanium TADDOL complexes like **41,** but with Cl replacing isopropoxy as a ligand, are stronger Lewis acids and are useful as asymmetric catalysts in Diels-Alder reactions (Scheme 3.19) [64]. The polymeric Lewis acids are kinetically as effective as their low molecular weight solution counterparts but generally give low enantioselectivities. A positive feature is the demonstrated recovery/recyclability, wherein nine batch reactions of a Diels-Alder reaction between 3-crotonoyloxazolidinone and cyclopentadiene were performed without decreasing yield or enantioselectivity. In this example, a nonlinear relationship between catalyst enantiomeric purity and product enantiomeric purity like that discussed by Noyori [63] was seen in a Diels-Alder reaction with a soluble TADDOL. However, this issue was not studied for the polymeric catalyst.

TADDOL complexes on polystyrene, like those described by Seebach, were also described by Altava [66]. While they, too, observed low enantioselectivity (the best result was 25% e.e. for the reaction shown in Scheme 3.19), they were able to obtain high conversions with less of an excess of the diene substrate.

Other chiral Lewis acid catalysts, bound as a polystyrene sulfonamide to polystyrene cross-linked with a mixture of divinylbenzene and a flexible bis(styryl)methyl poly(ethylene oxide) as cross-linking agents, are reportedly more effective in the Diels-Alder reaction of methacrolein and cyclopentadiene [67, 68]. For example, the chiral oxazaborolidinone **44** prepared in Scheme 3.20 was quite effective in Diels-Alder chemistry. Up to 95% e.e. (88% yield) was obtained in the reaction of methacrolein and cyclopentadiene using a polymer with an oligo(ethylene oxide) cross-linking

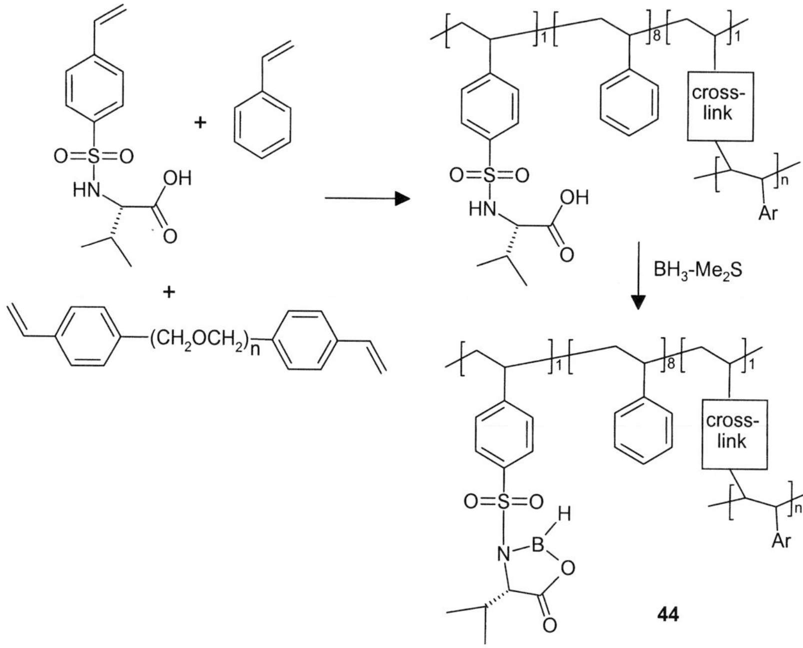

(20)

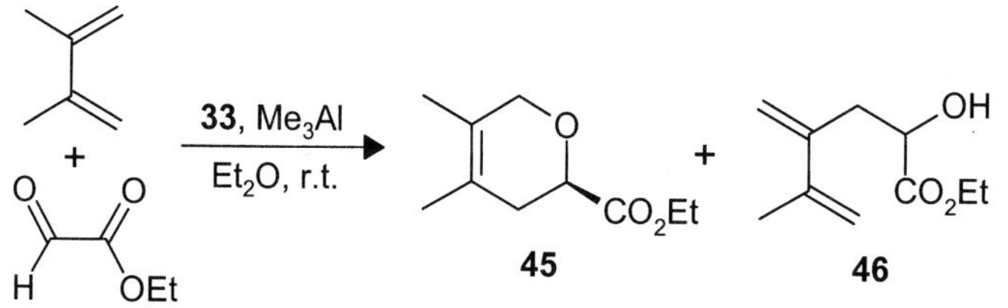

(21)

group containing ca. 10 oxygen atoms. Comparison of the enantioselectivities with the lower enantioselectivities seen with similar oxazaborolidinone catalysts on resins containing only divinylbenzene cross-links illustrates how resin structure can beneficially influence polymer-supported asymmetric catalysis.

A chiral polybinaphthyl-aluminum complex derived from the *in situ* reaction of **33** with Me_3Al is effective in the hetero-Diels Alder reaction of ethyl glyoxylate and 2,3-dimethylbutadiene (Scheme 3.21) [69]. While a mixture of Diels-Alder (**45**) and ene products (**46**) are produced, the Diels-Alder product is the predominant product (5:1) and is formed with 88% e.e. The ene product's e.e. is lower (38% e.e.). Recycling of this catalyst was analogous to the approach described earlier [56] and yielded a similar product mixture with similar high enantioselectivity for the Diels-Alder product.

The same chiral polymeric Lewis acid catalyst derived from Me_3Al and **33** has also been used in the 1,3-cycloaddition reaction of nitrones with alkyl vinyl ethers (Scheme 3.22) [70]. In this case, a variety of isoxazolidines were obtained in high synthetic yields with even higher diastereomeric (>98:2 *exo*:*endo*) and enantiomeric (>93% e.e.) purities. Four repetitive reactions where the catalyst was recycled were

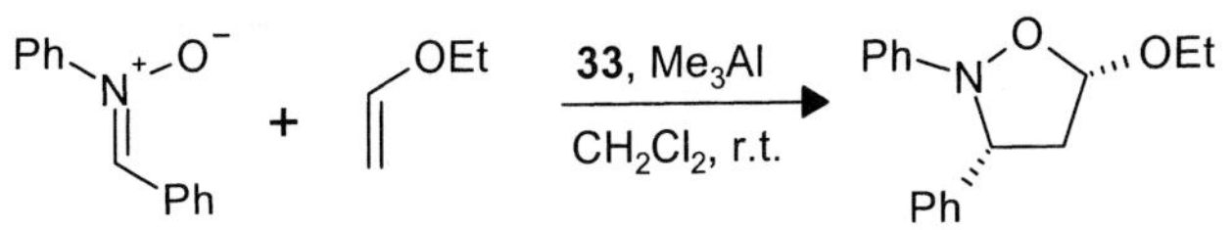

Ph, N, O, Ph + OEt — **33**, Me_3Al, CH_2Cl_2, r.t. → Ph, N, O, OEt, Ph (22)

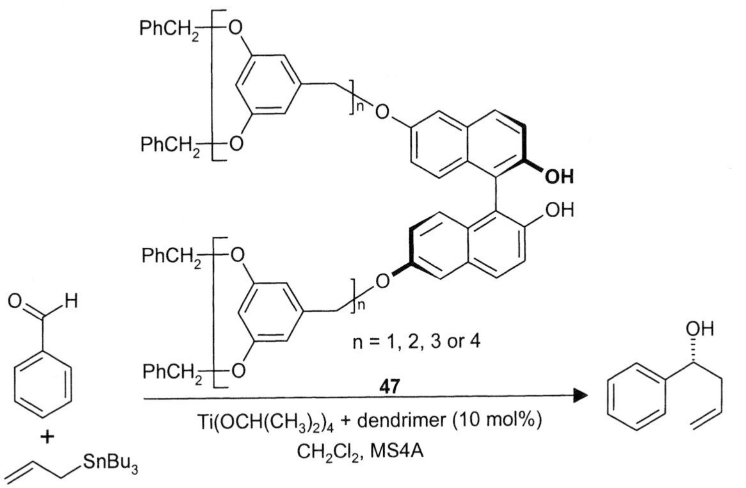

$PhCH_2$, O, n, OH, OH, n = 1, 2, 3 or 4

O, H + $SnBu_3$ — **47**, $Ti(OCH(CH_3)_2)_4$ + dendrimer (10 mol%), CH_2Cl_2, MS4A → OH (23)

described with the fourth cycle yielding the product with slightly lower enantioselectivity and a slightly lower synthetic yield. The gradual decrease in yield and enantioselectivity was ascribed to the small scale of the reactions and physical loss of catalyst. However, some catalyst decomposition may be occurring unless enantioselectivity is affected by small changes in catalyst concentration.

Dendrimers are a relatively new type of polymer and their use as supports for asymmetric catalysis has only just begun. TADDOL-containing dendrimers containing Lewis acids were discussed above as catalysts for diethylzinc additions to aldehydes and as Diels-Alder catalysts. Another example of this chemistry is found in the synthesis of a chiral dendritic binaphthol (**47**) and the use of its metal complexes in allylation of benzaldehyde (Scheme 3.23) [71]. This preliminary report suggests that very high enantioselectivities (88–92%) can be obtained in this chemistry though the reported synthetic yields are disappointing (18–36%). Recovery and reuse of the titanium-containing dendritic catalyst was also not described but could presumably be accomplished by membrane filtration – a technique that should be generally useful in recovery and separation of dendrimer supports.

3.7 Enolate Chemistry

Catalytic asymmetric enolate chemistry using polymeric enantioselective catalysts has received only limited attention. In early work, a polymer-bound proline was used in asymmetric Robinson cyclization chemistry [72]. This chemistry, developed in the early seventies is a classical asymmetric reaction used in steroid synthesis. In the early work, Takemoto attached *N,O*-protected *L*-hydroxyproline to DVB-cross-linked polystyrene to form the polymer-bound proline analog, which could reportedly be recovered and reused, presumably by filtration. However, the enantioselectivities seen (<40%) are far inferior to those obtained in solution chemistry.

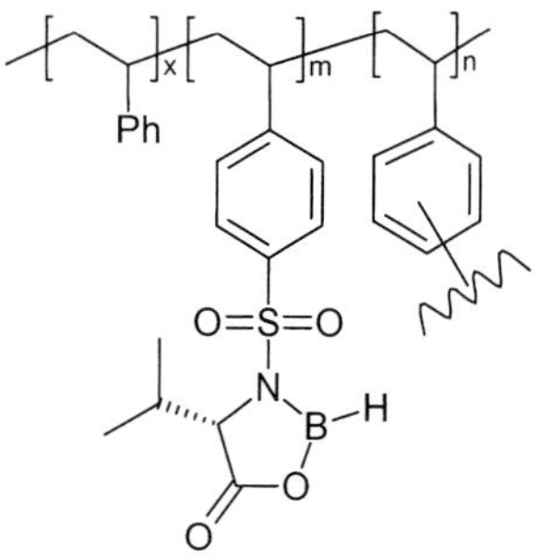

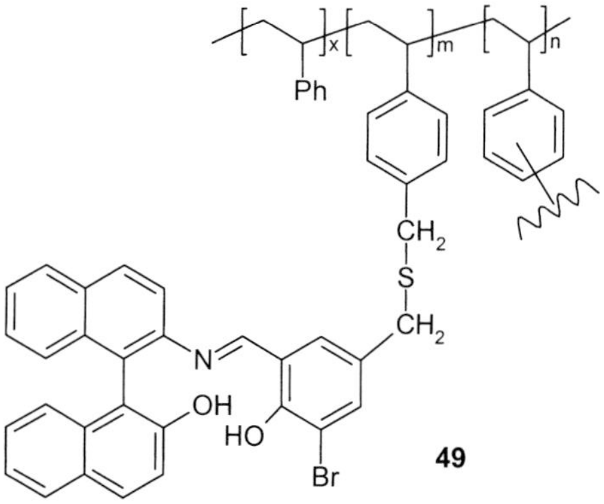

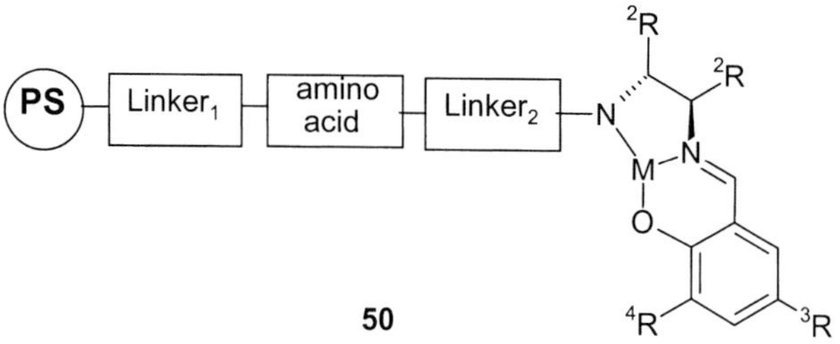

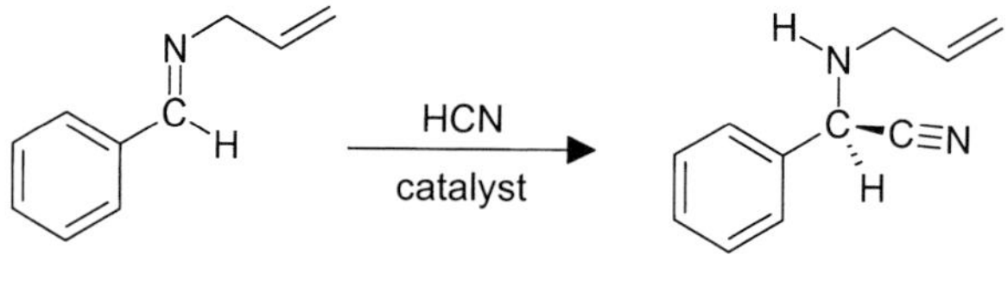

Other approaches to the development of useful polymeric catalysts for enolate chemistry have focused on polymeric Lewis acids but have not been significantly more successful in general. Typical enantioselectivities seen with various titanium and boron catalysts or reagents are less than 50% – a value that has to be increased substantially to be useful in synthesis [73, 74]. Low yields are also a problem in these cases. Nonetheless, this is an important area in organic catalysis, and it is likely that other polymer-supported catalysts can be devised that will be sufficiently reactive and selective to be useful. Polymers **48** and **49** are examples of the materials that have been used in stoichiometric and catalytic reactions of silyl ketene acetals with benzaldehyde.

3.8 Strecker Chemistry

Asymmetric addition of cyanide to carbonyl and imine groups is a reaction of considerable interest, since it leads to precursors for α-hydroxyl and α-aminoacids. This chemistry is especially attractive given that nature already provides us with examples of macromolecular catalysts (enzymes) that promote this reaction, and because simple synthetic dipeptides are also effective asymmetric heterogeneous catalysts for this chemistry [75]. However, while simple cyclic dipeptides like cyclo-Phe-His are effective catalysts for this chemistry [75], a cyclo-[(*S*)-Tyr-(*S*)-His] bound as a phenyl ether through the tyrosine phenolic group to chloromethylated polystyrene was an efficient, albeit not very enantioselective, catalyst for cyanohydrin formation from aromatic aldehydes [76].

A more successful approach to the development of polymer-bound catalysts for the asymmetric Strecker reaction has been described by Jacobsen [77]. In this work, they used combinatorial chemistry and polymer-supported catalysts to identify the best catalysts for this chemistry. The general idea was to design a tridentate polymer-bound Schiff base catalyst like **50** by using various tethering groups (Linker 1 and Linker 2), various amino acids, various chiral 1,2-diamines and various salicylaldehyde derivatives. The libraries of ligands were prepared using solid phase synthesis, and their metal complexes were analyzed for their ability to catalyze the Strecker reaction shown in Scheme 3.24. The best catalyst identified in these screens was the metal-free catalyst **51** ($^{1}R = -C(CH_3)_3$; $^{2}R = $cyclohexyl; $^{3}R = -OCH_3$) which gave the Strecker adduct in 80% e.e. To verify this result, the low molecular weight *N*-benzyl analog of this catalyst was synthesized independently and used as a soluble catalyst for Scheme 3.24, creating the product in 78% yield with 91% e.e. Since catalyst design was the emphasis of this work, catalyst reuse was not discussed. However, it seems likely that these insoluble enantioselective polymeric catalysts could be reused after filtration.

3.9 Asymmetric Dihydroxylation

Asymmetric dihydroxylation of olefins as described by Sharpless represents one of the most successful asymmetric catalytic reactions yet developed. It is not surprising

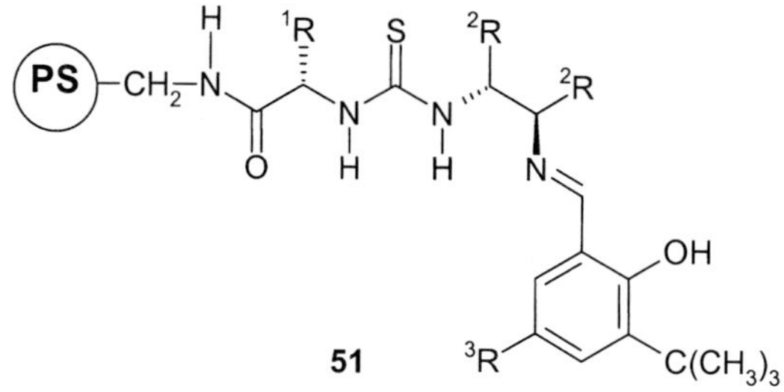

51

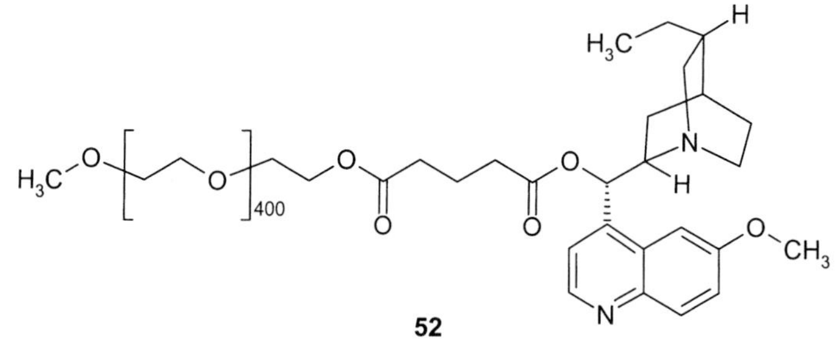

52

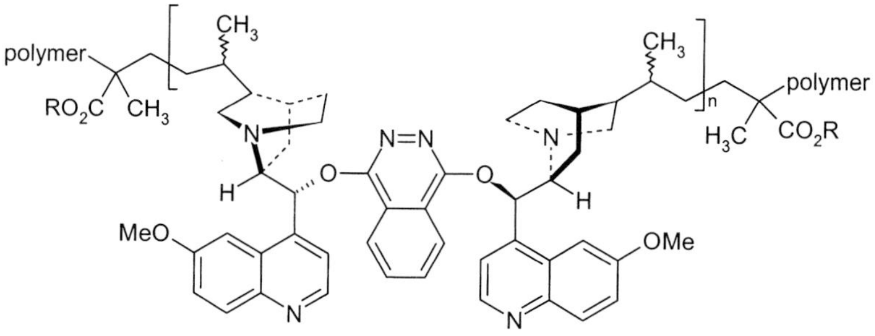

53; R = CH_2CH_2OH

54; R = CH_3

that this reaction should have attracted attention and that many groups should have explored techniques for catalyst/ligand immobilization and recovery. The work in this area has been summarized in a recent review [78].

Asymmetric dihydroxylation is also a particularly interesting case study for polymer-supported asymmetric catalysis because it is one example where several types of polymers have been promoted by different groups for catalyst recovery. The polymers used include insoluble polymers based on DVB-cross-linked polystyrene, insoluble polymers prepared from acrylonitrile and a soluble poly(ethylene glycol)-bound polymer. In the case of the PEG-bound catalysts like **52** [78, 79], fast catalysis, high yields and very high enantioselectivities of up to 99% e.e. were obtained. The PEG-bound catalyst **52** was also competent at oxidation of a *trans*-cinnamic acid substrate immobilized on a gel-like, DVB-cross-linked polystyrene polymer containing PEG grafts (Tentagel) [80]. This latter multipolymer system is of interest in asymmetric catalysis, since simultaneous multistep reactions generally remain relatively uncommon. Such multicomponent systems are widely used naturally in the cell. They could be advantageously used in asymmetric synthesis, when one product is formed in equilibrium and a second irreversible reaction drives that equilibrium. Polymer-supported asymmetric catalysts should be especially effective in such a system, because they intrinsically facilitate phase isolation.

The report by Song of a very enantioselective insoluble polymer-bound version of a dihydroxylation catalyst is interesting for several reasons [81]. First of all, the reactions generally proceeded at a reasonable rate with excellent enantioselectivity. For example, stilbene dihydroxylation proceeded within 15 h at $-10\,^{\circ}C$ giving a 93% yield and >99.9% e.e.. However, an equally instructive feature about polymer immobilization of catalysts is seen with polymer **53** which was so swollen that filtration was problematic. Polymer-bound catalyst **54** did not have this handling problem. This point is interesting in that handling of polymeric catalysts, their physical fragility or filterability is not always clearly addressed in publications. Nonetheless, polymer properties can change significantly with small changes in structure, and the polymer properties are a consideration when catalyst recovery with polymers is the ultimate goal.

Asymmetric dihydroxylation with insoluble polymer-supported ligands has also been discussed by Salvadori [82]. This work shows that a catalyst on an optimized cross-linked polymer can be as kinetically and stereochemically efficient as a catalyst with either low molecular weight or soluble polymeric ligands. However, while ligand recovery with the insoluble polymer is successful in this case, leaching of osmium is a problem. Small amounts of osmium have to be added in recycling experiments to maintain the original catalytic activity.

While there are differences between soluble and insoluble polymers in terms of their ease of development and use as supports for catalysis, results of studies by Janda's, Bolm's, Song's and Salvadori's groups show that highly enantioselective catalysis can be obtained by using either soluble or insoluble supports. These studies have also compared organic polymer supports with inorganic silica gel supports [83]. In the case of insoluble polymers, optimization of the polymer structure is clearly important in achieving high rates and high e.e.'s. Optimum ligand structure is clearly important in all cases.

A very different role for polymers in catalyst recovery is catalyst encapsulation. This idea has been demonstrated with several catalysts including dihydroxylation catalysts [84, 85]. In this approach to catalyst recovery, the polymer physically envelops the catalyst. This microencapsulation approach to catalyst recovery has some advantages. For example, in the case of a dihydroxylation, it avoids problems with osmium loss noted with other catalysts ligated by insoluble polymers [82]. This relatively new approach to catalyst recovery with polymers was demonstrated first with Lewis acids ($Sc(OTf)_3$), and it should find utility in asymmetric catalysis.

3.10 Epoxidation and Epoxide Ring Opening

A polymer-supported version of Sharpless' allylic alcohol epoxidation catalyst using the tartrate chiral auxiliary as part of the polymer main chain has been reported [86]. While mixed results were obtained, one of the better examples used the disodium salt of tartaric acid to form a polyester with a mixture of *o*- and *p*-bis(chloromethyl)benzene. The resulting polymeric tartrate ester gave 68% e.e. and 80% isolated yield in oxidation of *E*-2-hexenol.

Epoxidation using chiral salen Mn(III) complexes is another very well established catalytic asymmetric reaction in solution chemistry. Immobilization of these catalysts has been carried out in several ways, typically by attaching the phenyl ring of the salicylaldehyde to a resin either via a polymerizable styrene group or via an ether bond [87]. Various chiral diamines have then been used to prepare the chiral salen ligand. Examples from Laibinis (**55**) [88], Salvadori (**56**) [89] and Sivaram (**57**) [90] illustrate these approaches. Oxidations were then carried out with various substrates and with various oxidants. Enantioselectivities, yields and oxidants for asymmetric epoxidation of *Z*-1-phenylpropene are listed below.

Catalytic ring opening of *meso* epoxides and kinetic resolution of epoxides are also important reactions in synthesis. Polymer-supported enantioselective catalysts have recently been the focus of several studies in this area. As part of the development of new combinatorial syntheses of catalysts, Snapper and Hoveyda have described the development of polystyrene-bound peptidic Schiff bases that can complex $Ti(OCH(CH_3)_2)_4$ to form enantioselective catalysts for the addition of trimethylsilylcyanide to *meso* epoxides (Scheme 3.25) [91]. In this chemistry, synthesis of libraries of the Schiff base catalyst ligand **58** was simplified using solid phase chemistry. However, the initial approach to catalyst evaluation used solution-phase assays. This required cleavage of the catalyst from the polymer, a three-day reaction. To simplify further development of other dipeptide Schiff bases, Snapper and Hoveyda opted to test whether the polymer-bound catalysts reliably mimic their soluble analogs. To do this, they correlated the enantioselectivities for solution-phase and solid-phase *meso* epoxide ring-opening for a library of catalysts. This experiment showed that there was a reliable correlation between the e.e. for solid phase catalysts and the same catalyst after resin cleavage, although the enantioselectivities in each reaction were not the same. With this result, a more diverse library varying sections 1, 2 and 3 in **58** was screened in the solid phase with the interesting result that specific ligands optimal for individual substrates (cyclopentene oxide, cyclohexene oxide or cycloheptene oxide) could be identified.

In a second more recent example of this asymmetric catalysis involving epoxide ring-opening chemistry, Jacobsen's groups has attached chiral Co(III)(salen) complexes to polystyrene and silica gel and used these polymeric catalysts in hydrolytic kinetic resolution of terminal epoxides [92]. As noted above, the synthesis of polymer-bound ligands is a general limitation with the use of any sort of polymer-bound catalyst. To simplify the synthesis of **59**, Jacobsen's group used a novel strategy. While the known monophenolic salen ligand precursor could be coupled to the *p*-nitrophenyl carbonate derivative of a hydroxymethylpolystyrene, a simple route pre-

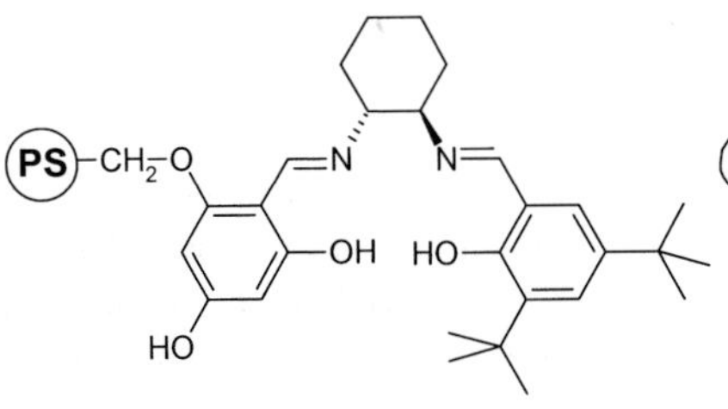

55, 79% e.e.; NaOCl

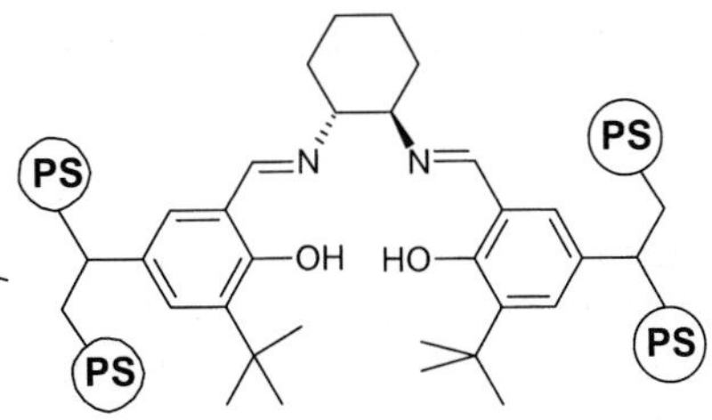

56, <41% e.e.; MCPBA/NMO

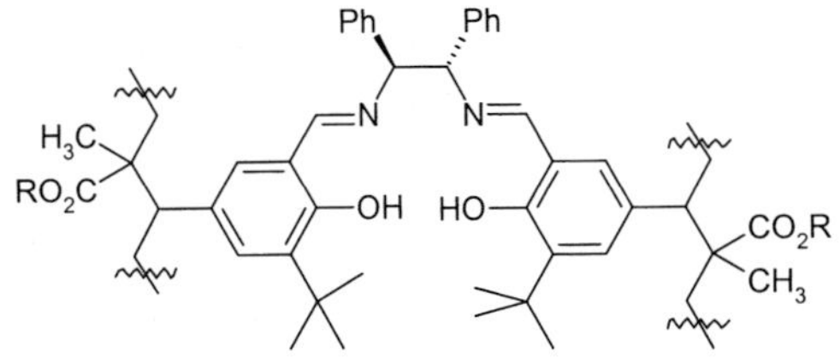

57, <30% e.e.; pyridine *N*-oxide

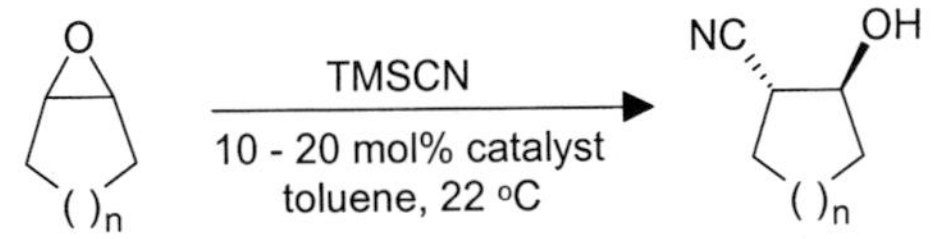

TMSCN
10 - 20 mol% catalyst
toluene, 22 °C
(25)

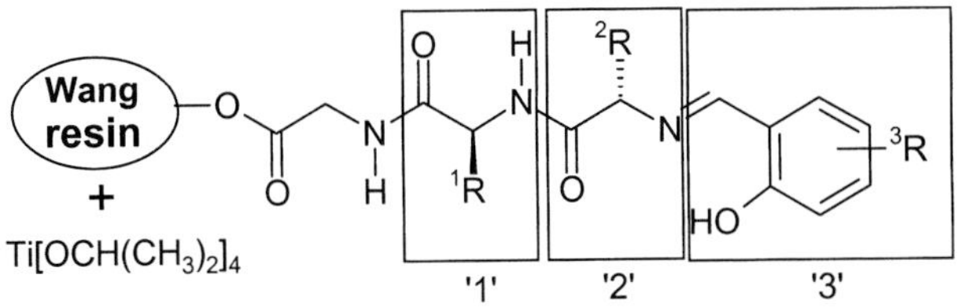

Wang resin
+
$Ti[OCH(CH_3)_2]_4$

'1' '2' '3'

58

pared a statistical mixture of salen ligand precursors. This mixture, derived from the reaction of *R*,*R*-1,2-diaminocyclohexane, excess di-*tert*-butylsalicylaldehyde and 2,5-dihydroxy-3-*tert*-butylbenzaldehyde, was then coupled to the *p*-nitrophenylcarbonate resin. Selective capture of the monophenolic precursor (38% of the mixture) avoided chromatographic separations in the synthesis of the salen ligand precursor. When the Co(III)(salen) complex derived from this synthesis was allowed to react with racemic epichlorohydrin, a mixture of **60** and **61** was obtained. At 51–52% conversion, the e.e. of the recovered **60** was greater than 99% throughout 5 cycles, and the e.e. of the hydrolysis product **61** was consistently 93–95%. Low catalyst loadings (0.25 mol%) and the insolubility of the catalyst made this kinetic resolution practical in that it avoided chloride-catalyzed racemization of **60** in the distillation required when a soluble catalyst was used (Scheme 3.26). The advantages of the insoluble catalyst **59** in dynamic kinetic resolution of epibromohydrin, in the hydrolytic kinetic resolution of

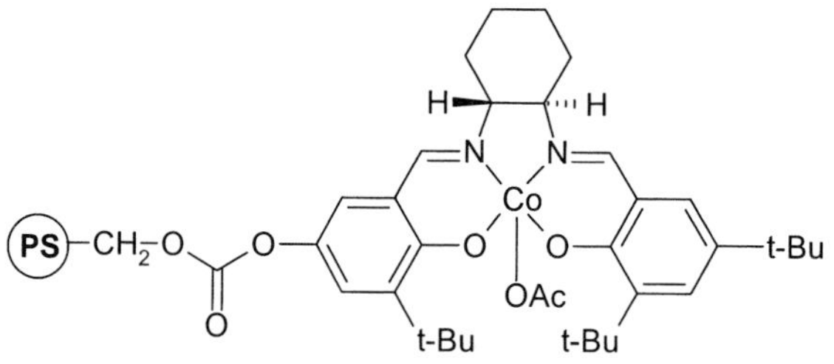

59

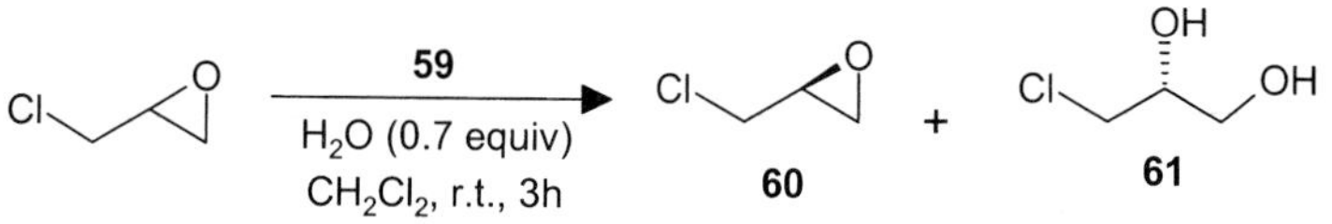

59, H_2O (0.7 equiv), CH_2Cl_2, r.t., 3h → 60 + 61 (26)

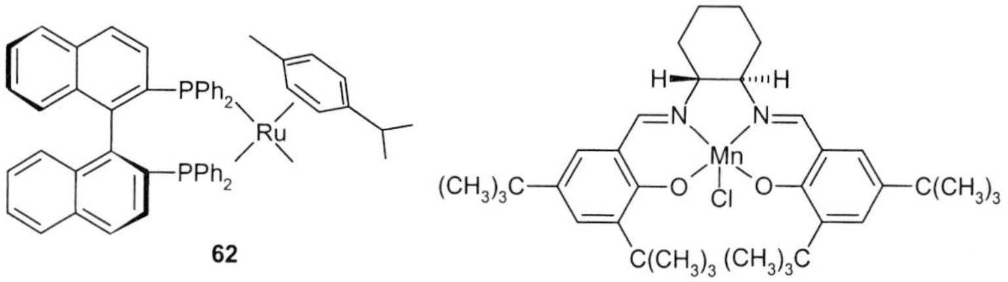

62

63

4-hydroxy-1-butene oxide (to form a triol precursor to a component of an HIV protease inhibitor) and in phenolic ring opening of a terminal epoxide to form the antihypertensive agent Propanolol are all noted by Jacobsen.

While the general strategies described in this chapter focus on chemical immobilization of a chiral catalyst by a polymer in order to facilitate catalyst recovery and reuse, alternative strategies using physical immobilization have also been reported. As noted above and in Chapter 2, this is most commonly accomplished by using a second liquid phase. However, Vankelecom has also described immobilization of chiral catalysts like **62** or **63** in a polydimethylsiloxane (PDMS) membrane [93, 94]. In these cases, the transition metal catalyst is simply mixed with the linear prepolymer polydimethylsiloxane chains containing terminal vinyl groups and a cross-linker. After removal of a solvent, a catalyst membrane with a 0.2 mm average thickness was obtained. The loading of catalysts was modest (0.108 μmol/cm^2 for **62** and 0.629 for **63**) but the immobilized catalysts had good selectivity in hydrogenation (**62**, typically 60–70% e.e. for methyl acetoacetate hydrogenation at 60 or 80 °C). The catalyst **63** immobilized in a PDMS membrane was successfully used in epoxidation. In a typical reaction, a terminal alkene, like styrene, was epoxidized by using NaOCl at 4 °C with a conversion of 78% to styrene oxide having 57% e.e. Leaching of catalysts in this system can be problematic, but this can be redressed either by changing the solvent or by linking the complexes covalently to the polymer.

3.11 Acylation Catalysts

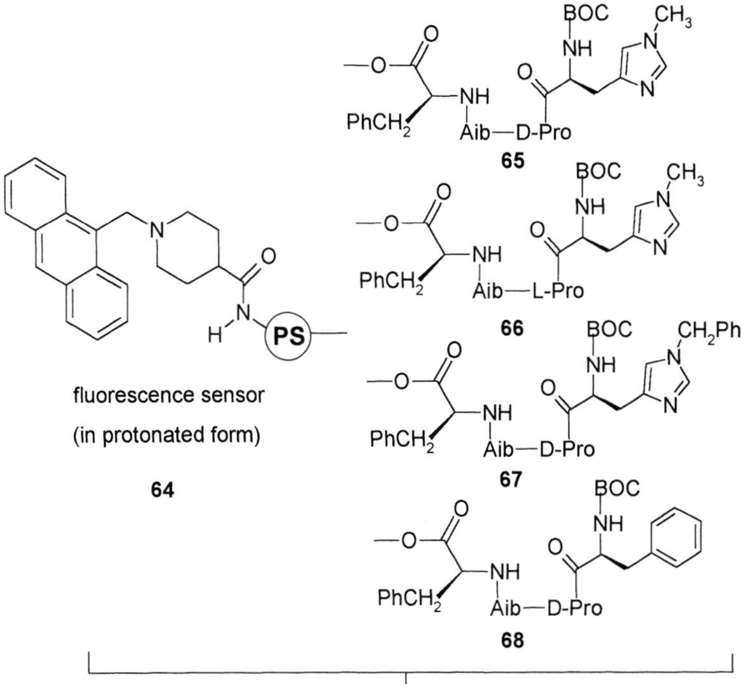

The use of polymers to immobilize ligands or catalysts as part of a combinatorial route to develop new asymmetric catalytic chemistry has developed in recent years into an increasingly important theme. The previous discussion pointed out several examples of this chemistry. A final example that incorporates a polymer-bound sensor and a polymer-bound enantioselective catalyst on the same resin bead shows the potential of this chemistry and the unique features of polymer supports in facilitating this chemistry [95]. In this example, a polymer like **69 a–d** was prepared which contains one of the chiral peptidic catalysts **65**, **66**, **67** or **68** for acylation as well as a fluorimetric sensor **64** that detects the release of acetic acid upon acylation. In this chemistry, a hydroxylmethyl containing polystyrene was functionalized first with the fluorescent probe aminomethylanthracene. This probe **64** is fluorescent on protonation and non-fluorescent otherwise. The resin mixture was subsequently divided in four portions, and a peptide catalyst, (**65**, **66**, or **67**) along with a non-catalyst **68**, was attached to each portion of aminomethylanthracene-containing beads. Subsequent acylation with acetic anhydride and racemic *trans*-1-acetamido-2-hydroxycyclohexane by using the beads individually, pair-wise or altogether successfully revealed the most active catalyst to be **65** based on the greater fluorescence of these beads due to a more effective kinetic resolution (more extensive acylation) which produced more acetic acid which was captured in an intraresin manner.

3.12 Conclusion

To summarize, the use of polymer supports in enantioselective catalysis continues to expand. Practical catalysis, novel separation strategies and new ways to develop catalysts are all demonstrated successes with polymer-supported enantioselective catalysts. There seems to be little doubt that the still untapped variety of polymer supports, the range of supporting chemistry and the still undisclosed ways to use polymers will further increase chemistry's repertoire of polymer-supported enantioselective catalysts.

References

[1] J.K. Stille, *J. Macromol. Sci., Chem.* **1984**, *A21*, 1689–93.
[2] E.C. Blossey and W.T. Ford in *Comprehensive Polymer Science. The Synthesis, Characterization, Reactions and Applications of Polymers*; D.C. Sherrington and P. Hodge, Eds.; Wiley, Chichester, 1988.
[3] S. Itsuno in *Polymeric Materials Encyclopedia; Synthesis, Properties and Applications*; J.C. Salamone, Ed., Vol. 10; CRC Press: Boca Raton, FL, 1996; p. 8078.
[4] S.J. Shuttleworth, S.M. Allin and P.K. Sharma, *Synthesis* **1997**, 1217–1238.
[5] H.U. Blaser and M. Studer, *Chirality* **1999**, *11*, 459–64.
[6] D.E. Bergbreiter *ACS Symposium Series* **1982**, *192*, 1–9. D.E. Bergbreiter *ACS Symposium Series* **1982**, *192*, 31–43.
[7] "Green Chemistry", P.T. Anastas and T.C. Williamson, Eds.; Oxford University Press, Oxford, UK, 1998.
[8] D.E. Bergbreiter *Catal. Today* **1998**, *42*, 389–97.
[9] D.E. Bergbreiter *Macromol. Symp.* **1996**, *105*, 9–16.
[10] D.E. Bergbreiter, B.L. Case, Y.S. Liu and J.W. Caraway *Macromolecules* **1998**, *31*, 6053–62.
[11] C.W. Harwig, D.J. Gravert and K.D. Janda *Chemtracts* **1999**, *12*, 1–26.
[12] "Aqueous-Phase Organometallic Catalysis: Concepts and Applications", B. Cornils and W.A. Herrmann, Eds.; Wiley-VCH, Weinheim, Germany, 1998.
[13] I.T. Horvath *Acc. Chem. Res.* **1998**, *31*, 641–50.
[14] E. de Wolf, G. van Koten and B.J. Deelman *Chem. Soc. Rev.* **1999**, *28*, 37–41.
[15] D.E. Bergbreiter, Y.S. Liu and P.L. Osburn *J. Am. Chem. Soc.* **1998**, *120*, 4250–1.
[16] C. Bianchini, P. Frediani and V. Sernau *Organometallics* **1995**, *14*, 5458–9.
[17] U. Nagel and J. Albrecht *Top. Catal.* **1998**, *5*, 3–23.
[18] N. Takaishi, H. Imai, C.A. Bertelo and J.K. Stille *J. Am. Chem. Soc.* **1978**, *100*, 264–8
[19] T. Masuda and J.K. Stille *J. Am. Chem. Soc.* **1978**, *100*, 268–72.
[20] A. Kazuo *Heterocycles* **1978**, *9*, 1539–43.

[21] G.L. Baker, S.J. Fritschel, J.R. Stille and J.K. Stille, *J. Org. Chem.* **1981**, *46*, 2954–60. G.L. Baker, S.J. Fritschel, and J.K. Stille, *J. Org. Chem.* **1981**, *46*, 2960–5.
[22] D.J. Bayston, J.L. Fraser, M.R. Ashton, A.D. Baxter, M.E.C. Polywka and E. Moses, *J. Org. Chem.*, **1998**, *63*, 3137–40.
[23] S.R. Gilbertson, X. Wang, G.S. Hoge, C.A. Klug and J. Schaefer, *Organometallics*, **1996**, *15*, 4678–80.
[24] C. Köllner, B. Pugin and A. Togni, *J. Am. Chem. Soc.*, **1998**, *120*, 10274–5.
[25] D.E. Bergbreiter and N. Koshti, submitted to *React. Polym.*
[26] P.E. Garrou *ACS Symposium Series* **1986**, *308*, 84–106.
[27] T. Malmstrom and C. Andersson *J. Mol. Catal. A: Chem.* **1999**, *139*, 259–70.
[28] K. Kaneda, H. Yamamoto T. Imanaka and S. Teranishi *J. Mol. Catal.* **1985**, *29*, 99–104.
[29] R. Selke, K. Haupke and H.W. Krause *J. Mol. Catal.* **1989**, *56*, 315–28
[30] I. Toth, B.E. Hanson and M.E. Davis *J. Organometal. Chem.*, **1990**, *397*, 109–17.
[31] J.J. Parlow, R.V. Devraj and M.S. South *Curr. Opin. Chem. Biol.* **1999**, *3*, 320–36.
[32] Q.H. Fan, C.Y. Ren, C.H. Yeung, W.H. Hu and A.S.C. Chen *J. Am. Chem. Soc.* **1999**, *121*, 7407–7408.
[33] S. Bhaduri, V.S. Darshane, K. Sharma and D. Mukesh *Chem. Commun.* **1992**, 1738–40.
[34] A. Baiker *J. Mol. Catal.* **1997**, *A115*, 473–93.
[35] X.B. Zuo, H.F. Liu, D.W. Guo and X.Z. Yang *Tetrahedron* **1999**, *55*, 7787–7804.
[36] A. Adima, J.J.E. Moreau and M.W.C. Man *J. Mater. Chem.* **1997**, *7*, 2331–3.
[37] G. Wulff and J. Vietmeier *Macromol. Chem.* **1989**, *190*, 1727–35.
[38] F. Locatelli, P. Gamez and M. Lamaire *J. Mol. Catal.* **1998**, *135A*, 89–98.
[39] P. Lecavalier, E. Bald, Y. Jiang, J.M.J. Frechet and P. Hodge *React. Polym.* **1985**, *3,* 315–26.
[40] S. Itsuno, Y. Sakurai, K. Ito, A. Hirao and S. Nakahama *Polymer* **1987**, *28*, 1005–8.
[41] S. Itsuno, K. Ito, T. Maruyama, N. Kanda, A. Hirao and S. Nakahama *Bull. Chem. Soc. Jpn.* **1986**, *59*, 3329–31.
[42] C. Caze, N. El Moualij, P. Hodge, C.J. Lock and J.B. Ma *J. Chem. Soc., Perkin Trans. I.* **1995**, 345–9.
[43] M. Felder, G. Giffels and C. Wandrey *Tetrahedron: Asymmetry* **1997**, *8*, 1975–7. S. Rissom, J. Beliczey, G. Giffels, U. Kragl and C. Wandrey *Tetrahedron: Asymmetry* **1999**, *10*, 923–8.
[44] J.C. Briggs, P. Hodge and Z.P. Zhang *React. Polym.* **1993**, *19*, 73–80.
[45] H.-U. Blaser, H.-P. Buser, K. Coers, R. Hanreich, H.-P. Jalett, E. Jelsch, B. Pugin, H.-D. Schneider, F. Spindler and A. Wegmann *Chimia* **1999**, *53*, 275–80.
[46] H. Doucet, T. Ohkuma, K. Murata, T. Yokozawa, M. Kozawa, E. Katayama, A.F. England, T. Ikariya and R. Noyori *Angew. Chem. Int. Ed.* **1998**, *37*, 1703–1707.
[47] T. Hayashi, N. Nagashima and M. Kumada, *Tetrahedron Lett.*, **1980**, *21*, 4623–6.
[48] S. Fritschel, J.J.H. Ackerman, T. Keyser and J.K. Stille, *J. Org. Chem.*, **1979**, *44*, 3152–7.
[49] K. Nozaki, Y. Itoi, F. Shibahara, E. Shirakawa, T. Ohta, H. Takaya and T. Hiyama, *J. Am. Chem. Soc.*, **1998**, *120*, 4051–2.
[50] P. Gamez, B. Dunjic, F. Fache and M. Lemaire, *J. Chem. Soc., Chem. Commun.* **1994**, 1417–8.
[51] P. Gamez, B. Dunjic, F. Fache and M. Lemaire, *Tetrahedron Asymmetry,* **1995**, *6*, 1109–16.
[52] Y. Uozumi, H. Danjo and T. Hayashi, *Tetrahedron Lett.* **1998**, *39*, 8303–6.
[53] M.P. Doyle, M.Y. Eismont, D.E. Bergbreiter and H.N. Gray, *J. Org. Chem.* **1992**, *57*, 6103–5.
[54] D.E. Bergbreiter, J. Poteat and L. Zhang *React. Polym.* **1993**, *20*, 99–109.
[55] L. Pu *Chem. Rev.* **1998**, *98*, 2405–94.
[56] W.-S. Huang, Q.-S. Hu, X.-F. Zheng, J. Anderson and L. Pu, *J. Am. Chem. Soc.*, **1997**, *119*, 4313–4.
[57] Q.-S. Hu, W.-S. Huang and L. Pu, *J. Org. Chem.*, **1998**, *63*, 2798–9.
[58] W.-S. Huang and L. Pu, *J. Org. Chem.*, **1999**, *64*, 4222–3.
[59] K. Soai, S. Niwa and M. Watanabe, *J. Org. Chem.*, **1988**, *53*, 927–8.
[60] K. Soai and M. Watanabe, *Tetrahedron Asymmetry,* **1991**, *2*, 97–100.
[61] S. Itsuno and J.M.J. Frechet, *J. Org. Chem.*, **1987**, *52*, 4140–2
[62] S. Itsuno, Y. Sakurai, K. Ito, T. Maruyama, S. Nakahama and J.M.J. Frechet, *J. Org. Chem.*, **1990**, *55*, 304–10.
[63] R. Noyori and M. Kitamura, *Angew. Chem. Int. Ed.* **1991**, *30*, 49–69.
[64] D. Seebach, R.E. Marti and T. Hintermann, *Helv. Chim. Acta*, **1996**, *79*, 1710–40.

[65] H. Sellner and D. Seebach, *Angew. Chem. Int. Ed.* **1999**, *38*, 1918–20.
[66] B. Altava, M.I. Burguete, B. Escuder, S.V. Luis, R.V. Salvador, J.M. Fraile, J.A. Mayoral and A.J. Royo, *J. Org. Chem.* **1997**, *62*, 3126–34.
[67] K. Kamahori, K. Ito and S. Itsuno, *J. Org. Chem.*, **1996**, *61*, 8321–4.
[68] K. Kamahori, S. Tada, K. Ito and S. Itsuno, *Tetrahedron: Asymmetry* **1995**, *6*, 2547–55.
[69] M. Johannsen, K.A. Jorgensen, X.F. Zheng, Q.S. Hu and L. Pu *J. Org. Chem.* **1999**, *64*, 299–301.
[70] K.B. Simonsen, K.A. Jorgensen, Q.S. Hu and L. Pu *Chem. Commun.* **1999**, 811–2.
[71] S. Yamago, M. Furukawa, A. Azuma and J. Yoshida, *Tetrahedron Lett.* **1998**, *39*, 3783–6.
[72] K. Kondo, T. Yamano, and K. Takemoto, *Makromol. Chem.* **1985**, *186*, 1781–5.
[73] S. Kiyooka, Y. Kido and Y. Kaneko, *Tetrahedron Lett.***1994**, *35*, 5243–6.
[74] A. Mandoli, D. Pini, S. Orlandi, F. Mazzini and P. Salvadori, *Tetrahedron: Asymmetry* **1998**, *9*, 1479–82.
[75] K. Tanaka, A. Mori and S. Inoue, *J. Org. Chem.* **1990**, *55*, 181–5.
[76] H.J. Kim and W.R. Jackson, *Tetrahedron: Asymmetry* **1992**, *3*, 1421–30.
[77] M.S. Sigman and E.N. Jacobsen, *J. Am. Chem. Soc.* **1998**, *120*, 4901–2.
[78] C. Bolm and G. Arne, *Eur. J. Org. Chem.* **1998**, *1*, 21–27.
[79] H. Han and K.D. Janda, *J. Am. Chem. Soc.* **1996**, *118*, 7632–3.
[80] H. Han and K.D. Janda, *Angew. Chem. Int. Ed. Engl.* **1997**, *36*, 1731–3.
[81] C.E. Song, J.W. Yang, H.J. Ha and S.G. Lee, *Tetrahedron: Asymmetry* **1996**, *7*, 645–8.
[82] P. Salvadori, D. Pini and A. Petri *J. Am. Chem. Soc.* **1997**, *119*, 6929–30.
[83] C. Bolm and A. Gerlach, *Angew. Chem. Int. Ed. Engl.* **1997**, *36*, 741–3.
[84] S. Nagayama, M. Endo and S. Kobayashi, *J. Org. Chem.* **1998**, *63*, 6094–5.
[85] S. Kobayashi and S. Nagayama, *J. Am. Chem. Soc.* **1998**, *120*, 2985–6.
[86] L. Canali, J.K. Karjalainen, D.C. Sherrington and O. Hormi, *Chem. Commun.*, **1997**, 123–4
[87] L. Canali and D.C. Sherrington, *Chem. Soc. Rev.* **1999**, *28*, 85–93
[88] M.D. Angelino and P.E. Laibinis *Macromolecules* **1998**, *31*, 7581–7.
[89] F. Minutolo, D. Pini, A. Petri and P. Salvadori *Tetrahedron: Asymmetry* **1996**, *7*, 2293–2302.
[90] B.B. De, B.B. Lohray, S. Sivaram and P, K. Dhal *J. Polym. Sci., Polym. Chem. Ed.* **1997**, *35*, 1809–18.
[91] K.D. Shimizu, B.M. Cole, C.A. Krueger, K.W. Kuntz, M.L. Snapper and A.H. Hoveyda *Angew. Chem. Int. Ed.* **1997**, *36*, 1703–7.
[92] D.A. Annis and E.N. Jacobsen *J. Am. Chem. Soc.* **1999**, *121* 4147–54.
[93] I.F.J. Vankelecom, D. Tas, R.F. Parton, V. Van de Vyver and P.A. Jacobs, *Angew. Chem. Int. Ed. Engl.* **1996**, *35*, 1346–8.
[94] I.F.J. Vankelecom, K. Vercruysse, P. Neys, D. Tas, K.B.M. Janssen, P.-P. Knops-Gerrits and P.A. Jacobs, *Topics in Catalysis* **1998**, *5*, 125–132.
[95] G.T. Copeland and S.J. Miller *J. Am. Chem. Soc.* **1999**, *121*, 4306–7.

4 Liquid Biphasic Enantioselective Catalysis

Brian E. Hanson

4.1 Introduction

Biphasic catalysis is now well established in the chemical community as a general solution to the problem of catalyst/product separation. Several widely recognized processes that use this methodology are the Shell Higher Olefin Process, the Ruhrchemie – Rhône Poulenc Process for the aqueous phase hydroformylation of propene, and the Rhône Poulenc myrcene alkylation step in vitamin E synthesis. Liquid biphasic catalysis is thus an attractive technology for potential commercial application of enantioselective homogeneous catalysis. In developing this chemistry, several unique aspects of biphasic catalysis, with regard to asymmetric transformations, become evident. For example, reaction rate and enantioselectivity may be influenced by the number of ionic groups in a water-soluble ligand or by addition of surfactants.

Just as ligands must be made water-soluble for application in aqueous biphasic catalysis, use of supercritical CO_2 or fluorous solvents requires specific ligand modifications. The preferred modification is usually the incorporation of perfluoroalkyl groups. As described in this review, the liquid biphasic approach has led to excellent results for a series of hydrogenation and biocatalytic reactions.

The practical value of asymmetric homogeneous catalysis for fine chemicals synthesis was first established with the process developed by Knowles at Monsanto for the enantioselective hydrogenation of olefins with a rhodium complex of the chelating chiral phosphine DIPAMP [1]. Monsanto used this process for a number of years to prepare L-DOPA. A closely related process was applied in the former GDR with a sugar-derived chiral phosphine, Ph-β-glup [2, 3]. Moreover, phenylalanine was apparently prepared commercially for a short period in Europe [4] via the asymmetric hydrogenation of α-acetamidocinnamic acid with Rh-PNNP catalysts [5]. The asymmetric hydrogenation of α-acetamidocinnamic acid derivatives is shown in Scheme 4.1 and the chelating phosphines DIPAMP, Ph-β-glup, and PNNP are shown in Fig. 4.1. Currently, asymmetric hydrogenation as a route to amino acids has been abandoned in favor of fermentation methods. The problem with the asymmetric catalytic route to chiral amino acids is not the asymmetric hydrogenation step but rather the cost of the precursor prochiral olefin. Although these processes have been abandoned, asymmetric hydrogenation is practiced for the preparation of other fine chemicals.

Interestingly, the process shown in Scheme 4.1 for the synthesis of L-DOPA is biphasic in a certain sense. The product amino acid is less soluble in the reaction medium (aqueous alcohol) than the precursor olefin. Thus, the product precipitates from the reaction mixture as a solid and the soluble catalyst-containing phase can be recovered or recycled.

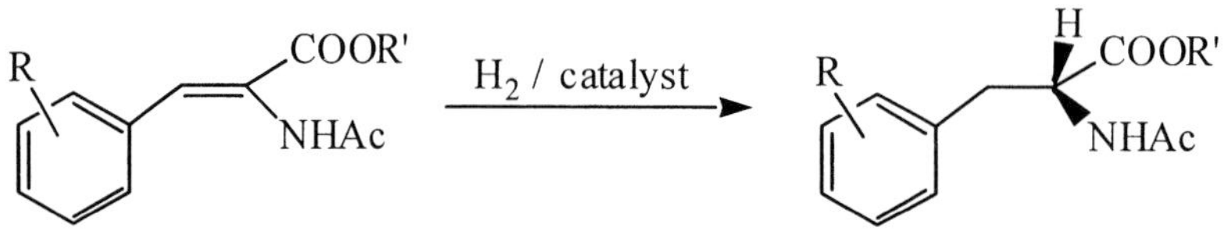

Scheme 4.1

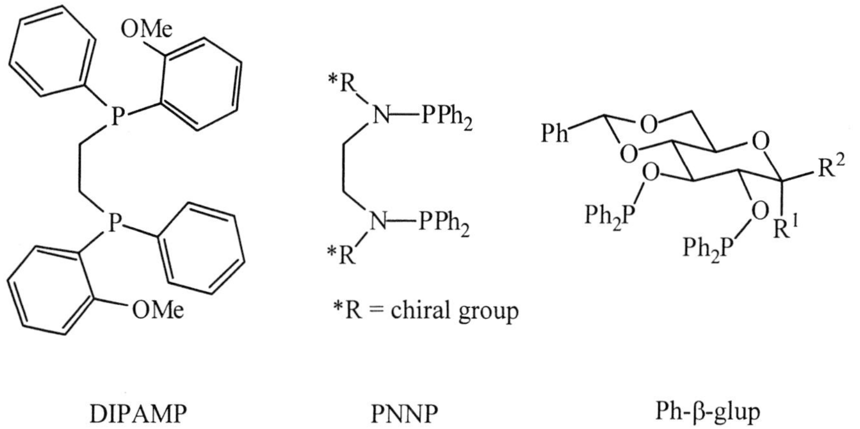

DIPAMP PNNP Ph-β-glup

Figure 4.1. Examples of chiral chelating phosphines for homogeneous asymmetric hydrogenation.

Substantial effort has been directed toward the preparation of water-soluble chiral catalysts for biphasic applications over the last 15 years. Chiral ligands have been made water-soluble by the incorporation of anionic groups such as sulfonate or carbonate; cationic groups such as quaternary ammonium ions or neutral hydrophilic groups such as polyethers. Descriptions of water-soluble ligands are available in recent reviews [6, 7]. Most water-soluble versions of chiral ligands have been tested for the asymmetric hydrogenation of cinnamic acid derivatives as described above. Some examples of water-soluble phosphines are shown in Fig. 4.2.

The purpose of this review is to summarize some of the recent results in the rapidly advancing area of biphasic enantioselective catalysis.

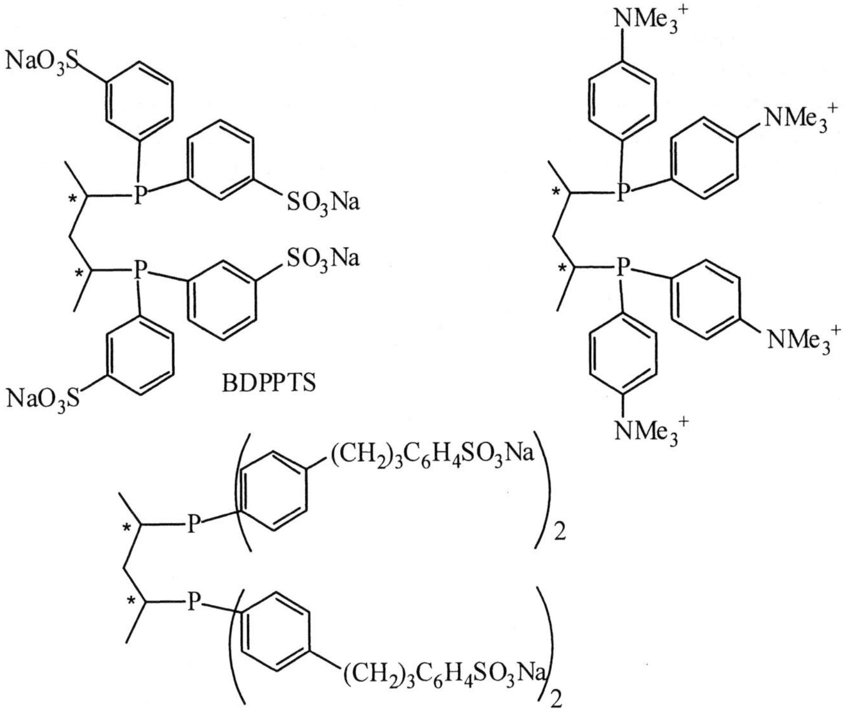

NaO_3S
SO_3Na
SO_3Na
NaO_3S
BDPPTS
NMe_3^+
NMe_3^+
NMe_3^+
NMe_3^+
$(CH_2)_3C_6H_4SO_3Na$
$(CH_2)_3C_6H_4SO_3Na$

Figure 4.2. Water-soluble versions of bisdiphenylphosphinopentane (BDPP): the tetrasulfonated ligand BDPPTS (left) [8], the tetrakis trimethylanilyl ligand (right) [9], and an amphiphilic version (bottom) [10].

4.2 Hydrogenation

Generally aqueous-phase asymmetric hydrogenation catalysts work well for α-acetamidocinnamic acid derivatives [6–10]. However none of the biphasic catalysts reported in the literature for this reaction are superior to the now abandoned homogeneous catalytic routes to amino acids mentioned above.

In addition to direct modification, phosphines may be made water-soluble by immobilization on a hydrophilic polymer. This has recently been achieved by chemically bonding the PPM ligand to polyacrylic acid via an amide linkage [11]. The resulting polymeric ligand is shown in Fig. 4.3. Under biphasic reaction conditions (water/ethyl acetate), Rh catalysts gave 100 turnovers (100% conversion) within five hours with an enantioselectivity of 89% for the biphasic hydrogenation of α-acetamidocinnamic acid. Compared to its homogeneous counterpart, the activity in the biphasic system is diminished and the enantioselectivity drops by nearly 10%.

Water-soluble catalysts for the asymmetric hydrogenation of propenoic acids have been described [12–14]. The best of these are ruthenium complexes of sulfonated bis-

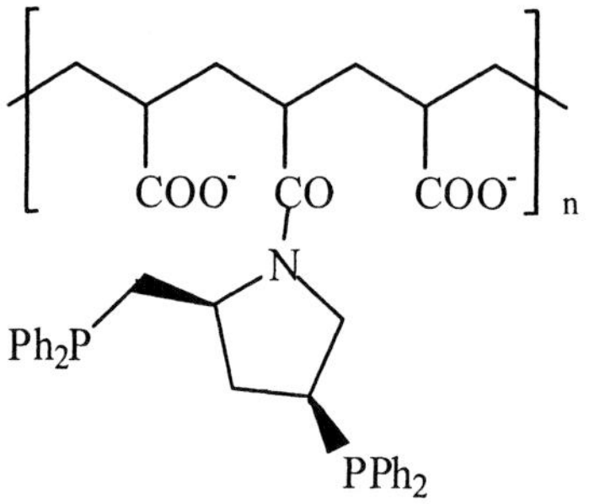

Figure 4.3. The PPM ligand bound to polyacrylic acid by an amide linkage.

diphenylphosphinobinaphthyl, BINAP*, which give selectivities comparable to the nonsulfonated catalysts [12, 13]. Activity is further improved in supported aqueous phase catalyst (SAPC) versions of the catalysts, but the enantioselectivity is diminished. A comparison of the selectivity under biphasic, SAPC, and homogenous conditions is given in Scheme 4.2 for the hydrogenation of a Naproxen precursor [12, 13].

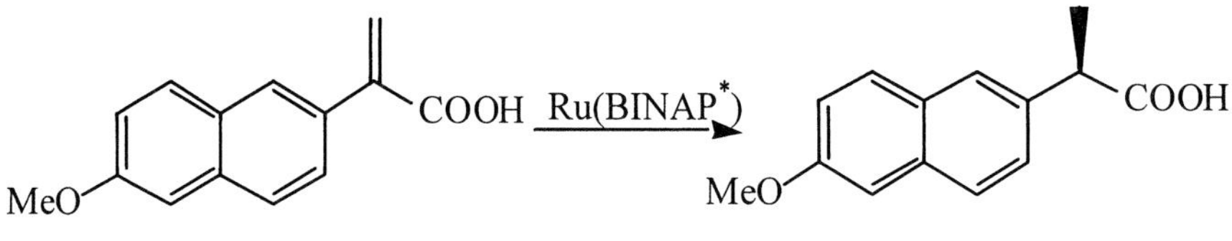

(BINAP* = sulfonated BINAP)

Phases	Solvent	E.e. (%)
L/L	EtOAc/H_2O	73–83
S/L	SAPC	29–77
L	MeOH	75–96

Scheme 4.2

An amphiphilic version of BINAP has been prepared by the introduction of long chains that are easily sulfonated. Curiously, this catalyst is not nearly as selective as BINAP* or BINAP for the hydrogenation of 2-arylpropenoic acids [14]. The ligand, designated BINAP–*, and results for asymmetric hydrogenation are shown in Scheme 4.3.

A group at Ciba Geigy, now Solvias, developed a homogeneous asymmetric imine hydrogenation catalyst for the synthesis of metolachlor [15, 16]. The reaction is shown in Scheme 4.4.

The biphasic approach to the asymmetric hydrogenation of imines may be of commercial interest as an alternative to the homogeneous hydrogenation step in the Solvias metolachlor synthesis. However, reaction rates and selectivities under biphasic conditions are currently not high enough to be economically viable [16, 17].

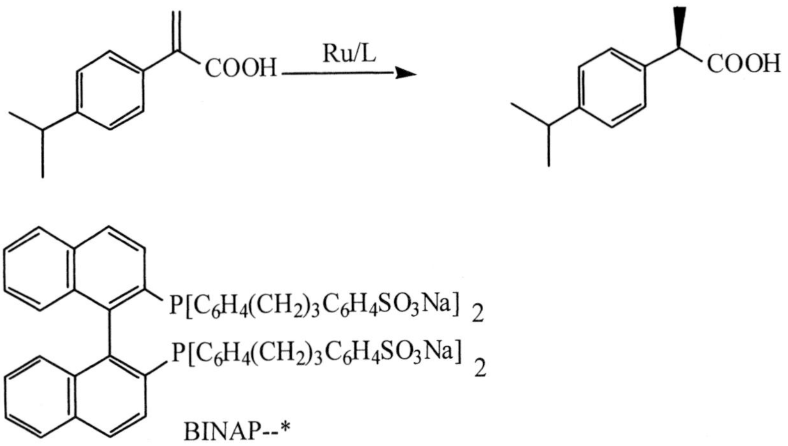

Ru/L

$P[C_6H_4(CH_2)_3C_6H_4SO_3Na]_2$
$P[C_6H_4(CH_2)_3C_6H_4SO_3Na]_2$

BINAP--*

Ligand	Solvent	E.e. (%)
BINAP–*	EtOAc/H_2O	19
BINAP–*	MeOH	56
BINAP	MeOH	83

Scheme 4.3

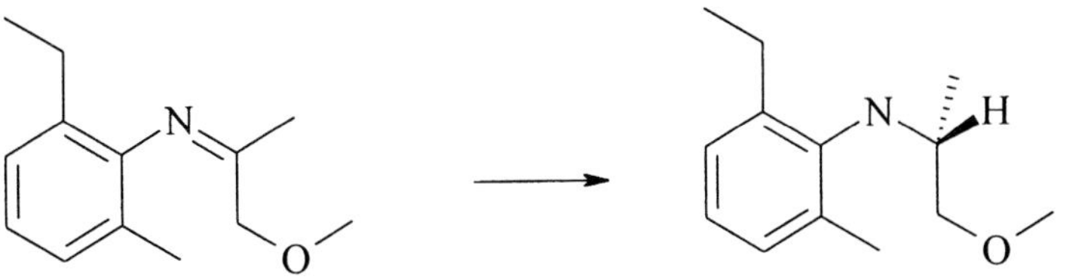

Metolachlor precursor

Scheme 4.4

The catalyst for the homogeneous process is an iridium(I) complex with a chiral ferrocenylphosphine plus acetic acid and iodine. The catalyst is exceptionally fast and very stable. Turnovers of up to 10^6 are reported.

In a series of patent applications to Ciba Geigy, Pugin et al. describe a variety of methods for the immobilization of chelating phosphines on solids [17]. Recently, Pugin described the preparation of water-soluble phosphines, in a manner similar to the example described above in Fig. 4.3 [17b]. Of particular interest are the chiral ferrocenyl phosphines that were immobilized on water-soluble polymers [17b]. When the phosphine is anchored to polyacrylic acid, either a biphasic catalyst or an extractable catalyst is obtained (Fig. 4.4).

It should be noted that the immobilized catalysts, including the biphasic catalysts, are not as active as their homogeneous counterparts for the hydrogenation [15b]. Consequently, the commercial process is performed homogeneously, and a distillation step is required for product purification.

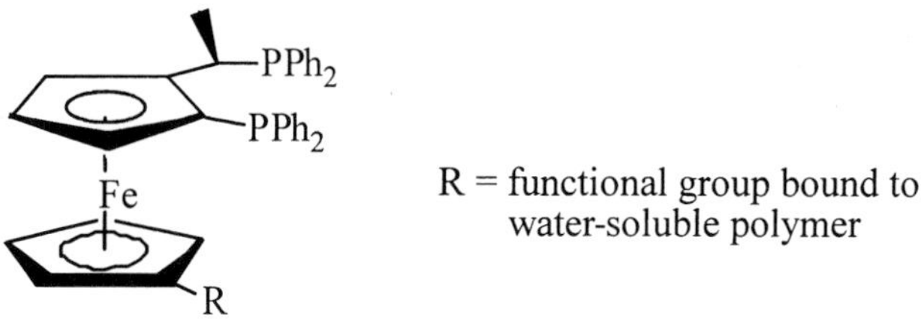

R = functional group bound to water-soluble polymer

Figure 4.4. Chiral ferrocenyl phosphines can be bound to polymers through bridging groups on the 2′ ring.

Several novel features are apparent in the rhodium-catalyzed hydrogenation of imines with sulfonated bis(diphenylphosphino)pentane [18–20]. Most notably, the enantioselectivity is strongly dependent on the degree of sulfonation with the greatest selectivity observed with the monosulfonated phosphine, BDPP-MS, Fig. 4.5. This seems counterintuitive, since monosulfonation introduces an additional stereogenic center which results in formation of a mixture of diastereomers. Some of the selectivity is regained when the tetrasulfonated derivative, which consists of a single isomer, is used in the hydrogenation. The origin of this unique effect is not clear at this time [18–20]. The results for the hydrogenation of N-benzylacetophenoneimine are summarized in Scheme 4.5, Table 4.1 and Fig. 4.6.

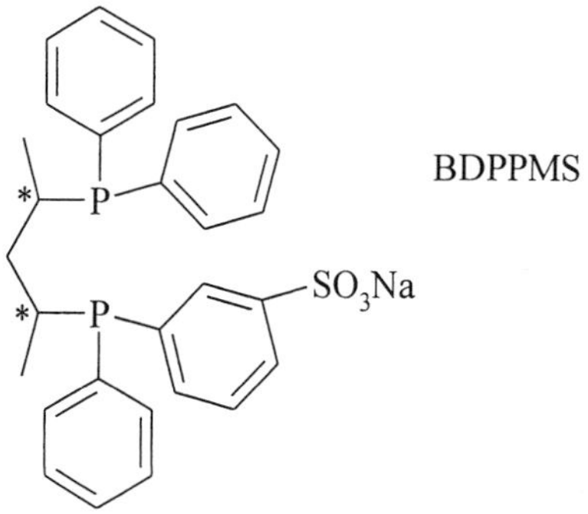

BDPPMS

Figure 4.5. Monosulfonated BDPP (BDPPMS).

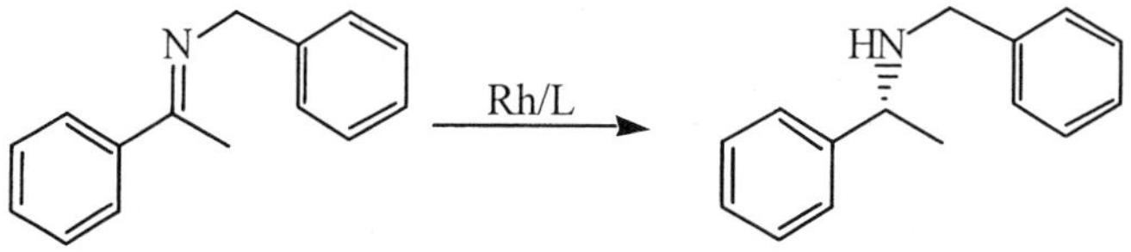

Rh/L

Scheme 4.5

Many subtle factors have been identified as significant in influencing the selectivity of imine hydrogenation with rhodium catalysts. These include reaction solvent and the nature of the anion [19–21]. Buriak and Osborn have observed that sulfonate anions improve reaction selectivity in reversed micelle systems in nonaqueous solvents [21].

Table 4.1. Results of the asymmetric hydrogenation of the imine in Scheme 4.5 where BDPP is R,R-2,4-bis(diphenylphosphino)pentane and -MS, -DS, -TrS, and -TS refer to monosulfonated, disulfonated, trisulfonated and tetrasulfonated, respectively [20].

Ligand	Solvent	Conversion (%)	E.e. (%)
BDPP	MeOH	100	68
BDPP	EtOAc/H_2O	0	–
BDPPMS	EtOAc/H_2O	100	94
BDPPDS	EtOAc/H_2O	100	2
BDPPTrS	EtOAc/H_2O	99	3
BDPPTS	EtOAc/H_2O	99	63

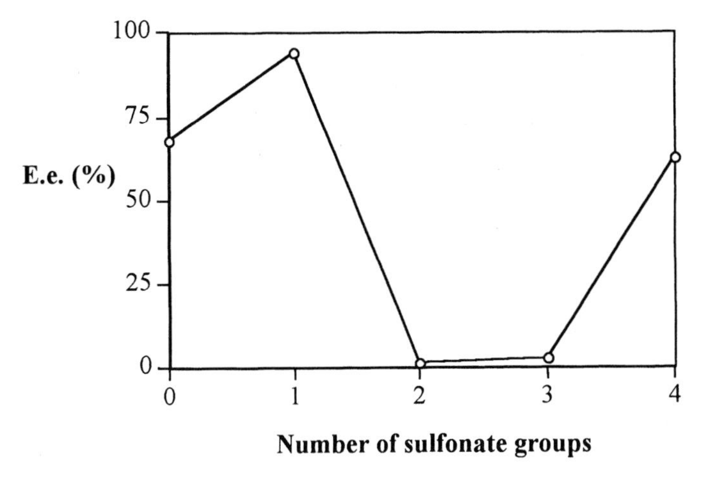

Figure 4.6. Enantioselectivity as a function of number of sulfonate groups for the asymmetric hydrogenation of N-benzylacetophenoneimine [20].

The Rh/BDPP-MS catalysts are relatively active for imine hydrogenation [15]. The problem is that the sulfonated ligands are relatively difficult to prepare, since the sulfonation step gives a mixture of products.

Ketone hydrogenation is a step in the Hoechst Celanese ibuprofen synthesis [22]. The reaction is practiced commercially over heterogeneous hydrogenation catalysts. Several catalysts for biphasic hydrogenation of ketones have been proposed [6, 7]. These, however, suffer from low reaction rates. The biphasic asymmetric hydrogenation of ketones has not yet been accomplished.

The addition of surfactants can have a remarkable influence on reaction activity and selectivity in biphasic asymmetric hydrogenation catalysis [23]. The effect is very pronounced with a series of new sulfonated phosphines prepared by the reaction of alcohol-substituted phosphines with sulfobenzoic anhydride as shown in Scheme 4.6. The synthetic method is general, and several new phosphines are reported.

When the ligand shown in Scheme 4.6 is used for the asymmetric hydrogenation of methyl α-acetamidocinnamate, the reaction half-life is 23 minutes, and the observed enantioselectivity is 38%. The half-life decreases to 5 minutes, and the enantioselectivity increases to 66% when sodium dodecyl sulfate (SDS) is added to the reaction [23]. A similar but smaller effect is observed when SDS is added to nonsulfonated versions of the ligand. Other examples of surfactant-modified asymmetric hydrogenation catalysts have been reported by Oehme [24].

The origin of the selectivity enhancement is not understood at this time. Most likely, the effect is attributed to local solvent effects and not to an ordering of the space at the active site. Nonetheless the addition of surfactants makes the biphasic approach to asymmetric hydrogenation more amenable to water-insoluble substrates and the development of future applications is expected.

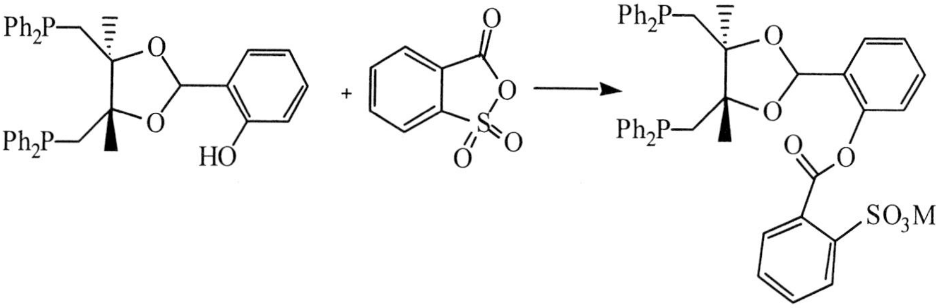

Scheme 4.6

The first example of a catalytic asymmetric hydrogenation reaction in supercritical (sc) CO_2, was the hydrogenation of enamides by Rh Duphos complexes [25]. For most substrates investigated, similar selectivities are observed in MeOH, hexane, and sc CO_2. However, in the case of β,β-disubstituted enamides, catalysis in sc CO_2 gives superior selectivities. An example of this phenomenon for the reaction in Scheme 4.7 is given in Table 4.2. The solubility of the Duphos ligand in sc CO_2 is due to the presence of aliphatic substituents on the phosphorus donor atom.

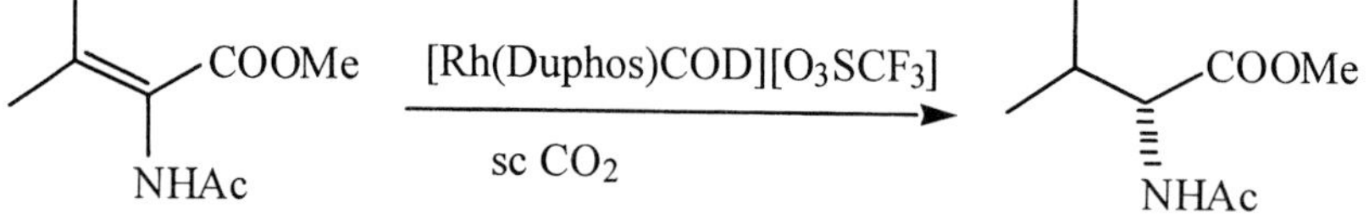

Scheme 4.7

Octahydro-BINAP (h_8-BINAP, Fig. 4.7) is a supercritical CO_2-soluble form of BINAP [26, 27]. Work from Noyori's group demonstrates the utility of this ligand in the sc CO_2 phase hydrogenation of tiglic acid [26]. Selected results from this study are presented in Table 4.3. It is interesting that selectivity and activity both improve upon the addition of $CF_3(CF_2)_6CH_2OH$ which can act as an amphiphile in sc CO_2.

Table 4.2. Enantioselectivity as a function of solvent with [Rh(Duphos)COD][O_3SCF_3] [23].

	MeOH	Hexane	Sc CO_2
E.e. (%)	67.4	70.4	88.4

T=40 °C; t=24 h; S/Rh=500; P_{to}=5000 psig; P_{H_2}=200 psig.

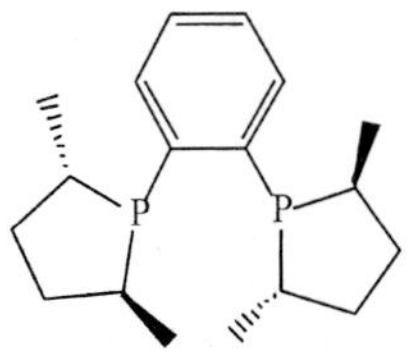

DuPhos

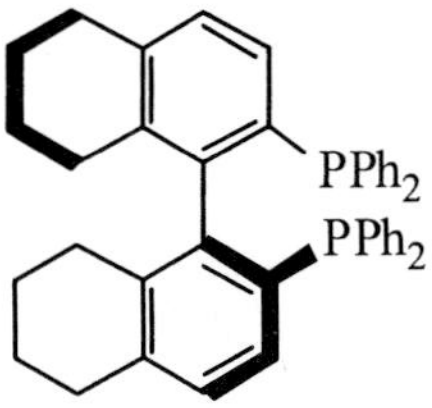

PPh_2
PPh_2

Figure 4.7. Hydrogenated form of BINAP for use in sc CO_2 phase catalytic reactions.

Table 4.3. Asymmetric hydrogenation of tiglic acid [26].

Phase	P_{H_2} (atm)	Yield (%)	E.e. (%)
sc CO_2	33	99	81 S
sc CO_2	7	23	71 S
sc CO_2 + $CF_3(CF_2)_6CH_2OH$	5	99	89 S
sc CO_2 + CD_3OD	6	81	78 S

t=12–15; T=50 °C; S/Rh=150–160; P_{CO_2}=170 atm.

The literature of homogeneous catalysis in supercritical fluids has recently been reviewed by Noyori [27]. However, the use of supercritical fluids as a reaction medium does not always lead to a biphasic catalytic system with respect to product workup and catalyst recycling. After the reaction is complete and the gases are vented, the catalyst and product are both left behind and must be separated. Biphasic reactions between water and sc CO_2 are of interest and are under investigation [28]. However, no work has been published yet on enantioselective versions of this type of biphasic catalysis.

4.3 Hydroformylation

The asymmetric hydroformylation of olefins continues to be an area of interest due to the potential for fine chemicals synthesis. Recent advances in single phase catalysis have emphasized binuclear catalysts [29], bisphosphites [30], and mixed phosphite –

phosphine ligands such as BINAPHOS [31]. The phosphite systems are inherently sensitive to water. Although there are reports of sulfonated versions of phosphites, they are not serious candidates for use in aqueous solvents [32]. On the other hand, BINAP has been directly sulfonated and the water-soluble version has been used for asymmetric hydrogenation (see above).

The large bite angle ligand NAPHOS is more appropriate for hydroformylation. Recently Herrmann et al. have shown that NAPHOS can be isolated in optically pure form and that it can be sulfonated to yield the axially chiral water-soluble BINAS (sulfonated NAPHOS) [33]. NAPHOS, however, is not sulfonated selectively. Typically two to eight sulfonate groups are incorporated per ligand [34]. The related ligand BIPHLOPHOS is disulfonated selectively, based on HPLC retention times [35]. The unsulfonated ligands NAPHOS and BIPHLOPHOS are shown in Fig. 4.8.

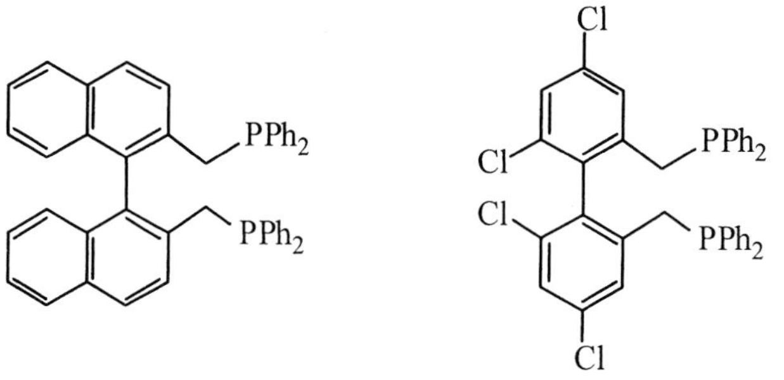

PPh_2 PPh_2 Cl Cl Cl PPh_2 PPh_2 Cl

NAPHOS BIPHLOPHOS

Figure 4.8. The NAPHOS and BIPHLOPHOS ligands have backbones similar to the bis-(diphenylphosphinomethyl)biphenyl ligand (BISBI) developed specifically for rhodium-catalyzed hydroformylation [36].

Although eight stereoisomers result from the disulfonation of BIPHLOPHOS, these could not be distinguished by NMR or HPLC [35]. A comparison of the performance of NAPHOS, BINAS, and BIPHLOPHOS for the hydroformylation of styrene is given in Table 4.4. The sulfonated version of BIPHLOPHOS has not yet been investigated in biphasic asymmetric hydroformylation. In the case of (–)-BINAS the enantioselectivity suffers compared to the monophasic hydroformylation of styrene with (–)-NAPHOS.

Table 4.4. Asymmetric hydroformylation of styrene with axially chiral phosphines including optically pure BINAS.

	T (°C)	Solvent	Conversion (%)	I (%)	E.e. (%)
(–)-BIPHLOPHOS [35]	40	toluene	49	82	15
(–)-BIPHLOPHOS [31]	80	toluene	68	81	0
(–)-NAPHOS [30]	40	toluene	89	96	32
(–)-BINAS [30]	40	toluene/H_2O-MeOH	92	93	18

I = branched/(branched + linear) aldehyde.

Rh-BINAPHOS-catalyzed asymmetric hydroformylation in the presence of dense and sc CO_2 phases has been recently reported [37]. Under the conditions used it is shown that at 60 °C increasing the CO_2 density diminishes the enantioselectivity of styrene hydroformylation. A plot of the observed selectivity as a function of CO_2 density is given in Fig. 4.9.

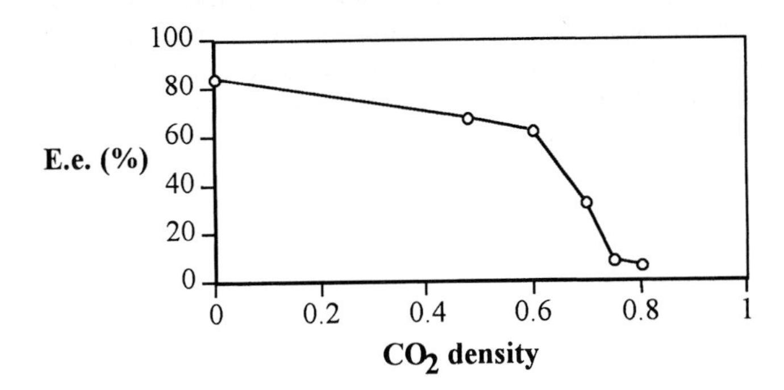

Figure 4.9. Enantioselectivity for Rh-BINAPHOS-catalyzed hydroformylation of styrene as a function of CO_2 density [37].

The sulfonated ligands BDPPTS, and sulfonated 1,2-bis(diphenylphosphinomethyl)cyclobutane have also been used for rhodium-catalyzed asymmetric hydroformylation of styrene with modest results. The ratio of branched to linear aldehyde product is 3:1, and the best e.e. value of 2-phenylpropanal is only 17% [38].

4.4 Oxidation

Wacker oxidation catalysts yield chlorine-containing products at high concentrations of chloride. In a novel modification of the Wacker process, Henry has shown that the catalysts can be made both water-soluble and chiral [39]. Thus, the oxidation of propene leads to chlorohydrin formation in e.e. values of up to 44%. Functionalized olefins give higher optical yields. The result shown in Scheme 4.8 is typical. The enantioselectivity is only reported for the predominant isomer. Up to 65 turnovers are possible, although details such as yield and substrate to catalyst ratio are not reported. This result and others from Henry's laboratory demonstrate that chlorohydrin synthesis is a versatile route to simple chiral building blocks for synthesis from relatively inexpensive starting materials.

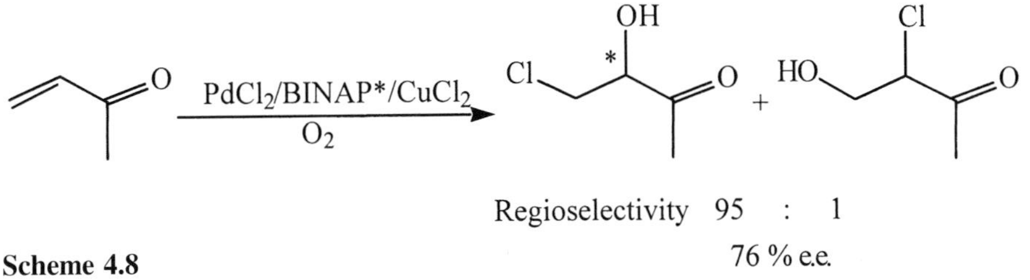

$PdCl_2$/BINAP*/$CuCl_2$, O_2

Regioselectivity 95 : 1

76 % e.e.

Scheme 4.8

Enantioselective manganese epoxidation catalysts have been made soluble in fluorocarbon solvents by adding fluorinated ponytails to chiral Schiff bases [40]. Depending on the substrate and the oxidant, e.e.'s of 5 to 92% are observed. In some cases, the e.e. is not reported. Generally, the enantiomeric excess is less than that obtained with a hydrocarbon-soluble homogeneous catalyst. The best result, an e.e. of 92%, was obtained for indene epoxidation with 1 atm O_2/pivalaldehyde as the oxidant and a manganese complex of the ligand shown in Fig. 4.10, where R-R = -$(CH_2)_4$-. The reaction proceeds to 100% conversion of indene, and the isolated yield of epoxide was 85% at a catalyst concentration of 1.5 mol%.

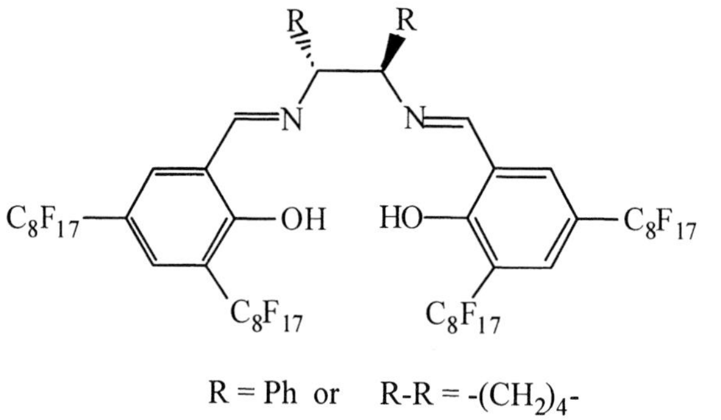

R = Ph or R-R = -$(CH_2)_4$-

Figure 4.10. A chiral Schiff base with fluorinated ponytails for enantioselective epoxidation.

4.5 Lewis Acid-Catalyzed Reactions

An example of fluorous biphasic separation in the catalytic application of Lewis acids is the enantioselective protonation of Samarium enolates [41] as shown in Scheme 4.9.

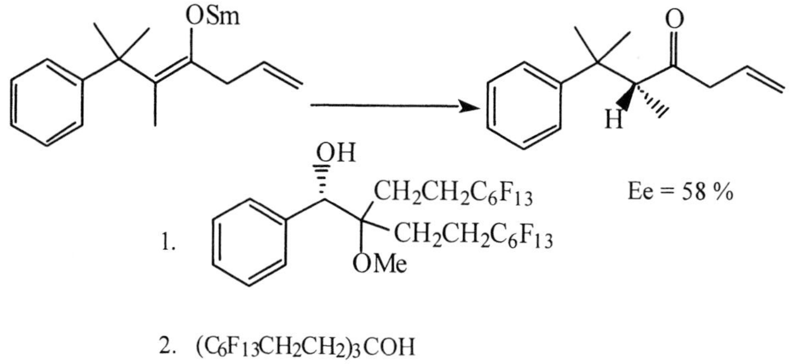

1. (chiral alcohol bearing OMe and two $CH_2CH_2C_6F_{13}$ groups)

2. $(C_6F_{13}CH_2CH_2)_3COH$

Ee = 58 %

Scheme 4.9

When the reaction is conducted in the presence of the chiral proton donor alone (i.e., with reagent 1 in Scheme 4.9), enantioselectivities of up to 66% are observed. When reagent 2 in Scheme 4.9 is present with the chiral proton donor in catalytic amounts, enantioselectivities of 43–58% are observed. Higher enantioselectivities are observed under biphasic reaction conditions compared to homogeneous reaction conditions in THF. This is most likely caused by the fact that the non-chiral proton donor partitions preferentially to the fluorous phase [41].

A novel pentafluorophenyl substituted chiral phosphinite has been used to generate an extractable Ru-Lewis acid catalyst for asymmetric Diels-Alder addition reactions [42]. The catalyst and typical results are given in Scheme 4.10. The catalysis is performed homogeneously in CH_2Cl_2, and workup is accomplished by adsorption of the ionic ruthenium catalyst onto celite followed by reextraction with tetraalkyl ammonium iodide. The efficient extraction procedure allows for convenient recycling of the catalyst.

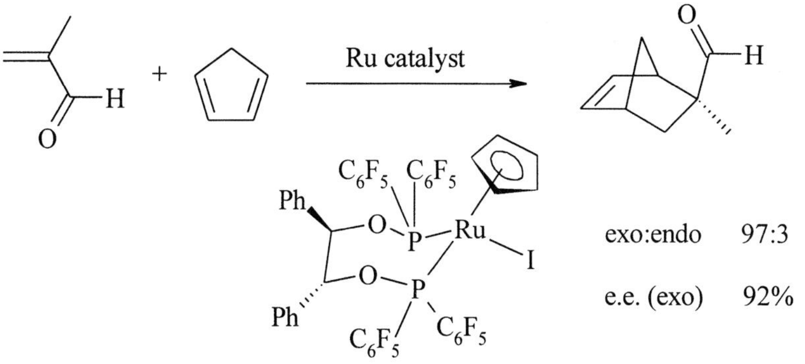

Ru catalyst

exo:endo 97:3

e.e. (exo) 92%

Scheme 4.10

4.6 Enzymatic Reactions

Biocatalytic reactions have been devised to operate under biphasic conditions. For example, *E. coli* JM109 cells modified to contain aldehyde reductase and glucose dehydrogenase activity, show good rates for the stereospecific reduction of certain aldehydes [42]. It was specifically shown that ethyl 4-chloro-3-oxobutanoate is reduced to ethyl 4-chloro-3-hydroxybutanoate in a biphasic reaction (Scheme 4.11). The aqueous, catalyst-containing phase, consists of phosphate buffer (pH 6), $NADP^+$, glucose, and modified *E. coli* cells. The nonaqueous phase is butyl acetate. In the best example, 13 500 turnovers were obtained (moles product/moles $NADP^+$) within 16 hours. The conversion was 94.1%, and the optical purity of the alcohol was 91.7%.

Scheme 4.11

The enzyme *S*-specific hydroxynitrile lyase, available from overexpression in *Pichia pastoris* [43], catalyzes the addition of HCN to aldehydes and ketones under biphasic reaction conditions [44, 45]. Reaction rates are improved in biphasic reaction conditions compared to reactions in water alone provided an emulsion is formed. Excellent enantioselectivity is observed with a wide range of substrates. Aldehydes react faster than ketones. The results shown in Scheme 4.12 are typical [44, 45]. Workup is achieved by allowing the emulsion to phase-separate, followed by extraction of the product with methyl *tert.*-butylether (MTBE).

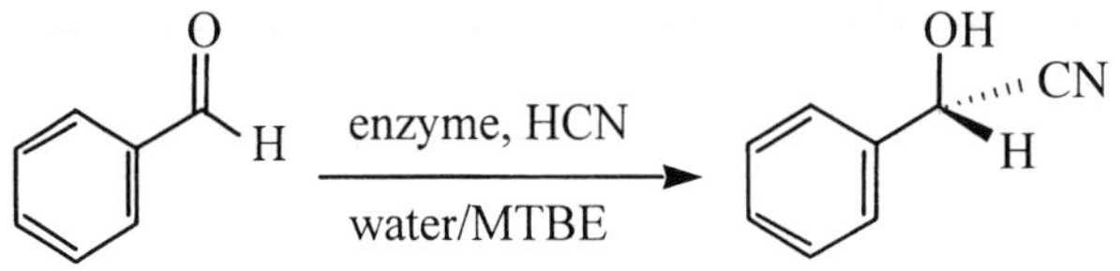

enzyme, HCN
water/MTBE

97 % conversion, 99 % e.e.
15 min, 15 °C

Scheme 4.12

4.7 Summary

It is evident that while biphasic catalysis has proven to be a general solution to the problem of catalyst/product separation, the value of the biphasic approach for a specific process, whether enantioselective or not, must be evaluated on a case-by-case basis. It is understood that in all small molecule applications, it is necessary to remove the catalyst from the product stream. Several factors dictate how this will be done. These include catalyst cost, product value, scale of reaction and reaction rate. For asymmetric applications the product value nearly always eclipses the cost of the catalyst. However, the catalyst must still be removed for product purity and environmental reasons.

The metolachlor process discussed above provides a good case study on the requirements for liquid biphasic enantioselective catalysis. The process is operated homogeneously rather than under biphasic conditions, primarily because the reaction is too slow when the catalyst is immobilized in an aqueous phase. Moreover, in this case, a distillation step is still required for product purification, thus diminishing the advantage of a phase separation step. Biphasic catalysis is still a worthy goal for this and other processes, however. The advantage will be evident when there is a clear improvement in selectivity upon performing the catalytic reaction under biphasic conditions. As recent results suggest, this condition will certainly be met in the near future for several enantioselective catalytic reactions.

References

[1] W.S. Knowles, *Acc. Chem. Res.,* 16 (1983) 106.
[2] R. Selke, H. Pracejus, *J. Mol. Catal.,* 37 (1986) 213.
[3] W. Vocke, R. Hänel, F.-U. Flöther, *Chem. Techn.,* 39 (1987) 123.
[4] G.W. Parshall, S.D. Ittel, *Homogeneous Catalysis*, Wiley Interscience, New York (1992).
[5] M. Fiorini, G.M. Giongo, *J. Mol. Catal.,* 5 (1979) 303.
[6] F. Joó, A. Kathó, in *Aqueous Phase Organometallic Catalysis: Concepts and Applications,* B. Cornils, W. A. Herrmann, eds., Wiley-VCH, Weinheim (1998).
[7] W.A. Herrmann, R.W. Eckel, in *Aqueous Phase Organometallic Catalysis: Concepts and Applications,* B. Cornils, W. A. Herrmann, eds., Wiley-VCH, Weinheim (1998).
[8] Y. Amrani, L. Lecomte, D. Sinou, F. Bakos, I. Toth, B. Heil, *Organometallics,* 8 (1989) 542.
[9] I. Toth, B.E. Hanson, *Tetrahedron Asymm.,* 1 (1990) 895.
[10] H. Ding, B.E. Hanson, J. Bakos, *Angew. Chem. Int. Ed. Eng.,* 34 (1995) 1645.
[11] T. Malström, C. Andersson, *J. Mol. Catal.,* 139 (1999) 259.
[12] (a) T. Wan, M.E. Davis, *J.C.S. Chem. Commun.,* (1993) 1262; (b) T. Wan, M.E. Davis; *Tetrahedron Asymmetry,* 4 (1993) 2461.
[13] M.E. Davis, T. Wan; *U.S. Patent: 5,827,794,* (1998).
[14] C.W. Kohlpaintner, B.E. Hanson, H. Ding; *U.S. Patent 5 777 087* (1998).
[15] F. Spindler, H.-U. Blaser, in *Transition Metals in Organic Synthesis*, M. Beller, C. Bolm, eds., Wiley-VCH, Weinheim (1998), p. 69.
[16] (a) H.-U. Blaser, H.-P. Buser, K. Coers, R. Hanreich, H.-P. Jalet, E. Jelsch, B. Pugin, H.-D. Schneider, F. Spindler, A. Wegmann, *Chimia,* 53 (1999) 275; (b) H.-U. Blaser, H.-P. Buser, H.-P. Jalett, B. Pugin, F. Spindler, *Synlett,* (1998) Special Issue, 867.
[17] (a) B. Pugin, H. Landert, *WO 98/01457* (1998); (b) B. Pugin, *U.S. Patent 5 432 289* (1998); (c) B. Pugin, F. Spindler, M. Muller; *U.S. Patent 5 382 729* (1995); (d) B. Pugin, F. Spindler, M. Muller, *U.S. Patent 5 432 289* (1995); (e) B. Pugin, F. Spindler, M. Muller, *U.S. Patent 5 308 819* (1994); (f) B. Pugin, F. Spindler, M. Muller, *U.S. Patent 5 252 751* (1993).
[18] J. Bakos, A. Orosz, B. Heil, M. Laghmar, P. Lhoste, D. Sinou, *J.C.S. Chem. Commun.* (1991) 1684.
[19] C. Lensink, J.G. DeVries, *Tetrahedron Asymmetry,* 3 (1992) 235.
[20] C. Lensink, E. Rijnberg, J.G. DeVries, *J. Mol. Catal. A,* 116 (1997) 199.
[21] J.M. Buriak, J.A. Osborn, *Organometallics,* 15 (1996) 3161.
[22] R.A. Sheldon, *Chemtech,* 3 (1994) 38.
[23] S. Trinkhaus, R. Kadyrov, R. Selke, J. Holz, L. Götze, A. Börner, *J. Mol. Catal. A* 144 (1999) 15.
[24] G. Oehme, I. Grassert, E. Paetzold, R. Meisel, K. Drexler, H. Fuhrmann, *Coord. Chem. Rev,* 186 (1998) 585.
[25] M.J. Burk, S. Feng, M.F. Gross, W. Tumas, *J. Am. Chem. Soc.,* 117 (1995) 8277.
[26] J. Xiao, S.C.A. Nefkins, P.G. Jessop, T. Ikariya, R. Noyori, *Tetrahedron Lett.,* 32 (1996) 2813.
[27] P.G. Jessop, T. Ikariya, R. Noyori, *Chem. Rev.,* 99 (1999) 475.
[28] G.B. Jacobson, J. Brady, J. Watkin, W. Tumas, *Abs. Papers, American Chemical Society,* 217 (1999) 244.
[29] G.G. Stanley, in *Advances in Catalytic Processes, Asymmetric Catalysis*, vol. 2, M.P. Doyle, ed., JAI Press Greenwich, CN, (1997), p. 221.
[30] J.E. Babin, G.T. Whiteker, *U.S. Patent 5 491 266* (1996).
[31] K. Nozaki, W.G. Li, T. Horiuchi, H. Takaya, *Tetrahedron Lett.* 38 (1997) 4611.
[32] A.G. Abatjoglou, D.R. Bryant, R.R. Peterson, *U.S. Patent 5 180 854* (1993).
[33] R.W. Eckl, T. Priermeier, W.A. Herrmann, *J. Organomet. Chem.* 532 (1997) 243.
[34] W.A. Herrmann, C.W. Kohlpaintner, R.B. Manetsberger, H. Bahrmann, H. Kottmann, *J. Mol. Catal.,* 97 (1995) 65.
[35] F. Rampf, M. Spiegler, W.A. Herrmann, *J. Organomet. Chem.,* 582 (1999) 204.
[36] C.P. Casey, G.T. Whiteker, M.G. Melville, L.M. Petrovich, J.A. Gavney, D.R. Powell, *J. Am. Chem. Soc.* 114 (1992) 5535.

[37] S. Kainz, W. Leitner, *Catal. Lett.*, 55 (1998) 223.
[38] M.D. Miquel-Serrano, A.M. Masdeu-Bulto, C. Claver, D. Sinou, *J. Mol. Catal. A*, 143 (1999) 49.
[39] O. Hamed, P.M. Henry, *Organometallics,* 17 (1998) 5184.
[40] G. Pozzi, F. Cinato, F. Montanari, S. Quici, *J.C.S. Chem. Commun.* (1998) 877.
[41] S. Takeuchi, Y. Nakamura, Y. Ohgo, D.P. Curran, *Tetrahedron Lett.*, 39 (1998) 8691.
[42] E.P. Kundig, C.M. Saudan, G. Bernarinelli, *Angew. Chem. Int Ed.,* 38 (1999) 1220.
[43] M. Kataoka, K. Yamamoto, H. Kawabata, M. Wada, K. Kita, H. Yanase, S. Shimizu, *App. Microbiol. Biotech.*, 51 (1999) 486.
[44] M.Hasslacher, M. Schall, M. Hayn, H. Griengl, S.D. Kohlwein, H.J. Schwab, *J. Biol. Chem.* 271 (1996) 5884.
[45] H. Griengl, N. Klempier, P. Pöchlauer, M. Schmidt, N. Shi, A.A. Zabelinskaja-Mackova, *Tetrahedron*, 54 (1998) 14477.

5 Immobilized Enzymes in Enantioselective Organic Synthesis

Peter Rasor

5.1 Introduction

In a book on heterogeneous enantioselective catalysis, a chapter on immobilized enzymes stands out. Most chemocatalysts presented in this book have a metal atom in the catalytic center. It is worth mentioning that many enzymes feature a metal atom in their catalytic center as well, although it is enveloped by a highly complex chiral ligand – designed by nature, not by man. These "*bio*catalysts" (in contrast to "*chemo*-catalysts") are, with few exceptions, soluble in aqueous systems and thus differ from organometallic catalysts which degrade rapidly in water.

The notion that enzymes are unstable and too costly to be used in organic synthesis is changing rapidly. This is partly due to efficient methods to immobilize enzymes, thus stabilizing them and reducing catalyst costs through repeated use of the catalyst. Additionally, enzymes have become considerably cheaper because of the progress made in genetic engineering.

This chapter will cover general aspects of enzyme immobilization, some applications and some recent developments. Specific issues have been discussed elsewhere in detail and are mentioned only briefly, such as mass transfer effects [1] and the use of enzymes in membrane reactors [2]. Since immobilized enzymes allow repeated use, which immediately translates into lower costs, this issue is more important to industry than to academia. Emphasis will be placed on technologies applicable to industry rather than on the latest development in pure research.

Enzymes have gained a place in the toolbox of the organic chemist, because they are selective catalysts. They feature chemo-, regio- and stereoselectivity and thus are able to perform tasks not easily accomplished otherwise (Fig. 5.1).

In addition, they have a number of advantages over other catalysts, most notably the ability to perform reactions without the need of protecting labile groups. Unfortunately, enzymes are not the solution to all synthetic challenges, and as with any other chemical method or catalyst, there are limitations that may prevent their use (Table 5.1).

Not long ago, enzymes were employed in organic chemistry only on research scale, simply because they were too expensive to use in industrial processes. Progress

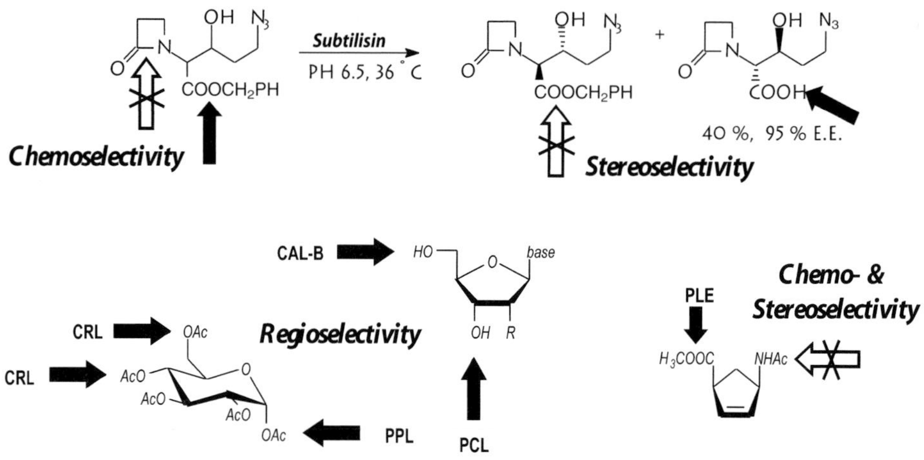

Figure 5.1. Enzymes catalyze chemo-, regio- and stereoselective reactions. The full arrow indicates the group that reacts preferentially; the open arrow with a cross indicates groups that do not react. Abbreviations: CAL-B, lipase from *Candida antarctica*, type B; CRL, lipase from *Candida rugosa*; PCL, lipase from *Pseudomonas cepacia*; PLE, esterase from pig liver; PPL, lipase from porcine pancreas.

Table 5.1. Advantages and limitations of enzymes in organic synthesis.

Advantages	Limitations
• Mild conditions	• Availability and price
• No organic solvent needed in most cases	• Sensitivity/operational stability
• Wide range of activity	• Lack of flexibility
• High selectivity (chemo, regio, stereo)	• Not all reaction types accessible
• Often fewer reaction steps (no protecting groups needed)	• Cofactor regeneration sometimes needed
• High process optimization potential	• Solubility of substrates in water

in two areas has changed the situation completely: genetic engineering has drastically reduced the production costs of enzymes, and the ability to immobilize enzymes allows their repeated use. Together, these technologies have reduced the cost so that even commodities can be produced by biocatalysis. Enzymes find applications in many industrial processes, most notably in the pharmaceutical and fine chemicals industry [3]. The synthesis of β-lactam antibiotics is a good example: the most important β-lactam nuclei and some side chains are produced on a yearly scale of thousands of tons by using immobilized enzymes. The earliest example on an industrial scale is the resolution of amino acids using an immobilized aminoacylase, as pioneered by Tanabe et al. [4].

5.2 Immobilization

This section will emphasize aspects of enzyme immobilization (Section 5.2.1). It will describe types of carriers (Section 5.2.1.1), enzymes (Section 5.2.1.2) and how they are bound to carriers (Section 5.2.1.3). Special attention will be paid to the emergence of a new technology, the cross-linked enzyme crystals (Section 5.2.1.4). Quite often, the aspect of evaluating the performance of a specific immobilization method is neglected. Therefore, two sections will deal with the activity assay of immobilized enzymes (Section 5.2.2) and the activity balance of an immobilization experiment (Section 5.2.3).

The immobilization of enzymes serves three purposes. First of all, it transforms the enzyme from a rather large and elusive molecule to a catalyst particle which can be handled through simple mechanical operations ("carrier-fixation" or cross-linkage). Alternatively, the entrapment within a membrane or encapsulation restricts the motion of the enzyme to a defined space ("localization"). In this case, the enzyme can be used repeatedly or continuously. Secondly, certain methods of immobilization lead to a catalyst with a stability under reaction conditions which is superior to that of the free enzyme ("stabilization"). Last but not least, the workup of reaction mixtures is greatly simplified.

When evaluating advantages and disadvantages of free enzymes (enzyme solution, powder or lyophilizate) vs. immobilized enzymes (Table 5.2), the reason for using immobilized enzymes has to be clear before deciding on the enzyme technology to be adopted in a reaction or process. Questions need to be asked such as:
"Do I want to pay for immobilization in order to get better workup?"
"Do I really need a highly stable catalyst if I only have to run a few batches?"
"Do I want to endure lengthy experiments to determine the operational stability?"

Certainly, questions and answers wildly differ in academia and industry. This is the reason why so many methods of immobilization have been published but so few have been adopted by industry.

Table 5.2. Comparison of free vs. immobilized enzyme.

Free enzyme	Immobilized enzymes
• Often crude mixtures of enzymes	• High stability
• Low activity with crude mixtures	• Easy workup
• Limited stability	• Expensive but ...
• Workup difficult (*e.g.* phase separation)	• ... favorable process economics through repeated use
• No costly development of immobilized enzyme necessary	• Need special reactor equipment
• Use of existing process equipment	• Costly process development (operational stability?)
• Favorable for a few large-scale batches	• Need to run many batches (high volume products)
• Repeated use limited to organic solvents, enzyme membrane reactor	

5.2.1 Methods of Immobilization

Enzymes have been immobilized onto almost everything, from sawdust [5] to crushed chicken bones [6], from agarose [7] to porous glass beads [8].

In principle, enzymes are either adsorbed on or covalently bound to a carrier, or they are entrapped in a membrane or by a phase barrier. Enzymes can be attached to prefabricated carriers, cross-linked without a carrier or during polymerization, or encapsulated/included within a certain space (Table 5.3).

Table 5.3. Classification of immobilization methods (adapted from [9])

Attachment to prefabricated carriers	Cross-linkage	Encapsulation/inclusion
• Covalent binding • Ionic binding • Adsorption • Metal binding • (Bio)affinity binding	• Direct cross-linking • Co-cross-linking • Crystallization & cross-linking (cross-linked enzyme crystals) • Copolymerization after chemical modification	• Membrane devices • Inclusion by polymerization • Microcapsules • Liposomes/reversed micelles • (Organic solvents)

With respect to biocatalysis on a preparative to industrial scale, adsorptive and covalent binding to carriers plays a dominant role (Fig. 5.2), while metal and affinity binding are negligible due to high costs of the carrier.

It should be pointed out that when the bond of the carrier to the enzyme is stronger (covalent vs. non-covalent binding), the interference with the enzyme structure is greater, leading to reduced enzyme activity. Picking the optimal carrier and binding type will make the enzyme more stable. In contrast, the incorporation into a polymeric network, a membrane device, or a liposome/reversed micelle results in little or no interference with the enzyme structure (Fig. 5.2). In those cases, the enzyme can only be stabilized by adapting the reaction conditions to the specific needs of the enzyme. However, these conditions are not necessarily optimal for substrate or product stability.

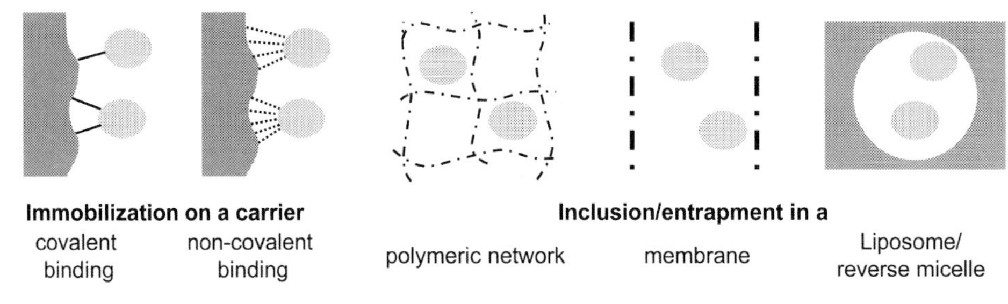

Figure 5.2. An enzyme (gray ellipse) can be bound to a carrier by a single or a few covalent bonds, by adsorption, or by entrapment.

Direct cross-linking or cross-linked enzyme crystals pose a viable alternative, since immobilized enzymes can be prepared without the need of developing and producing a suitable carrier up-front (Fig. 5.3). Typical cross-linking agents are bifunctional molecules such as glutaraldehyde. Just as in the case of covalent binding to a carrier, a higher degree of cross-linking may stabilize the structure of the agglomerate or crystal, but it also disturbs the enzyme structure and may lead to considerable deactivation. It is possible to cross-link an enzyme in the presence of another, possibly inert protein. This reduces overall activity but may stabilize the enzyme. If the enzyme is crystallized before cross-linking, particles of defined shape are generated. Covalent incorporation into a polymeric network by copolymerization can be considered as a special case of cross-linking (Fig. 5.4). Thus, the enzyme is chemically modified with polymerizable groups and then covalently incorporated into a polymeric network. Using this principle, enzymes can be incorporated into plastics [10].

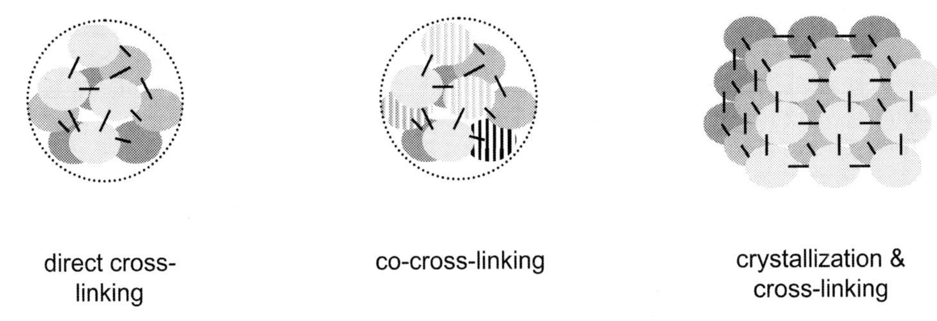

Figure 5.3. Enzymes (gray ellipses) can be directly cross-linked as such, or co-cross-linked in the presence of an inert protein (striped ellipses). Alternatively, the enzyme can be crystallized before cross-linking.

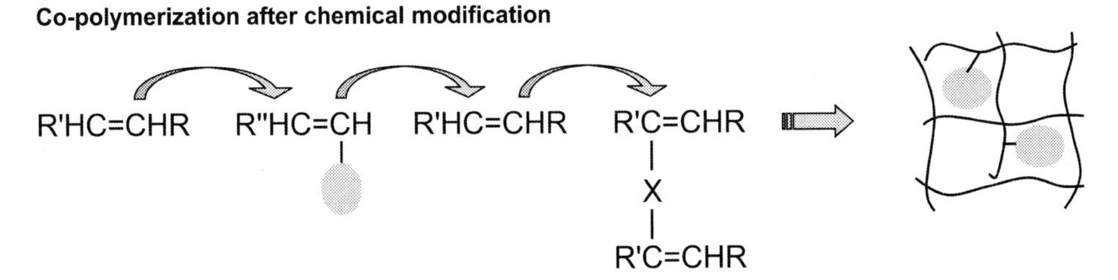

Figure 5.4. A special case of cross-linking is the copolymerization of a chemically modified enzyme (ellipses).

5.2.1.1 Enzymes

An enzyme is a catalytically active macromolecule having a backbone of L-amino acids. It may consist of one or several peptide chains, either linked covalently by disulfide bonds or held together by non-covalent binding forces. After it is produced by translation in an eukaryotic cell, it may be modified, e.g. by glycosylation.

While many chemical methods can be envisaged to modify certain amino acids for cross-linking or covalent binding to a carrier, most methods deactivate the enzyme and/or use very expensive reagents, which are unsuitable for immobilization beyond a small laboratory scale. Out of the 20 proteinogenic amino acids, only few are actually useful for covalent immobilization. Lysine is by far the most important amino acid. Its ε-amino group can be acylated, alkylated or arylated [11]. The same applies to the N-terminus of the peptide chain. The modification of carboxylate groups (aspartic or glutamic acid, C-terminus) requires their activation, e.g. by water-soluble carbodiimides. This first step and the following steps usually lead to a strong loss of activity. Tyrosine can play a certain role in cross-linking reactions.

The sugar groups of glycosylated enzymes can be converted into dialdehydes by periodate cleavage [12]. Thereafter, the enzyme can be bound to carriers containing amino groups via a Schiff base.

For adsorption onto a carrier, all amino acids on the enzyme surface can contribute to non-covalent binding. The 3D-structure of the enzyme is little disturbed by this process.

In most cases, semi-pure to pure enzymes are used for immobilization. The purer the enzyme, the higher the activity of the resulting catalyst is and the less likely side reactions will be caused by contaminating activities. In some fortunate cases, a purification step and carrier fixation can be combined by selective binding of the preferred enzyme to the carrier [13]. Similarly, the crystallization before cross-linking is a purification step in itself [14].

5.2.1.2 Carriers

Besides enzyme characteristics and the choice of a process technology, the chemical and physical properties of a carrier determine the success or failure in developing an immobilized enzyme. The most important factors are listed in Table 5.4. It is easily recognized that there is no 'ideal' carrier for all enzymes and applications.

As already mentioned in the introduction, there are hundreds of possible carriers for enzyme immobilization to choose from. In the following, only some of the most important carriers have been selected (Table 5.5).

Besides using prefabricated carriers, the particle can be generated in the presence of the enzyme. As mentioned before (see Section 6.2.1), this applies to cross-linking of enzymes in amorphous or crystalline form and to the incorporation into polymeric networks. For good process characteristics, the particle size distribution should be as homogeneous as possible. Usually, a sieving step eliminates fines and aggregates of particles. If this step is done before binding the enzyme, as for prefabricated carriers, no enzyme is lost (Fig. 5.5). If the particle is generated in the presence of the enzyme (polymerization, cross-linking), part of the activity is lost in the sieving step. This does not apply to the cross-linked enzyme crystal technology, since sieving can be done before cross-linking and lost enzyme recycled.

Immobilized enzymes can have many different shapes, depending on the manufacturing process and the desired features of the particles. Ideally, the particles should be spherical, with a high porosity and mechanical stability. However, other shapes are possible as well, as can be seen for the immobilized D-amino acid oxidase (Fig. 5.6) or the cross-linked enzyme crystal (Fig. 5.9).

Table 5.4. Properties of the carrier and influence on the eventual immobilized enzyme and its use.

Properties of the carrier	Comments
Active groups	For covalent binding, the carrier must have reactive groups on the surface. Epoxides, iminocarbonates, aldehydes or amines are very effective (Fig. 5.7). Density and distribution influence the amount of activity bound to the carrier.
Particle size and form	The particles must be large enough to be handled by simple mechanical operations, like sieving. The larger the particles, the easier they are to handle but the lower the activity is based on weight. For good flow (column) or filtration characteristics (stirred tank), the particles should be uniform, and preferably spherical. A particle-size distribution maximum between 0.2 to 0.6 mm seems to be the optimum. A high percentage of fines below 0.1 mm diameter will clog filters and may lead to enzyme loss.
Inertness to solvents	Especially for column processes, the carrier should not swell or shrink in different solvents; this may lead to enzyme deactivation and blockage of the column.
Mechanical stability	For stirred-tank reactors, the carrier should be stable against abrasion and breakage by the stirrer. This can be achieved by using flexible or even soft carriers, although the latter will increase filtration time. For packed-bed reactors, rigid carriers with little or no compressibility are required.
Surface area and pore size	A large surface area (100 m^2/g and larger) allows high loading with enzyme. The pore size should be above 30–50 nm to make the pores available to the enzyme.
Polarity	The polarity of a carrier influences the non-bonding interactions between carrier and enzyme; for example, hydrophobic enzymes like lipases bind strongly to hydrophobic carriers, while esterase from pig liver does not. Polarity also affects the adsorption and availability of the substrate.

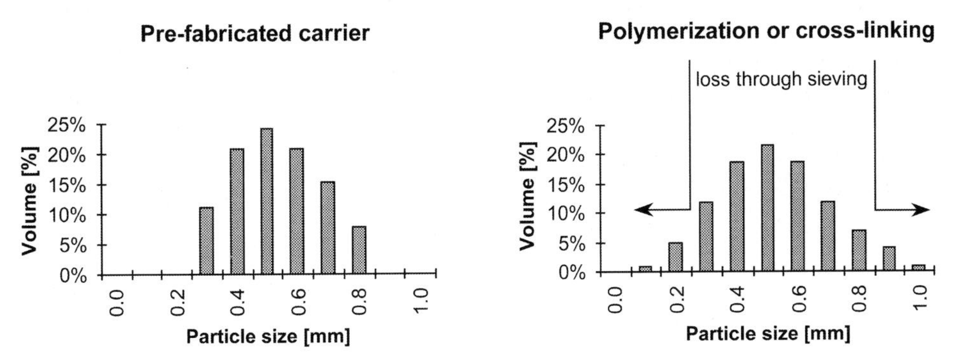

Figure 5.5. A particle size distribution without any fines or aggregates is preferred to ensure good filtration characteristics.

Table 5.5. Types of carriers (selection).

Carrier	Comments
Inorganic carriers	*Inorganic carriers have the advantage that they are inert to organic solvents (no swelling), are mechanically stable and resistant to high temperature. They may have limited binding surface. Activation/modification is needed for covalent binding.*
Diatomaceous earth, Celite, alumina, bentonite	Cheap carriers, often used for the adsorption of lipases [15].
Silica gel, controlled pore glass (CPG), glass beads	Often low activity per weight, expensive (CPG). Activation/modification is needed for covalent binding.
Organopolysiloxanes	Deloxan (Degussa AG): inorganic core, modified to carry organic groups for covalent enzyme binding. Good overall properties for enzyme immobilization [16].
Kaolin	Toyonite (Toyo Denka Kogyo Corp.): a chemically modified kaolin carrying either amino or methacrylo groups.
Organic carriers	*Activated carriers are available. Swelling may occur in organic solvents.*
Ion exchangers: Duolite, Amberlite, Lewatit, Relite	Often cheap carriers with good immobilization yield. Mostly spherically formed, good filtration characteristics. If the enzyme is adsorbed, enzyme bleeding is possible. Industrial scale carriers.
Epoxy-activated carriers	Eupergit (Röhm AG): no activation necessary, easy modification of surface properties and/or binding chemistry, e.g. by introducing primary amino groups; industrial scale carriers.
Chitosan	Chitopearl (Fuji Spinning): spherical cross-linked chitosan.
Polysaccharides	Dextran or agarose in beaded form: Sephadex, Sepharose. Often good yields by adsorption but very expensive carriers.

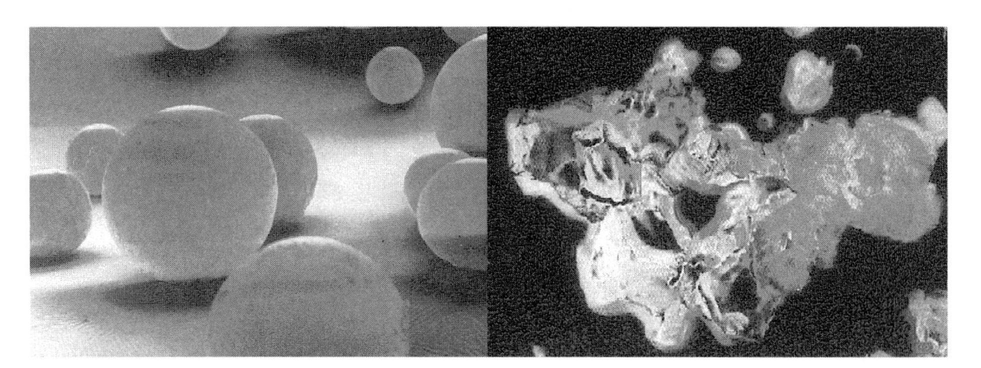

Figure 5.6. Immobilized enzymes can come in a variety of shapes: (left) CHIRAZYME L-2, Carrier 2 (Novozym 435), a lipase from *Candida antarctica*, type B, immobilized on a macroporous acrylic resin has spherical particles. (right) D-amino acid oxidase, immobilized on a crushed polymerizate, displays irregularly formed particles (photographs by M. Kage, Germany, on behalf of Roche Molecular Biochemicals).

5.2.1.3 Binding Enzymes to Carriers

The amino group of lysine is the group of choice to bind an enzyme covalently to a carrier (Fig. 5.7). In some cases, the support needs to be activated to carry appropriate functional groups. Although binding is strong, considerable activity loss can occur during immobilization.

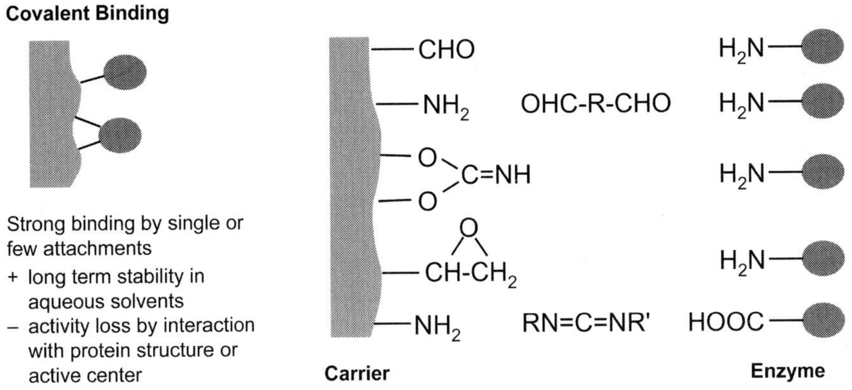

Covalent Binding

Strong binding by single or few attachments
+ long term stability in aqueous solvents
– activity loss by interaction with protein structure or active center

—CHO | H₂N—
—NH₂ | OHC-R-CHO | H₂N—
—O, —O >C=NH | H₂N—
—CH-CH₂ (epoxide, O) | H₂N—
—NH₂ | RN=C=NR' | HOOC—

Carrier | Enzyme

Figure 5.7. Covalent binding of an enzyme to a carrier.

Glutaraldehyde is often used to attach enzymes to carriers having functional amino groups. The binding is not as straightforward as it seems. Although formation of a Schiff base may play a role, carrier and enzyme react with a variety of different species present in a glutaraldehyde solution [17]. At acidic to neutral pH, glutaraldehyde exists as a monomer (free aldehyde or hydrate) or polymeric hemiacetal, while under basic conditions it undergoes aldol condensation to form *α,β*-unsaturated multimeric aldehydes, which can react with proteins or carriers to form stabilized Schiff bases or Michael adducts. This explains why the often recommended reduction of the Schiff base with sodium borohydride does not seem to increase stability.

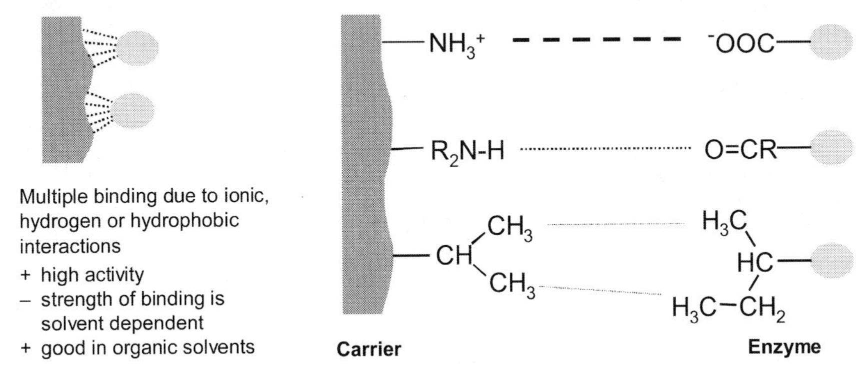

Figure 5.8. Non-covalent binding of an enzyme to a carrier.

Non-covalent binding of enzymes, on the other hand (Fig. 5.8), is often performed in a simple manner: the enzyme is mixed with the carrier in a buffer with a defined pH and polarity and then filtered off. In most cases, the immobilized enzyme is dried for better storage and subsequent use in organic solvents. The retained activity is high, but the enzyme desorbs under certain reaction conditions, especially in the presence of detergents or emulsifying agents.

5.2.1.4 Cross-Linked Enzyme Crystals

Like numerous chemical substances, proteins can also be crystallized. Since binding forces between protein molecules are weak, it may take months to find the conditions in which a protein crystallizes, and the crystal itself often possesses limited mechanical stability. If protein molecules in a crystal are inter- and intramolecularly cross-linked, the crystal becomes insoluble in aqueous media and mechanically more stable. Since proteins usually crystallize in a stable conformation, the protein of a cross-linked protein or enzyme crystal is even more stable than its free counterpart. The enzyme conformation is locked, and partial denaturation by the influence of temperature, pH or co-solvents is reduced. The pores of a crystal are big enough (2 nm diameter and greater) to allow easy diffusion of compounds with a molecular weight of up to 2000 Da (Fig. 5.9) [18, 19].

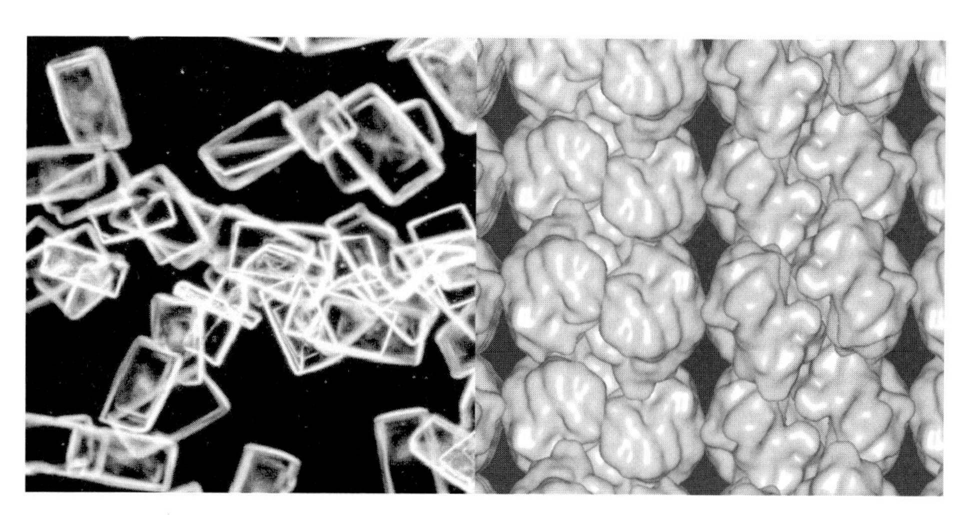

Figure 5.9. (left) Cross-linked enzyme crystals of Penicillin G amidase (Penicillin acylase). The average length of a crystal is 0.14 mm. On the right, a computer-generated image of the crystal structure of lipase from *Candida rugosa* shows large pores (20 Å) running through the crystal (Images kindly provided by J. Lalonde, Altus Biologics, Cambridge, MA, USA; right picture taken from [19], copyright American Chemical Society).

Crystallization has been used for a long time for purifying proteins. In this respect, it is noteworthy that the quality of the crystals need not be as high as that required for X-ray crystallography. A distinct advantage of the cross-linked enzyme crystal (CLEC) technology is certainly that generating the crystals is simultaneously a final purification step towards very high purity [20, 21].

A cross-linked enzyme crystal has the highest possible density of active enzyme sites, and thus, a very high activity per weight. Since the catalytic activity is distributed homogeneously through the crystal, considerable diffusion limitation may occur (Table 5.6). Additionally, in hydrolysis reactions of esters or amides, the generated acid leads to a pH-gradient inside the crystal. This means that the enzyme inside the crystal has to operate at lower pH. This phenomenon also applies to any other immobilized enzyme in which the enzyme is distributed homogeneously through the particle, but it may be less pronounced due to the lower enzyme density (see also Section 5.3.2, Fig. 5.20.).

Table 5.6. Dependence of the hydrolytic activity of SynthaCLEC-PA (Penicillin G amidase/acylase) on the crystal size (J. Lalonde, unpublished results).

Formulation	Crystal size	Activity* (dry weight)
Free enzyme	–	25 U/mg
SynthaCLEC-PA	approx. 0.04 mm	12–18 U/mg
SynthaCLEC-PA	approx. 0.11 mm	3–5 U/mg

* Conditions: 1.2% (w/v) Penicillin G potassium salt, 28 °C, pH 8.0, 2 mM phosphate buffer.

The CLEC technology, as pioneered by Altus Biologics, yields stable biocatalysts, capable of operating under enzyme-hostile conditions. However, learning how to crystallize and cross-link enzymes is difficult. This is not the method of choice for an afternoon experiment for making an immobilized catalyst but rather should be left to experts in the field. It is too early to tell whether this technology can and will be adopted for many industrial applications.

5.2.2 Activity Assay

The activity of free or immobilized enzymes is expressed in *Units*. In general, the following definition is used: 1 Unit converts 1 μmol of substrate per minute under defined conditions (concentration, pH, temperature, etc.). When comparing activities, one has to be careful whether the definition is comparable.

In order to get a reliable value for the reaction rate, the reaction is followed at regular intervals up to a conversion of 10–20%. Depending on the reaction type, the conversion can be followed spectrophotometrically, by titration of liberated acid or base (for hydrolysis reactions), by the consumption of oxygen or the generation of hydrogen peroxide (oxidation reactions), or by quantitative chromatography (HPLC, GC). The assays preferred by enzymologists fail in most cases with carrier-fixed enzymes, because floating particles inside a magnetically stirred cuvette may disturb the measurement.

Measuring the activity of immobilized enzymes seems straightforward but there are many pitfalls. An important factor is the size of the samples used in the test. The smaller the sample size required to do the testing, the less likely it is to draw a sample for testing with uniform particle-size distribution. Since in most cases activity located on the outer shell of the particle is measured, the activity depends also on the size of the particles. The smaller the particles, the larger the surface per weight is and, thus, the activity measured (Fig. 5.10). Sieving out the small particles reduces the apparent activity.

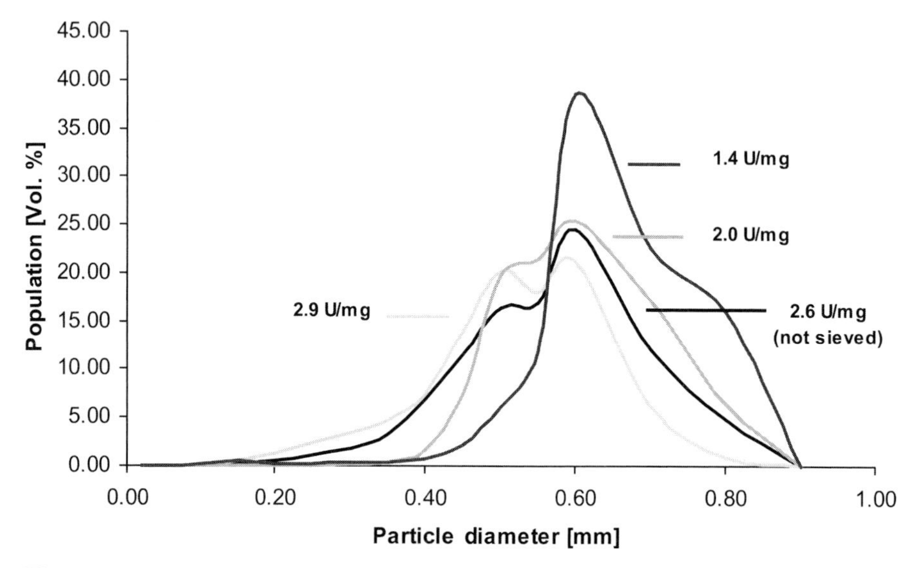

Figure 5.10. Dependence of activity on particle size distribution for CHIRAZYME L-2 Carrier 2 (a lipase from *Candida antarctica*, type B, immobilized on a macroporous acrylic resin). Activity test: hydrolysis of 100 mM tributyrin, 10 mM K phosphate, pH 7.0, 25 °C, automatic burette [22].

Another issue is the activity assay of immobilized enzymes containing water or buffer. For example, variations can be expected from condensation inside the storage vessel, reducing the amount of water or buffer in the catalyst particles. In order to get an accurate result, the following procedure should be strictly followed:

- A representative sample (mixing!) is taken from the storage vessel and weighed.
- The sample is washed using water or a very weak buffer and weighed again.
- A portion of the washed sample is used in the activity assay.
- A portion of the washed sample is dried to remove all moisture and then weighed.
- By using the weight ratio of washed sample to the sample in storage conditions, the activity of the material (storage conditions) can be calculated.
- By using the weight ratio of washed sample to the dried sample, the activity of the material based on dry weight can be calculated.

Since the moisture content of a preparation can vary during storage or by usage, the activity based on dry weight is the reference used to determine whether a catalyst has lost activity during storage or under operational conditions.

5.2.3 Activity Balance

After an immobilization experiment, it is important to assess the chosen method by determining the activity balance. Only measuring the activity of the immobilized enzyme and comparing it with the total amount of activity used in the experiment is not enough. It is necessary to understand where the activity remains in order to effectively optimize an immobilization procedure. Figure 5.11 gives an example in which the immobilization yield would be 45%. The supernatant contained 10% at the end of the immobilization procedure. Due to diffusion limitation, some enzyme may be "trapped" inside the catalyst particles; in this example, 20%. This so-called hidden activity can be determined by crushing the catalyst particles without denaturing the enzyme and measuring the activity again. In most cases, some enzyme activity is lost due to denaturation of the enzyme under the immobilization conditions (25%).

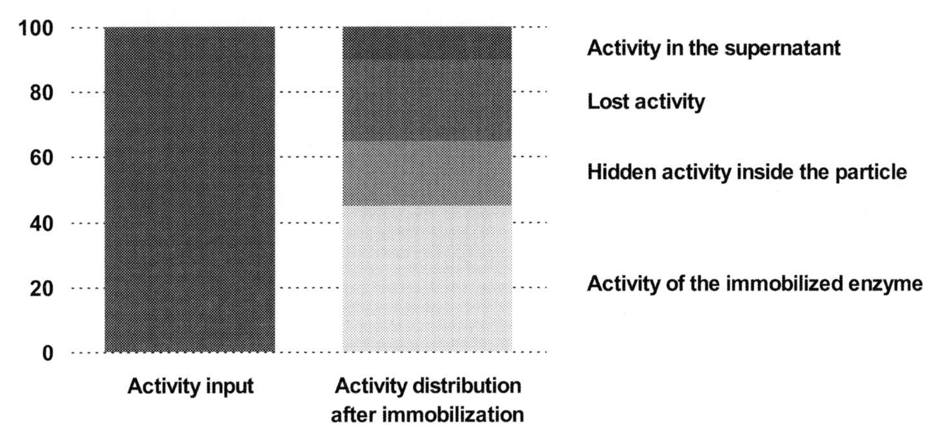

Figure 5.11. The activity balance of an immobilization method (example).

With lipases, it has been observed that the activity of the immobilized lipase can be higher than the activity input (free enzyme) [23–25]. This is due to a specific feature of lipases, namely the interfacial activation. Most lipases contain a lid over the active site which is opened when the lipase binds to a lipid droplet. The lipase can be immobilized in this activated state if highly hydrophobic carriers are used as a lipid substitute.

On the other hand, one has to be careful in interpreting activities observed for reactions in organic solvents. Here, the accessible surface plays an important role: an enzyme powder may be less active than the immobilized form, because only part of the powder may be accessible to the reactants.

5.2.4 Cost of Immobilization

It is obvious that the immobilization of enzymes is a considerable experimental effort which does not come inexpensively. How does carrier fixation affect the price of the enzyme per activity?

When comparing the price of various commercially available enzyme formulations based on activity, it turns out that a powder or lyophilizate is generally twice as expensive as the corresponding enzyme solution. The enzyme immobilized on a carrier is 10–25 times more expensive than the solution. More extreme cases are also known with costs ranging from 3 up to 500 times higher than that of the enzyme solution.

Table 5.7 explains what cost advantage can be expected from a carrier-fixed enzyme, and how the various effects add up in the end. As already explained, carrier fixation can lead to a stabilization of the enzyme under the reaction conditions which translates into lower enzyme input. However, during the immobilization procedure itself, enzyme is lost (immobilization yield). Together with labor and equipment costs, the price for the carrier, etc., the carrier-fixed enzyme is approximately 10 times more expensive than the free enzyme. If the immobilized enzyme can be used for 200 cycles, it will be 40 times cheaper than the free enzyme. In other words, 5 cycles need to be run before the carrier-fixed enzyme becomes cheaper than the free enzyme. This estimation does not take into account the equipment costs for the actual process to be run and must be regarded as a rough generalization.

Table 5.7. Cost comparison of carrier-fixed vs. free enzyme.

Parameter	Averaged factor
Free enzyme (calculation basis)	1
Stabilization by immobilization under reaction conditions	2
Cost based on activity (free vs. immobilized enzyme)	1/10
Repeated use/number of cycles	200
Cost comparison (advantage of immobilized vs. free enzyme)	40

5.3 Operation

It is a distinct advantage of biocatalytic processes that their technology does not differ very much from other chemical processes. For processes using free enzymes, standard equipment such as a stirred-tank can be used. Since the predominant purpose of using carrier-fixed enzymes is the reuse of the catalyst, some modifications are required to allow a straightforward operation and a long lifetime of the biocatalyst. This section will focus on the type of reactor appropriate for the application and the parameters influencing the biocatalysts' operational stability. Several industrial processes are also mentioned here. For better understanding, these processes are briefly introduced.

Especially in the area of β-lactam antibiotics, biocatalysis has found widespread use. Since the seventies, the enzymatic hydrolysis of Penicillin G to 6-APA and Cephalosporin G to 7-ADCA has replaced the chemical process (Scheme 5.1). The phenylacetic acid that is generated in such hydrolysis reactions must be neutralized by addition of a base, preferably ammonia.

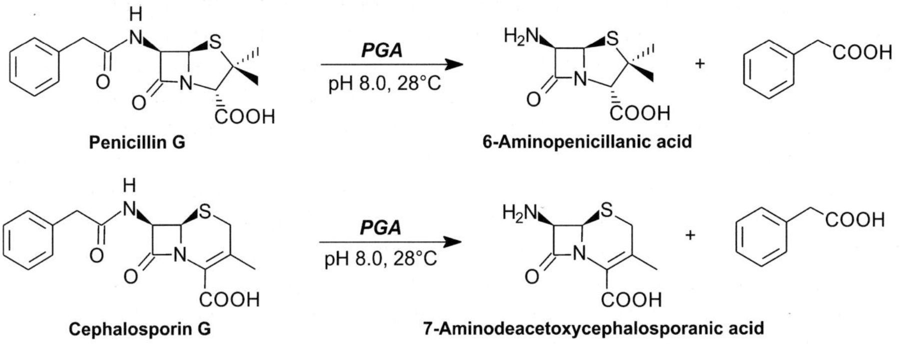

Penicillin G → PGA, pH 8.0, 28°C → 6-Aminopenicillanic acid + phenylacetic acid (COOH)

Cephalosporin G → PGA, pH 8.0, 28°C → 7-Aminodeacetoxycephalosporanic acid + phenylacetic acid (COOH)

Scheme 5.1. The enzyme Penicillin G amidase (PGA; also called Penicillin acylase) hydrolyzes Penicillin G or Cephalosporin G to 6-APA or 7-ADCA.

A related application is the enzymatic synthesis of unnatural, amino acid side chains for semi-synthetic penicillins and cephalosporins. A D-hydantoinase is used to make the important side chain D-p-hydroxyphenylglycine (PHPG) (Scheme 5.2). The process is used to make thousands of tons of D-p-hydroxyphenylglycine per year. Under the reaction conditions the hydantoin racemizes rapidly. This dynamic resolution allows a theoretical yield of 100% D-enantiomer.

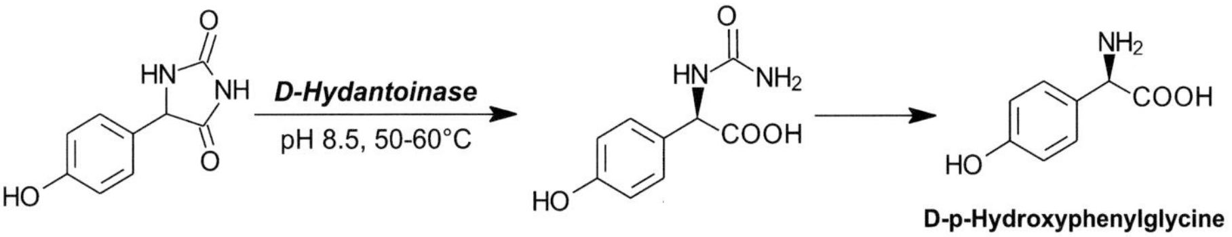

D-Hydantoinase, pH 8.5, 50-60°C → D-p-Hydroxyphenylglycine

Scheme 5.2. D-Hydantoinase hydrolyzes hydantoins to the corresponding carbamoylic acids which can be converted to the free D-amino acid either by diazotization or by enzymatic hydrolysis using a decarbamoylase.

Up to now, most biocatalytic processes on industrial scale are hydrolysis reactions involving the hydrolysis of an ester or amide. An example of an oxidation reaction is the first step of the enzymatic synthesis of (7-ACA) from Cephalosporin C (Scheme 5.3). The hydrogen peroxide generated in the oxidase reaction is consumed in the decarboxylation from the keto acid to glutaryl-7-ACA [26].

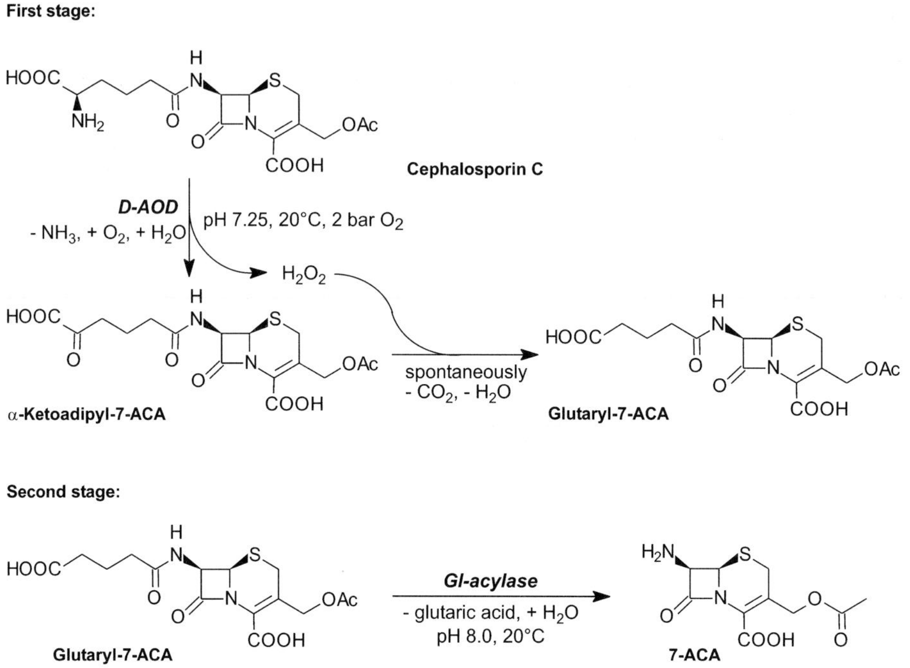

First stage:

Cephalosporin C

D-AOD
$- NH_3$, $+ O_2$, $+ H_2O$
pH 7.25, 20°C, 2 bar O_2
H_2O_2

α-Ketoadipyl-7-ACA

spontaneously
$- CO_2$, $- H_2O$

Glutaryl-7-ACA

Second stage:

Glutaryl-7-ACA

Gl-acylase
- glutaric acid, $+ H_2O$
pH 8.0, 20°C

7-ACA

Scheme 5.3. Enzymatic transformation of Cephalosporin C to 7-ACA via oxidation by D-amino acid oxidase (D-AOD) and hydrolysis of the intermediate by glutaric acid acylase (Gl-acylase).

For lipases, the hydrolysis of isosorbide diacetate may serve as an example (Scheme 5.4). The process was developed by Roche Diagnostics [27] based on work of Seemayer [28]. However, it was never commercialized, although it is highly competitive to the chemical process. In this process, the diacetate is hydrolyzed to the 2-acetate which can serve as a precursor for isosorbide-5-nitrate, a coronary vasodilator.

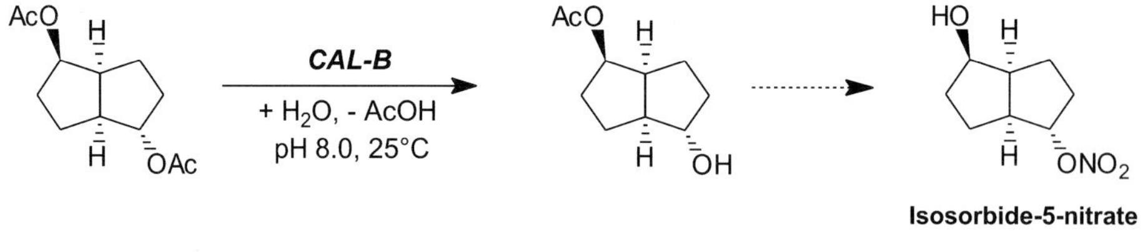

CAL-B
$+ H_2O$, $-$ AcOH
pH 8.0, 25°C

Isosorbide-5-nitrate

Scheme 5.4. Hydrolysis of isosorbide diacetate by a lipase from *Candida antarctica*, type B.

Many other industrial processes are known but very few data have been published on the stability of the enzymes. Since the processes mentioned above have been extensively studied by Roche Diagnostics, these data will serve as a basis for the following statements.

5.3.1 Reactors

Three classes of reactors have shown to be technically robust and reliable for biocatalytic reactions. Figure 5.12 shows the reactor types, and Table 5.8 lists some characteristics. The choice of the best reactor greatly depends on the type of reaction (e.g. hydrolysis or transesterification), the reaction media (e.g. aqueous or organic phase, biphasic systems, cosolvents and detergents etc.), and the properties of substrate and product (e.g. solubility, stability).

The stirred-tank reactor has proven to be simple and extremely efficient for hydrolysis reactions which require efficient mixing, e.g. upon addition of acid or base, or

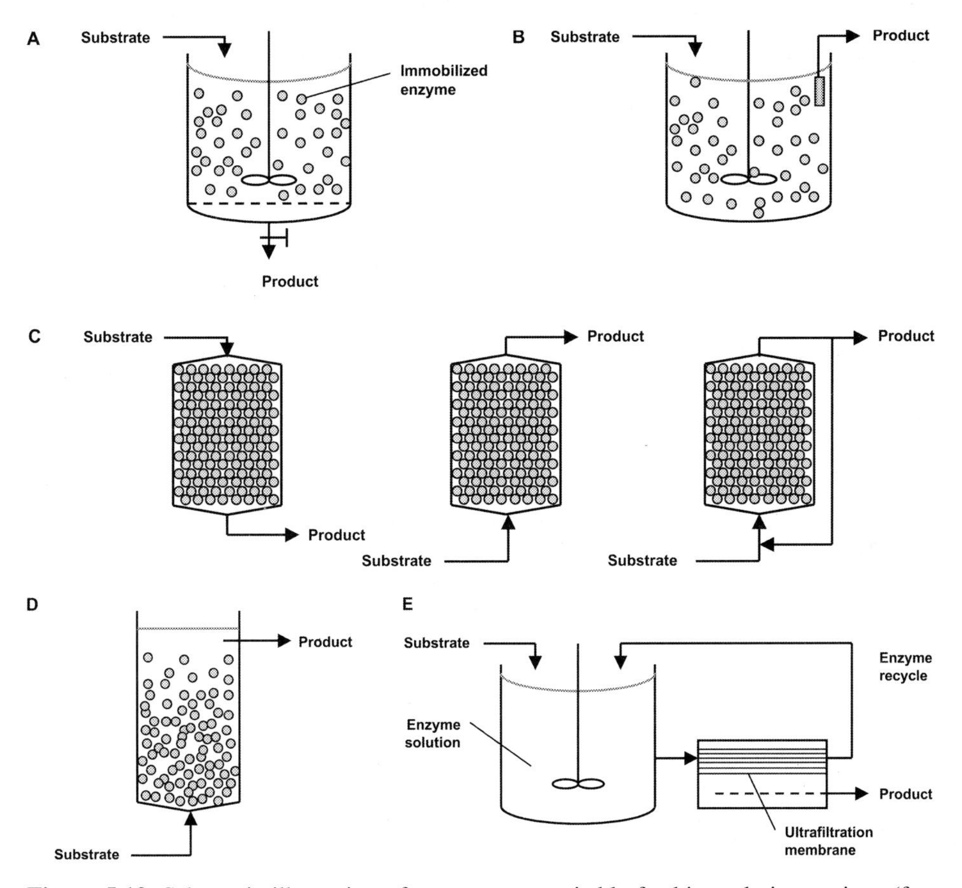

Figure 5.12. Schematic illustration of reactor types suitable for biocatalytic reactions (from [29]). The fed-batch stirred-tank reactor (A) has a sieve in the bottom to retain the catalyst particles within the reactor. The continuous fed-batch stirred-tank reactor (B) has an overflow with a sieve. In the continuous packed-bed or column reactor (C), the immobilized enzyme is trapped inside a column, while the immobilized enzyme moves freely in the continuous fluidized-bed reactor (D). The enzyme membrane reactor, EMR (E) uses hollow fibers to separate the non-immobilized enzyme from the reaction mixture.

Table 5.8. Characteristics of reactor types suitable for biocatalytic reactions.

Stirred-tank reactor	Packed-bed reactor	Enzyme membrane reactor

• Simple equipment but discontinuous operation (fed-batch)
• Easy to operate
• Efficient mixing, e.g. for neutralization or oxygen supply
• Catalyst loading up to 30% of reactor volume
• Operated down to 30% of residual enzyme activity
• Enzyme addition (topping) for constant reaction time
• Non-abrasive, elastic particles required
• Particle size preferably within 0.1–1 mm range
• Filter screen width <0.04–0.07 mm
	• High space-time-yield
	• Simple operation
	• Column cascades provide high flexibility
	• Less recommended for reactions requiring
	  • Neutralization
	  • Cofactors
	  • Gaseous substrates
	• Need for pressure resistant particles
		• Continuously operated
		• High space-time-yield
		• Molecular weight limits for substrates & products
		• In-situ solvent extraction is easy
		• Enzyme not modified
		• Enzyme stabilized only by reaction conditions
		• Actual enzyme activity easily controlled

when oxygen is consumed. Column processes show advantages for reactions in organic media or for reactions with uncharged reaction products. If a base needs to be added to keep the pH constant, as during the hydrolysis of esters or amides (Scheme 5.1 to 5.4) for example, a pH-gradient builds up inside the column as acid is generated. In the downward part of column, the pH decreases one to three units below the optimal pH. This means that a considerable portion of the catalyst is operating in sub-optimal conditions, leading to reduced overall activity and stability.

This can be illustrated by comparing the splitting of Penicillin G to 6-aminopenicillanic acid (6-APA) in a fed-batch stirred-tank and in a fed-batch column reactor (Scheme 5.1, Figs. 5.13 and 5.14). For the latter system, the operational stability under comparable conditions is 2.5 times lower in the loop than in the stirred-tank reactor [30].

For reactions that require the addition of gaseous reactants, like oxygen for the oxidation of Cephalosporin C (Scheme 5.3), again the stirred-tank reactor is highly favored due to better gas transfer.

The enzyme membrane reactor differs in many aspects from other reactor types. The predominant advantage is its use of free enzyme and the fact that enzyme activity can easily be controlled. On the other hand, the benefit of enzyme stabilization by immobilization is missing and, thus, the enzyme stability can only be controlled by adapting the reaction conditions or changing the enzyme properties chemically or through genetic engineering.

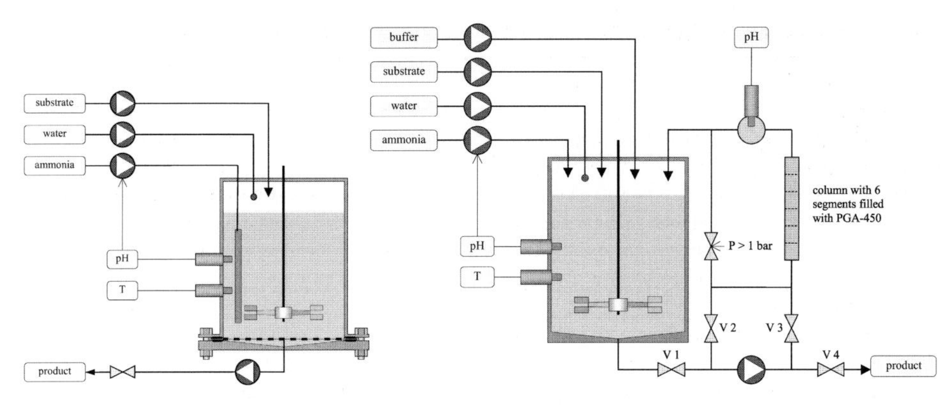

Figure 5.13. Schematic drawing of the reactor setup for hydrolysis of Penicillin G to 6-APA using Penicillin G amidase 460 (PGA-450) in a fed-batch stirred-tank (left) and in a fed-batch packed-bed loop reactor (right) [30].

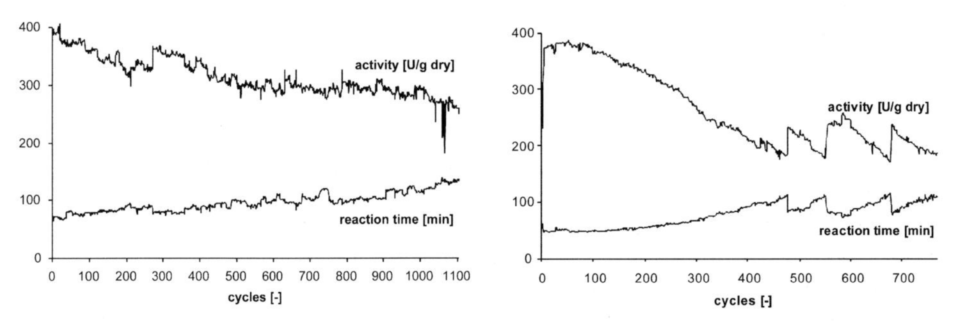

Figure 5.14. Hydrolysis of Penicillin G to 6-APA using Penicillin G amidase 450 (PGA-450) in a fed-batch stirred-tank (left) and in a fed-batch packed-bed loop reactor (right) [30]. The initial activity decreases more slowly in the stirred-tank reactor than in the loop reactor. During the run, part of the column was refilled three times with fresh enzyme to keep the reaction time at around 100 minutes. Reaction conditions are comparable: Stirred-tank reactor 7 kU/l, 8% Penicillin G, pH 8, 28 °C, 250 rpm, catalyst consumption 0.16 kU/kg 6-APA; loop reactor 10 kU/l, 1 volume/volume/minute circulating rate, catalyst consumption 0.42 kU/kg 6-APA.

5.3.2 Operational Stability

The term 'operational stability' describes the stability of the catalyst under given reaction conditions. As already mentioned, the predominant feature of immobilized enzymes is their ability to be reused many times and, thus, reducing the cost of catalyst per kg of produced product. Great care has to be taken to create a balance of reaction conditions and the long-term stability of the catalyst to maximize the operational stability. How some opposing parameters influence the hydrolysis of Penicillin G to 6-APA is shown in Table 5.9 [30].

Table 5.9. Influences of changing parameters on the enzymatic hydrolysis of Penicillin G to 6-APA by immobilized Penicillin G amidase. Optimal reaction conditions: stirred-tank reactor 4.8 kU/l, 8% Pen G, pH 8, 28 °C.

Change with respect to optimum	Positive consequences	Negative consequences
Higher pH	Higher enzyme activity Lower enzyme input	Lower enzyme stability Lower substrate and product stability
Higher temperature	Higher enzyme activity Lower enzyme input	Lower enzyme stability Lower substrate and product stability
Higher substrate concentration	Lower reaction volume Better product isolation yield	Higher substrate and product inhibition Less favorable equilibrium Longer reaction time
Lower enzyme input	Lower catalyst costs	Longer reaction times Higher equipment costs Lower substrate and product stability
Use of buffer	Higher enzyme activity	Increased salt load Additional costs for buffer

While the influence of some parameters like temperature and substrate concentration is fairly obvious or characteristic in each reaction studied, a few comments need to be made concerning the use of buffer in hydrolysis reactions.

The conversion of Penicillin G to 6-APA again serves as an example. If the enzyme loading on and inside the carrier particle is high, diffusion limitation can be observed. The phenylacetic acid (PAA) generated within the catalyst particle lowers the pH inside the particle which then leads to a pH-gradient inside the particle (Fig. 5.15) and, thus, to a lower activity of the enzyme inside the particle [31]. Adding a buffer to the system reduces this effect and therefore increases the overall activity of the catalyst per weight (Fig. 5.16).

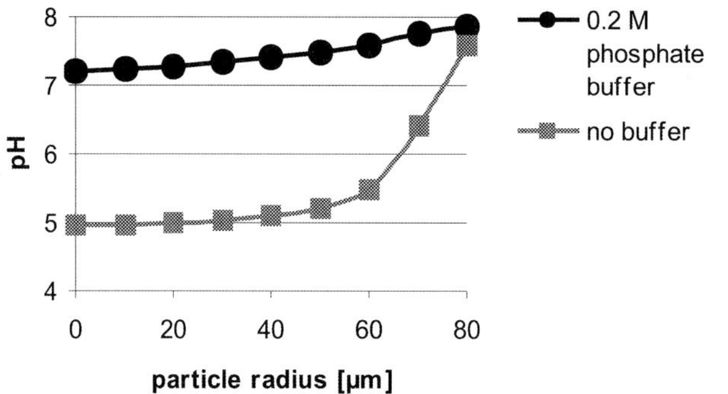

Figure 5.15. Hydrolysis of Penicillin G by PGA: Calculated pH-profile with or without buffer (37 °C, $pH_{bulk} = 7.8$) [31].

When establishing a biocatalytic process, the issue of catalyst cost needs to be addressed early on in the development. After a few laboratory-scale experiments, it is al-

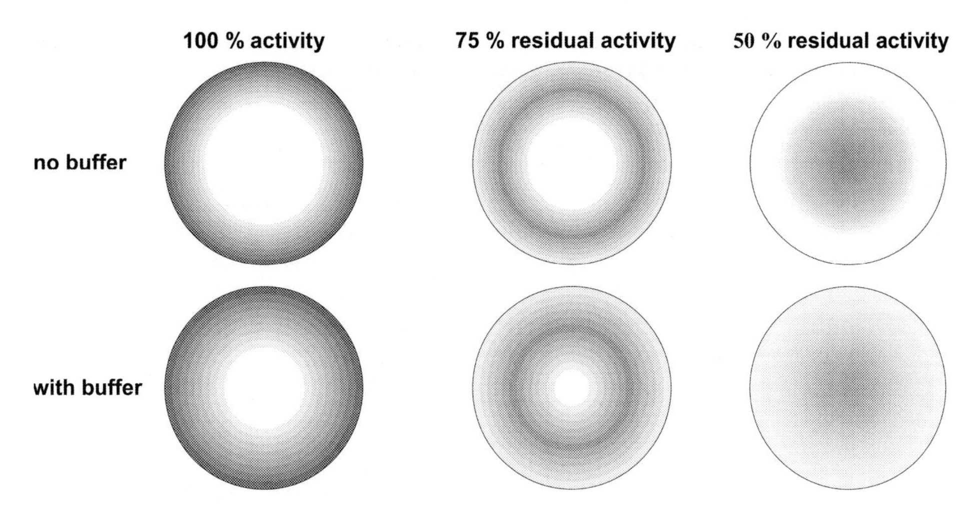

Figure 5.16. The scheme explains how hydrolysis of an ester or amide under diffusion limitation may evolve in a long-term operation [30]. The dark zone represents actively catalyzing enzyme. As the enzyme gradually deactivates during operation, the active zone migrates to the centre of the particle. If the reaction system is buffered, the pH-gradient is less pronounced and more enzyme is active (the active zone is broader).

ready possible to get a rough estimate of the amount of enzyme needed to do the reaction within a reasonable time. Once the figure is known, one can decide whether or not recycling is necessary, and one may calculate from the cost of the enzyme the number of reuses that must be reached to meet the economic target. The important parameter for cost estimation is the **enzyme consumption** expressed in catalyst activity per kg of product, or its reciprocal, the **productivity**, expressed in kg product per catalyst unit (activity or weight). This is exemplified for the hydrolysis of Penicillin G to 6-APA (Scheme 5.1 and Fig. 5.14, Table 5.10).

Table 5.10. How to calculate enzyme consumption and productivity of a catalyst, e.g. in the hydrolysis of Penicillin G to 6-APA. Data are based on laboratory-scale results for a fed-batch stirred-tank [30].

Substrate concentration:		8%	(w/v) Penicillin G, potassium salt
Chemical yield:		87%	(isolated)
		50%	based on weight
Enzyme input:		4.8	kU/l
		60	kU/kg Penicillin G
		~0.45	kg PGA-450 (dry weight)/kg Penicillin G
	60/0.5 =	120	kU/kg 6-APA
Enzyme cost:		Y	DM/kU
Recycles:		900	
Enzyme consumption:	120/900 =	0.13	kU/kg 6-APA
	Y×0.13 =	X	DM/kg 6-APA
Productivity:	1/0.13 =	7.6	kg 6-APA/kU PGA-450

The determination of the number of achievable cycles (batches) with one enzyme load is a cumbersome series of experiments, especially if the number to be reached is high. Varying reaction parameters (pH, temperature, concentration etc.) multiplies the number of necessary experiments. In the end, the results are summarized in graphs showing the decrease of activity over the cycle number, and the associated increase of reaction time (Fig. 5.14).

In analogy with the calculation in Table 5.10, data are given for other processes (Table 5.11). The table indicates typical values for industrial processes. It must be pointed out that the number of attainable cycles is given for half activity, meaning that the catalyst has lost 50% of its initial activity. By then, the reaction time has at least doubled. Before developing an immobilized enzyme, emphasis should be placed on the activity of the free enzyme towards the substrate. If the specific activity (activity per amount of protein) is below 1 U/mg protein, the enzyme is just not a good catalyst for the intended reaction. It is unlikely that a commercially viable process can be developed by using the enzyme repeatedly in its immobilized form.

As an example of a resolution using a cross-linked enzyme crystal, the hydrolysis of 1-phenylethyl acetate by ChiroCLEC-PC, a cross-linked crystalline lipase from *Pseudomonas cepacia* was investigated and optimized [32]. The reaction was upscaled to 100 l with the following conditions: 30% substrate concentration, pH 6.0, 30 °C, 0.5% (w/v) ChiroCLEC-PC, corresponding to an activity input of 8 kU/l, 18 h reaction time (first cycle). Although the reaction proceeded with extremely high stereoselectivity, a strong loss of activity after each cycle was observed. Leakage through the reactor sieve was eliminated by using a fine sieve (27 μm together with a 5 μm cloth) and adding a filtration aid. The lipase had lost 60% of its activity after nine batches (not counting for physical loss). This shows that a sufficiently large crystal size is a very important factor for a good performance on a large scale.

The authors estimate that the lifetime of the catalyst could reach approximately 20 cycles producing 223 kg (*R*)-1-phenylethanol although this means using the catalyst far below 50% of its initial activity. This would correspond to a productivity of 0.28 kg (*R*)-alcohol per kU ChiroCLEC-PC (catalyst consumption: 3.6 kU/kg or 440 U/mol).

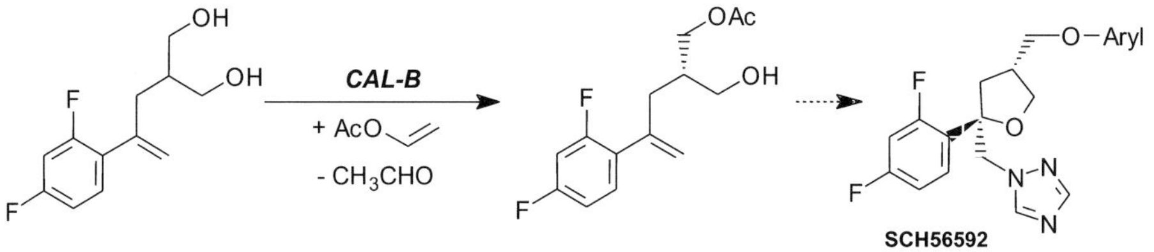

Scheme 5.5. Enzymatic process by Schering-Plough for the synthesis of an enantiomerically pure drug intermediate using CHIRAZYME L-2, Carrier 2 (Novozym 435), an immobilized lipase from *Candida antarctica*, type B. The reaction conditions on pilot scale: scale 150 liters, solvent acetonitrile, diol concentration 20% (w/v), 2 equivalents of vinyl acetate, enzyme load 5% (w/w), temperature 0 °C, reaction time 6 h, yield (not isolated) 74% monoacetate, 26% diacetate [33]. On a laboratory scale, the number of possible reuses was at least 10. This would give a productivity of approx. 30 kg monoacetate per kg catalyst.

Table 5.11. Comparison of various processes. Numbers are based on laboratory results and may vary on an industrial scale. Depending on the price of the catalyst, catalyst costs per kg of product range from DM 2 to 25.

Catalyst	PGA-450	D-AOD	Gl-Acylase	D-Hydantoinase	CAL-B
Substrate	Penicillin G	Cephalosporin C	Gl-7-ACA	5-(*p*-Hydroxyphenyl)-hydantoin	Isosorbide diacetate
Reaction: refer to	Scheme 5.1	Scheme 5.3	Scheme 5.3	Scheme 5.2	Scheme 5.4
Reference	[30]	[30]	[30]	[30]	[27]
Catalyst activity					
soluble enzyme (U/mg)	30	25	5	100	600
immob. enzyme (dry, U/g)	330	>85	>29	>1000	>250
immob. enzyme (moist, U/g)	135	>20	>85	>150	>100
Catalyst input					
input per volume (kU/l)	4.8	2.5	5	6	5
input per kg substrate (U/g)	60	79	165	60	62
Substrate concentration (%)	8	3	3	10	8
Other conditions	50 mM borate buffer, pH 8.0, 28 °C	pH 7.25, 20 °C, 2 bar oxygen	pH 8.0, 20 °C	2 mM $MnCl_2$, 5 mM $Na_2S_2O_3$, pH 8.5, 50 °C	pH 7, 25 °C
Reaction time (initial, min)	90	60	60	65	60
Reuse (until half activity)	900×	130×	210×	180×	200×
Theoretical yield (%)	87 (isolated)		overall 85 (isolated)	>96 (HPLC)	>96 (HPLC)
Productivity (kg/kU)	7.6	0.83	0.67	2.1[1]	2.5
(based on dry weight, kg/kg)	2500	71	193	2000	625
Catalyst consumption (U/kg)	130	1.2	1.5	0.5	0.4
(U/mol)	28	492	408	85	75

[1] Productivity based on total conversion from 5-(p-hydroxyphenyl)hydantoin to D-p-hydroxyphenyglycine (the second step is a diazotization; assumed overall isolated yield 80%).

Most large-scale examples mentioned in literature are hydrolysis reactions. Little has been published on the operational stability of e.g. lipase-catalyzed reactions in organic solvents. Schering-Plough gives conditions for the pilot-scale synthesis of an intermediate for an antifungal drug, currently in clinical trials (Scheme 5.5) [33].

From experience with the processes mentioned here, some guidelines can be drawn for the use of immobilized enzymes in stirred-tank reactors:

- Avoid high stirrer speed,
- Remove all sharp edges from baffles, stirrer blades, probes etc.,
- Avoid close contact between moving parts (stirrer vs. probes/baffles),
- Do not stir when there is little or no solvent,
- Check regularly for enzyme leakage indicating a punctured sieve.

For all reactions in aqueous media, microbial contamination must strictly be avoided, since it will quickly lead to loss of activity and filtration problems.

5.4 Summary

It took organic chemists and process engineers some time to acknowledge that biocatalysts are a viable alternative to chemocatalysts. A strong increase in the development of biocatalytic processes has been seen in recent years. Many new single stereoisomer drugs will be produced via routes in which at least one biocatalytic step has been tested. However, the success of any chemoenzymatic process ultimately depends on the performance of the drug candidate in clinical trials.

Because an increasing number of enzymes have become commercially available at lower prices, biocatalysts belong in the catalyst toolbox of every chemist involved in enantioselective synthesis. The immobilization of enzymes greatly enhances their potential on a large scale. No new equipment needs to be designed or developed for their use. However, before using immobilized enzymes, one should evaluate whether or not free enzymes perform just as well at lower costs.

Acknowledgements

I wish to thank my colleagues at Roche Diagnostics, Ulrich Giesecke, Wilhelm Tischer and Frank Wedekind for carefully reviewing the article. My appreciation is also extended to Padma Rao for editing the English and to Jim Lalonde from Altus Biologics for making available unpublished data on cross-linked enzyme crystals.

References

[1] Tischer W., Kasche V., Immobilized enzymes: crystals or carriers? *Trends Biotechnol* 17 (1999) 326-35.

[2] Biselli A., Kragl U., Wandrey C., in Drauz K., Waldmann H. (eds), Enzyme catalysis in organic synthesis, Verlag Chemie, Weinheim, New York, Basel, Cambridge, Tokyo (1992), pp. 132–52.

[3] Rasor J. P., Voss E., Enzyme-catalyzed processes in pharmaceutical industry, *Applied Catalysis A: General* (2000) accepted.

[4] Chibata I., Tosa T., Shibatani T., in Collins A. N., Sheldrake G. N. and Cosby J. (eds), Chirality in Industry, John Wiley & Sons Ltd, Chichester, New York, Brisbane, Toronto, Singapore (1992), pp. 351–370.

[5] Schafhauser D. Y., Storey K. B., Immobilization of amyloglucosidase onto granular chicken bone, *Appl Biochem Biotech* 32 (1992) 89–109.

[6] Kerscher V., Plainer H., Sproessler B., Immobilization of enzymes for use in organic media, German Patent DE 4029374 (1992).

[7] Nilsson K., Mosbach K., Immobilization of enzymes and affinity ligands to various hydroxyl group carrying supports using highly reactive sulfonyl chlorides, *Biochem Biophys Res Commun* 102 (1986) 449–457.

[8] Miller D. A., Blanch H. W., Prausnitz J. M., Enzymic interesterification of triglycerides in supercritical carbon dioxide, *Ann NY Acad Sci* 613 (Enzyme Eng 10) 1990) 534–7.

[9] Tischer W., Wedekind F., Immobilized enzymes: Methods and application, *Top Curr Chem* 200 (1998) 95-126.

[10] Wang P., Sergeeva M. V., Lim L., Dordick J. S., Biocatalytic plastics as active and stable material for biotransformations, *Nature Biotechnology* 15 (1997) 789–93.

[11] Wong S. S., Chemistry of protein conjugation and cross-linking, CRC Press Inc. (1991)

[12] Roxer G. P., *Methods Enzymol* 135 (1987) 143.

[13] Bastida A., Sabuquillo P., Armisen P., Fernandez-Lafuente R., Huguet J., Guisan J. M., A single-step purification, immobilization and hyperactivation of lipases via interfacial adsorption on strongly hydrophobic supports, *Biotech Bioeng* 58 (1998) 486–93.

[14] Margolin A. L., Novel crystalline catalysts, *Trends Biotechnol* 14 (1996) 223–230.

[15] Balcao V. M., Paiva A. L., Malcata F. X., Bioreactors with immobilized lipases: state of the art, *Enzyme Microb Technol* 18 (1996) 392–416.

[16] Wedekind F., Daser A., Tischer W., Penicillin G amidase, glutaryl-7-ACA acylase or D-amino acid oxidase immobilized on amino-substituted siloxane polymers, WO 9516773 (1994).

[17] Walt D. R., Agayn V. I., The chemistry of enzyme and protein immobilization with glutaraldehyde, *Trends Anal Chem* 13 (1994) 425–30.

[18] Persichetti R. A., Clair N. L. S., Griffith J. P., Navia M. A., Margolin A. L., Cross-linked enzyme crystals (CLECs) of thermolysin in the synthesis of peptides, *J Am Chem Soc* 117 (1995) 2732–7.

[19] Lalonde J., Practical catalysis with enzyme crystals, *Chemtech* 27 (1997) 38–45.

[20] Lalonde J., Govardhan C., Khalaf N., Martinez A. G., Visuri K., Margolin A. L., Cross-linked crystals of Candida rugosa lipase: Highly efficient catalysts for the resolution of chiral esters, *J Am Chem Soc* 117 (1995) 6845–52.

[21] Wang Y.-F., Yakovlevsky K., Zhang B., Margolin A. L., Cross-linked crystals of subtilisin: Versatile catalyst for organic synthesis, *J Org Chem* 62 (1997) 3488–95.

[22] Bacher W., Rasor P., unpublished results, Roche Diagnostics GmbH (1999).

[23] Reetz M. T., Zonta A., Simpelkamp J., Efficient heterogeneous biocatalysts by entrapment of lipases in hydrophobic sol-gel materials, *Angew Chem Int Ed Engl* 34 (1995) 301–3.

[24] Reetz M. T., Zonta A., Simpelkamp J., Efficient immobilization of lipases by entrapment in hydrophobic sol-gel materials, *Biotechnol Bioeng* 49 (1996) 527–34.

[25] Fernandez-Lafuente R., Armisen P., Sabuquillo P., Fernandez-Lorente G., Guisan J. M. Immobilization of lipases by selective adsorption on hydrophobic supports, *Chem Phys Lipids* 93 (1998) 185–197.

[26] Tischer W., Giesecke U., Lang G., Röder A., Wedekind F., Biocatalytic 7-aminocephalosporanic acid production, *Ann NY Acad Sci* 612 (1992) 502–9.
[27] Wedekind F., Rasor P., Tischer W., unpublished results, Roche Diagnostics GmbH (1995).
[28] Seemayer R., Bar N., Schneider M. P., Enzymic preparation of isomerically pure 1,4:3,6-dianhydro-D-glucitol monoacetates – precursors for isoglucitol 2- and 5-mononitrates, *Tetrahedron: Asymmetry* 3 (1992) 1123–6.
[29] Katchalski-Katzir E., Immobilized enzymes – learning from past successes and failures, *Trends Biotechnol* 11 (1993) 471–8.
[30] Giesecke U., Product description PGA-450, http://indbio.roche.com/indbio/Ind/PGA/f_p0.htm, Product description CC2, .../Ind/CC2/f_cp0.htm, Product description D-hydantoinase, .../Ind/Hyd/f_hyd0.htm, Roche Diagnostics GmbH (1999).
[31] Spiess A., Schlothauer R.-C., Hinrichs J., Scheidat B., Kasche V. pH Gradients in immobilized amidases and their influence on rates and yields of -lactam hydrolysis, *Biotechnol Bioeng* 62 (1999) 267–277.
[32] Collins A. M., Maslin C., Davies R. J., Scale-up of a chiral resolution using cross-linked enzyme crystals, *Organic Process Research & Development* 2 (1998) 400–6.
[33] Morgan B., Dodds D. R., Zaks A., Andrews D. R., Klesse R., Enzymatic desymmetrization of prochiral 2-substituted-1,3-propanediols: A practical chemoenzymatic synthesis of a key precursor of SCH51048, a broad-spectrum orally active antifungal agent, *J Org Chem* 62 (1997) 7736–7743.

6 Enantioselective Hydrogenation Catalyzed by Platinum Group Metals Modified by Natural Alkaloids

Peter B. Wells and Richard P. K. Wells

6.1 Historical Perspective

A new era in heterogeneous catalysis dawned when it was discovered that conventionally supported metal catalysts, modified by the adsorption of certain chiral organic compounds onto their surfaces, exhibited enantioselectivity in the hydrogenation of prochiral reactants.

This era began with a report by Erlenmeyer in 1930 that the hydrocinchonine salt of cinnamic acid underwent homogeneous asymmetric bromination in chloroform solution [1]. This observation prompted Lipkin and Stewart to investigate whether a hydrocinchonine salt would interact with a platinum catalyst to induce enantioselective hydrogenation, and in 1939 they reported that hydrocinchonine β-methylcinnamate in ethanol at room temperature and pressure was hydrogenated over Adams-Pt to give β-phenylbutyric acid which showed a rotation corresponding to an optical yield of about 8% [2]. Some twenty years later, Akabori and coworkers reported that Pd/silk-fibroin was enantioselective for compounds containing the >C=N- and >C=C< functions with optical yields being as high as 66% [3], but the origin of the enantioselectivity was not clear. After another twenty years, a series of papers from Orito, Iwai and Niwa [4–6] established unequivocally that the pre-adsorption of cinchona alkaloids onto conventionally supported Pt catalysts rendered them enantioselective for the hydrogenation of α-keto esters, particularly methyl pyruvate, MeCOCOOMe, and ethyl benzoylformate, PhCOCOOEt. Optical yields of 75–80% were achieved at room temperature using proprietary catalysts, common solvents and elevated hydrogen pressures. While there is a clear conceptual connection between this investigation and that of Lipkin and Stewart, there is no indication that the Japanese group was aware of the earlier work. Their papers initiated immense interest which has resulted in the discovery of several families of alkaloids and a variety of synthetic analogs which are effective as modifiers of platinum and palladium giving catalysts which facilitate the enantioselective reduction of compounds containing the functions: >C=O, >C=N-, >C=C<, and the pyrazine ring system. The subject has been much reviewed [7–13]. Consequently, the objective of this chapter is not to set out a complete factual account of the field but rather to describe and assess the processes whereby cinchona and other natural alkaloids direct enantioselectivity in metal-catalyzed reactions, and to point to possible future developments.

6.2 Enantioselective Hydrogenation of Activated Ketones Over Platinum

The general structures of the cinchona alkaloids are shown in Fig. 6.1; these molecules contain an aromatic quinoline ring system and a saturated quinuclidine ring system separated by the carbon atom numbered C_9. The molecules contain chiral centers at C_3, C_4, C_8, and C_9, wherein C_8 and C_9 have the *S*- and *R*-configuration in cinchonidine and quinine, respectively, but the *R*- and *S*-configuration in cinchonine and quinidine, respectively. Thus, cinchonidine and cinchonine are related as 'near enantiomers' and their enantiodirecting effects are, indeed, opposite. Simple derivatives have been investigated, the most important being the 10,11-dihydroalkaloids, compounds in which the substituent at C_9 has been varied, and compounds in which the quinuclidine-N atom has been quaternized.

Hydrogenation of pyruvate esters

Hydrogenations of methyl and ethyl pyruvate to the corresponding lactates (Scheme 6.1) have received much attention both from an exploratory practical viewpoint and from a theoretical and interpretative standpoint. Representative results are found in Table 6.1. Performance in these reactions is quantified in terms of the *enantiomeric excess,* e.e., which is defined as $[R]-[S]/([R]+[S])$. It is immediately clear that modi-

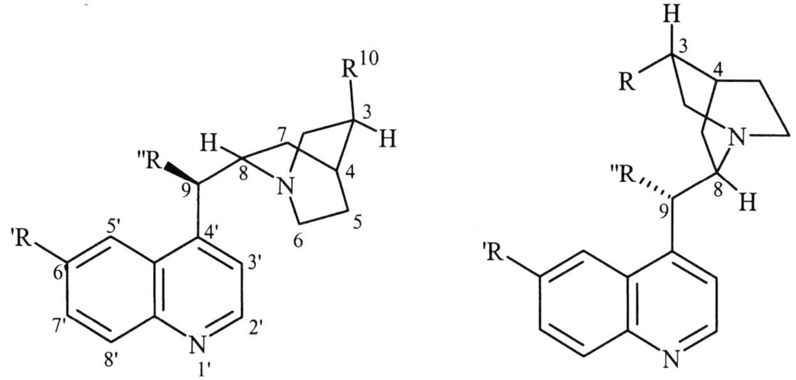

Figure 6.1. Structures of the cinchona alkaloids:

cinchonidine (CD, left), cinchonine (CN, right):	$R = C_2H_3$	$R' = H$	$R'' = OH$
quinine (QN, left), quinidine (QD, right):	$R = C_2H_3$	$R' = OMe$	$R'' = OH$
dihydroderivatives (e.g. HCD):	$R = C_2H_5$		
O-methyl derivatives (e.g. O-MeHCD):	$R'' = OMe$		

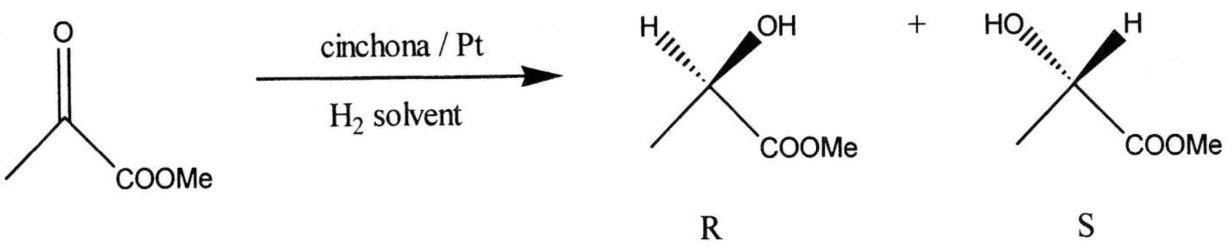

O COOMe — cinchona / Pt, H_2 solvent → H OH COOMe (R) + HO H COOMe (S)

Scheme 6.1

fiers having the cinchonidine configuration [*S*- at C_8 and *R*- at C_9] give an enantiomeric excess in favor of the *R*-product, whereas modifiers having the cinchonine configuration [*R*- at C_8 and *S*- at C_9] give the *S*-product in excess.

Substantial enantioselectivity (e.e. = 60 to 80%) is achieved by using the natural alkaloids as modifiers of reactions at elevated hydrogen pressure and ambient temperature without system optimization (entries 1–8, Table 6.1). The 10,11-dihydroderivatives provide slightly higher enantioselectivities (entries 5, 7, 8, Table 6.1). To validate comparisons, most reactions presented in Table 6.1 were conducted in ethanolic solution. However, reactions occur in a variety of solvents, with enantiomeric excess being reduced by use of solvents having a high dielectric constant. Thus, by employing cinchonidine as a modifier, Baiker, Blaser and coworkers have observed e.e. values of 79.9 to 81.4% in solvents with a low dielectric constant (cyclohexane, $\varepsilon=2.02$; toluene, 2.38; tetrahydrofuran, 7.6; dichloromethane, 9.08), lower values of 75.0 to 78.0% in alkanols, (pentanol, $\varepsilon=13.9$; ethanol, 24.3; methanol, 33.6) and inferior values of 54.1% in water ($\varepsilon=80.4$) and 48.0% in formamide ($\varepsilon=109$) [21]. The exception to this generalisation is that carboxylic acid solvents lead to better enantioselectivities than expected from their polarity [18]. Thus, for ethyl pyruvate hydrogenation in acetic acid, dihydrocinchonidine-modified Pt gave an e.e.=88% and dihydro-O-methyl-cinchonidine an e.e.=95% (entry 12, Table 6.1). Acetic acid may also be used advantageously as a co-solvent, for example with toluene [18]. In comparison, the use of bases as solvents is very disadvantageous (entry 13, Table 6.1). The effectiveness of cinchonidine, cinchonine, and their dihydroderivatives as modifiers is decreased above 315 K [17].

A synthetic alkaloid, resembling cinchonidine but in which C_9 was not chiral, yielded an enantioselective reaction but with a reduced enantiomeric excess (entry 16, Table 6.1). Epiquinidine, in which both C_8 and C_9 have the *R*-configuration, is not an effective modifier (entry 17, Table 6.1).

Modification of cinchona alkaloids by quaternization at the quinuclidine-N atom, e.g. by conversion to benzyl cinchonidinium chloride, substantially destroys enantioselectivity (entries 18, 19, Table 6.1). However, hydrocinchonidine sulfate, in which the quinuclidine-N is simply protonated, behaves as cinchonidine [16].

The dependence of enantiomeric excess on conversion and hydrogen pressure may vary with the procedure used for catalyst preparation and modification (see below). With separate modification there may be little variation with conversion [17], whereas with *in situ* modification, reactions have been reported in which values are low initially, rise steeply over the first 20% of conversion [22,23], and remain fairly steady thereafter [21, 24]. 5′,6′,7′,8′,10,11-Hexahydrocinchonidine, a modifier having a par-

Table 6.1. Representative values of the enantiomeric excess (e.e.) observed in the hydrogenation of methyl pyruvate (MePy) and of ethyl pyruvate (EtPy) catalyzed by cinchona-modified Pt/silica or Pt/alumina[a]

Entry	Modifiers[b]	R	R′	R″	Reactant	Solvent	e.e./%	Ref.
1	CD	C_2H_3	H	OH	MePy	ethanol	64–77(*R*)	14
2	CD	C_2H_3	H	OH	EtPy	ethanol	76(*R*)	15
3	CN	C_2H_3	H	OH	EtPy	ethanol	56(*S*)	15
4	QN	C_2H_3	OMe	OH	MePy	ethanol	61(*R*)	9
5	HQN	C_2H_5	OMe	OH	MePy	ethanol	65(*R*)	15
6	QD	C_2H_3	OMe	OH	MePy	ethanol	55(*S*)	16
7	HCD	C_2H_5	H	OH	MePy	ethanol	80(*R*)	17
8	HCD	C_2H_5	H	OH	EtPy	ethanol	82(*R*)	18
9	O-MeHCD	C_2H_5	H	OMe	EtPy	ethanol	82(*R*)	18
10	O-MeHCD	C_2H_5	H	OMe	EtPy	toluene	88(*R*)	18
11	O-MeHCD	C_2H_5	H	OMe	EtPy	propionic acid	91(*R*)	18
12	O-MeHCD	C_2H_5	H	OMe	EtPy	acetic acid	95(*R*)	18
13	O-MeHCD	C_2H_5	H	OMe	EtPy	triethylamine	3(*R*)	18
14	O-AcHCD	C_2H_5	H	OAc	MePy	ethanol	44(*R*)	9
15	O-AcHCD	C_2H_5	H	OAc	EtPy	ethanol	20(*R*)	15
16	Deoxy-CD	C_2H_3	H	H	EtPy	ethanol	44(*R*)	15
17	Epiquinidine	C_2H_3	OMe	H	MePy	ethanol	0	9,19
18	Benzyl(CD)Cl	C_2H_3	H	OH	EtPy	ethanol	0	15
19	Benzyl(CD)Cl	C_2H_3	H	OH	MePy	ethanol	2(*R*)	9
20[c]	CD	C_2H_3	H	OH	MePy	ethanol	97(*R*)	20

[a] Reactions conducted at ca. 293 K and analysed at high conversions. Refs. 9, 14, 16 and 20, pressure of 10 bar, catalysts premodified before admission to reactor. Refs. 15 and 18, pressure of 70 to 100 bar, catalysts not premodified, modifier added to reactor with reactant.
[b] Modifier structures and nomenclatures are defined in Fig. 6.1.
[c] Reaction over stabilized colloidal Pt (see text).

tially hydrogenated quinoline ring, gave an initial enantiomeric excess of 45% which decreased steadily to 23% at 80% conversion [24].

High performances have been reported for modified Pt catalysts in which the active metallic phase has been supported on silica, alumina, and carbon [9, 18, 25] or deposited in the mesopores of MCM-41 [26]. The dependence of enantiomeric excess on catalyst preparation procedure and Pt particle size has been studied [25] as well as the broader theme of particle morphology [12], but a detailed performance/structure correlation has yet to be established. Very small Pt particles (ca. 0.7 nm in diameter) do not promote enantioselectivity [27]. Particles of average size (1.8 nm particles such as are present in the 6.3% Pt/silica reference catalyst EUROPT-1) provide average enantioselectivities and the sintering of this particular catalyst, so that the mean Pt particle size increased to 4 nm, improved the enantiomeric excess to 80% [12]. This concurs with the general conclusions of a broader study [25]. Modification of catalyst grain size by use of ultrasonication may also lead to an improvement in enantioselectivity

[28]. In contrast, the highest enantiomeric excess of 97.5% yet reported for pyruvate ester hydrogenation (entry 20, Table 6.1) has been achieved by using a polyvinylpyrrolidone-stabilized Pt colloid having a narrow particle size distribution centered at 1.4 nm [20]. A colloid having a mean particle size of 1.8 nm gave an e.e. = 81.4% in ethanol (a Pt particle size and e.e. closely similar to those provided by the 6.3% Pt/silica, EUROPT-1), and this performance was improved to e.e. = 95.9% by applying acetic acid as solvent. Exceptional performance (e.e. = 91.3%) was attained when the 1.4 nm colloidal particles were deposited onto alumina to provide a supported catalyst. This suggests that optimum Pt particle configuration is not simply determined by size and that other factors contribute to optimum performance. For example, the polyvinylpyrrolidone protecting agent may control Pt-particle morphology in some appropriate manner or it may restrict the available Pt surface and thereby reduce the number of unmodified sites which catalyze racemic reaction.

Enhanced rates in enantioselective pyruvate hydrogenation

In the absence of a modifier, supported Pt catalysts show modest activities for the racemic hydrogenation of pyruvate to lactate esters. However, when minute quantities of a cinchona alkaloid are introduced into the system, the reaction not only shows an enantiomeric excess but also an enhanced rate [17, 29, 30] (Fig. 6.2).

The rates of production of *both* enantiomers are enhanced. The enantiomeric excess and the enhanced rate reach maximum values at alkaloid concentrations which provide high surface coverage. In general, the rate is increased by a factor of 10 to 50. Variation of any experimental parameter which causes improved enantiomeric excess generally also causes an increased reaction rate and *vice versa*. Thus, reactions at tem-

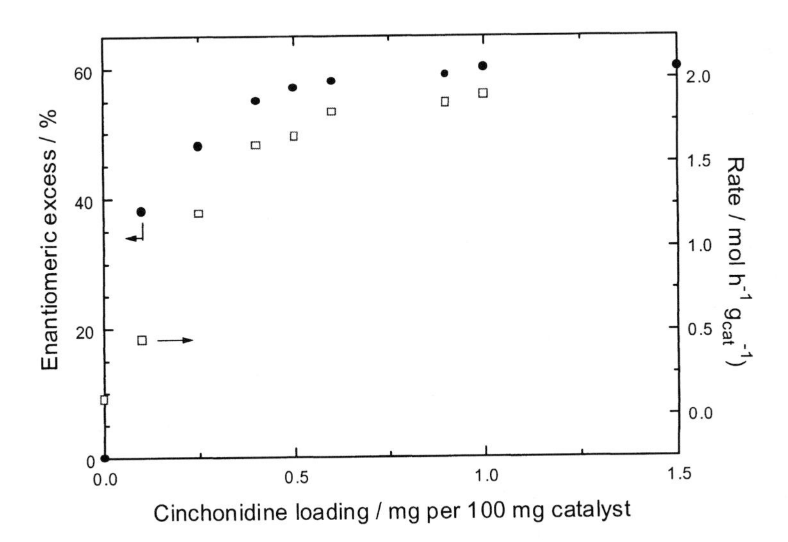

Figure 6.2. Pyruvate ester hydrogenation over 6.3% Pt/silica at a pressure of 10 bar and at 293 K. Dependence of enantiomeric excess and initial reaction rate on the mass of cinchonidine used to modify 100 mg catalyst (D_{Pt} = 60%).

peratures above 315 K or in solvents having a high dielectric constant in which enantiomeric excess is adversely affected, show lower rate enhancement factors [17, 21]. Reactions over catalysts modified by quaternized alkaloid or by epiquinidine, which show no enantioselectivity, proceed at rates only slightly above the rate of racemic reaction in the absence of alkaloid. It is therefore clear that mechanisms that interpret the origin of enantioselectivity in these systems must simultaneously provide an understanding of enhanced rate.

Hydrogenation of other prochiral ketones

The preceding account demonstrates the intensity with which the enantioselective hydrogenation of α-ketoesters over cinchona-modified Pt has been studied. However, other families of prochiral ketonic compounds have been investigated. Their general structures and performances are presented in Fig. 6.3 [7, 31–39]. Conversion of α-alkanediones to the corresponding hydroxyketones provides only modest enantioselectivities, but kinetic differentiation during the second stage of reaction (conversion to alkanediol) results in an increasing enantiomeric excess in the diminishing hydroxyketone remaining (Table 6.2) [32]. By the same token, an enantiomeric excess spontaneously appears in 3-hydroxybutan-2-one when a racemic mixture is hydrogenated over cinchona-modified Pt. There are recent reports on the enantioselective hydrogenation of α-ketoacetals, such as pyruvaldehyde dimethyl acetal [33, 34], and detailed interpretations of the high enantioselectivities achieved have yet to be presented (see

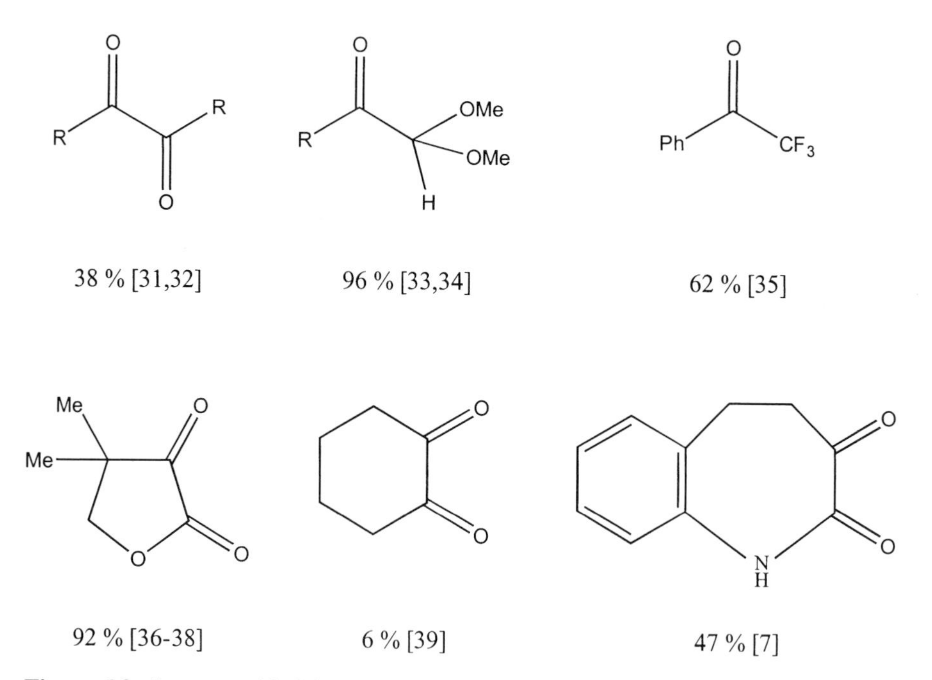

Figure 6.3. Some prochiral ketones, which are hydrogenated enantioselectively over cinchonidine-modified Pt, and observed e.e. values.

Table 6.2. Enantiomeric composition of 3-hydroxybutan-2-one as it is formed and subsequently removed in butane-2,3-dione hydrogenation to butane-2,3-diol at 273 K (entries 1–6) or removed during hydrogenation of the initially racemic hydroxyketone at 293 K (entries 7–9)[a]. Catalyst=6.3% Pt/silica (EUROPT-1).

Entry	Modifier[b]	Reactant	Conversion[c]/%	Hydroxybutanone		
				R/%	*S*/%	e.e./%
1	none	dione	100	50	50	0
2	CD	dione	20	60	40	20(*R*)
3	CD	dione	50	61	39	22(*R*)
4	CD	dione	95	62	38	24(*R*)
5	CD	dione	160	81	19	62(*R*)
6	CD	dione	180	90	10	80(*R*)
7	none	hydroxyketone[d]	40	50	50	0
8	CN	hydroxyketone[d]	40	36	64	28(*S*)
9	CN	hydroxyketone[d]	80	20	80	60(*S*)

[a] Pressure=10 bar.
[b] CD=cinchonidine; CN=cinchonine.
[c] 100%=completion of uptake of one mole of H_2; 200%=completion of uptake of two moles of H_2.
[d] Racemic mixture at zero conversion.

Chapter 1). Cyclic conjugated diones represent an interesting class of reactants with reactants having five-, six-, and seven-membered rings providing very different selective outcomes.

Methods of catalyst modification: a cautionary note

Orito and coworkers exposed their Pt catalysts to alkaloid in a prereaction modification step and subsequently transferred the alkaloid-treated Pt to the reactor where further solvent and reactants were added [4–6]. This procedure has been followed by Wells and coworkers [9, 12, 14, 17, 19, 32], but in other laboratories it has been more common to admit solvent, modifier, and reactants to the reactor in one operation. Some of the highest enantioselectivities have been obtained by the latter method (see entry 12, Table 6.1). It was recently shown that successful modification of Pt by cinchona alkaloids requires aerobic conditions [12, 17, 19, 40], i.e. effective alkaloid adsorption must occur in competition with oxygen, a competition that is crucial to the formation of surface conditions necessary for subsequent enantioselectivity. In the separate modification method, oxygen treatment is achieved by a specific step in which the slurry of catalyst and alkaloid solution are stirred in air for several hours before transfer of the catalyst to the vessel. The *in situ* modification method relies on the air normally dissolved in the solvent and ester to provide the necessary oxygen. Because of these differences in catalyst modification procedure, it is necessary to exercise caution when comparing results from different laboratories and when attempting to reproduce work published in the literature.

6.3 Mechanisms of Enantioselective Pyruvate Hydrogenation Over Platinum

Interpretations of the Orito reaction and related processes have adopted one of two starting points. Most investigators have supposed that molecules of alkaloid modifier are adsorbed onto the metal catalyst surface, thereby creating a chiral environment at certain adjacent unoccupied sites. This is consistent with the well-known strong adsorption of N-heterocyclic compounds onto platinum group metals and arises naturally from the original Orito methodology of prereaction modification described above. In an alternative approach, it has been proposed that intermolecular complexation of alkaloid with reactant occurs in solution to achieve chemical shielding in such a way that, as the complex approaches the surface, only one enantioface of the reactant is exposed to the catalyst surface and susceptible to hydrogen atom addition. In the *adsorption model,* the catalyst is conceived as acquiring enantioselective sites by the action of the alkaloid. In the *chemical shielding model,* the catalyst is not viewed as acquiring any chiral quality but simply acts as a source of activated hydrogen. These models will now be discussed separately.

6.3.1 The Adsorption Model

Adsorbed states of cinchona alkaloids

The adsorbed state of cinchonidine and of 10,11-dihydrocinchonidine on Pt has been examined in some detail. Adsorption isotherms indicate that alkaloid adsorbs on Pt/silica at room temperature both on the Pt active phase and on the support [14]. Pt-catalyzed hydrogen-deuterium exchange in 10,11-dihydrocinchonidine occurs at all positions in the quinoline moiety and in the OH-group located at C_9, but not in the quinuclidine ring system, indicating that the alkaloid is chemisorbed to the Pt surface via the aromatic ring system [41]. The exchange shows this to be a very reactive adsorption, although the mechanism of the exchange is unclear. Adsorption of 10,11-dihydrocinchonidine on the (111)-face of a Pt single crystal has been examined by X-ray photoelectron and NEXAFS spectroscopies and by LEED [42–44]. XPS and LEED showed that high coverage is achieved and that adsorption is not ordered (as had been proposed in early work [9, 14]). The expected stoichiometry was observed and retained up to 550 K [43]. The angular dependence of the N *K*-edge spectra in the near-edge region demonstrated that the quinoline ring is oriented parallel to the Pt surface at 293 K but inclined at about 60° to the surface at 323 K [44]. It is clear, therefore, that catalysts exposed to solutions of cinchonidine contain alkaloid molecules chemisorbed by the interaction of the π-electrons of the quinoline moiety with the Pt active phase. This is consistent with the aforementioned observation that partial hydrogenation of the quinoline ring, as in 5′,6′,7′,8′,10,11-hexahydrocinchonidine, reduces the effectiveness of the alkaloid as a modifier [24]. Adsorption from ethanolic solution is partially irreversible; a fraction of cinchonidine and a smaller fraction of quinine can be removed by repeatedly washing modified Pt/silica with solvent [45]. Adsorption has also been proposed to occur via the quinuclidine-N atom [46] and, since dihydrocinchonidine is formed in reactions over catalysts modified by cinchonidine [47], it is evident that adsorption involving the vinyl group can also occur. Models show that the vinyl and quinoline groups in cinchonine and quinidine are able to interact simultaneously with a flat Pt surface, whereas this is not the case for cinchonidine or quinine. However, cinchonine or quinidine as modifiers show no special properties attributable to this property.

Conformations of cinchona alkaloids

Rotations leading to changes in conformation in cinchona alkaloids can occur around all single bonds, i.e. the bonds between carbon atoms $C_{4'}$ and C_9, C_8 and C_9, C_3 and C_{10}, and between C_9 and O. For each alkaloid, molecular energy varies with rotation around these bonds (and particularly about the first two listed). Possible molecular conformation(s) are represented by the low energy states identified from maps of potential energy as each torsion angle is varied. The methods of molecular mechanics have been used to construct such potential energy maps for dihydro- and variously substituted quinidines and epicinchona alkaloids [48–51] and, more recently, for cinchonidine and cinchonine [38, 52]. The outcome is that, discounting minor perturbations induced by rotation around the C_9-O bond, cinchonidine, cinchonine, quinine and quinidine possess four states of minimum energy. Those for cinchonidine are shown in Fig. 6.4, where closed-1, closed-2, open-3, and open-4 are descriptors that have been widely adopted. States 1 and 2 are described as 'closed' because the lone pair on the quinuclidine-N atom points towards the quinoline ring, whereas states 3 and 4 are 'open' because the lone pair points away from the quinoline ring. In the open-3 conformation, the lone pair points into the cleft formed by the quinoline and quinuclidine ring systems, whereas in the open-4 state, it points away from both ring systems. These calculations relate to the molecule *in vacuo,* i.e. they do not take the environment of the molecule into account. For present purposes it is necessary to determine experimentally, by using NMR, the relative occupations of these lowest energy states for molecules in solution, and to make the assumption that the conformational balance so determined is largely retained upon adsorption of the alkaloid at the Pt surface. This involves, first, the complete assignment of the ^{1}H NMR spectrum of each alkaloid in the solvent of choice, and then the deduction of conformation by observation of inter-ring NOEs and quantification of certain JJ coupling constants. These methods quickly revealed a very rich conformational chemistry. Some alkaloids predominantly adopted a closed conformation (e.g. (O-*p*-chlorobenzoyl)-dihydroquinine), some an open conformation (e.g. quinine and dihydroquinine), and in others an intermediate situation existed (e.g. acetyldihydroquinidine showed only a small energy difference between closed-2 and open-3). In the most recent substantial study [38], Burgi and Baiker have calculated (by employing a reaction field model in combination with Hartree Fock and density functional calculations) the electronic energies, Gibbs free energies and dipole moments for the four conformations of cinchonidine relative to the open-3 conformation for the molecules in the free state (dielectric constant, $\varepsilon = 1.0$) and in solvents having $\varepsilon = 2.0$, 4.8, 20.7 and 78.5. It is clear that: (i) open-4 is

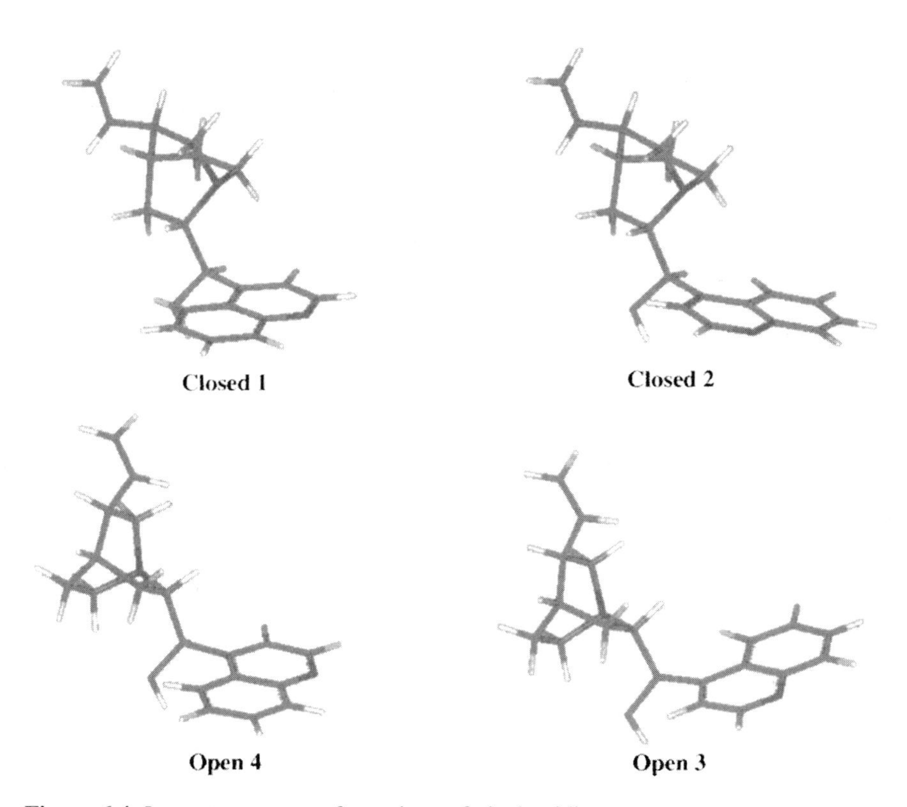

Figure 6.4. Lowest energy conformations of cinchonidine.

a conformation of high energy relative to the others under all conditions and can be neglected, (ii) the energies of the closed-1 and closed-2 state approach that of open-3 as the dielectric constant is increased. This was verified by experimental measurement of the fraction of molecules in the open-3 conformation from observation of JJ coupling constants. Thus, for toluene ($\varepsilon=2.34$) the fraction of open-3 was 70%, for diethyl ether ($\varepsilon=4.3$) 71%, for tetrahydrofuran ($\varepsilon=7.6$) 62%, for acetone ($\varepsilon=20.7$) 40%, dimethylformamide ($\varepsilon=36.7$) 33%, for dimethylsulfoxide ($\varepsilon=40$) 27%, and for water ($\varepsilon=78.5$) 30%. One solvent, ethanol ($\varepsilon=24.3$), failed to conform to this correlation, giving [open-3]=77%; this additional stabilization of the open-3 confromation was attributed to hydrogen bonding between the ethanol and the quinuclidine-N of the alkaloid.

To summarize, both calculation and experiment show that open-3 is the most stable conformation of cinchonidine in apolar solvents but that, as solvent polarity is increased, the high dipole moments of the closed-1 and closed-2 conformations lead to their relative stabilization and to open-3 becoming the minority conformation in polar solvents at room temperature. It is reasonable to suppose that corresponding statements hold true for the four minimum energy states of cinchonine.

Entry 17 of Table 6.1 shows that epiquinidine is not an effective modifier; its adsorption onto Pt provides neither enantioselectivity nor a greatly enhanced rate. In epiquinidine, C_8 and C_9 both have the *R*-configuration. NMR spectroscopy and molecular mechanics calculations have shown that this molecule exists almost completely as an open conformation (resembling open-4 for cinchonidine) in which the lone pair on the quinuclidine-N is directed away from both the quinoline and quinuclidine ring systems [19]. This conformation is also predicted to be the most stable by molecular mechanics calculations [52]. Docking procedures identify no paired configurations which are sensitive to the substrate ester geometry, and hence, no selective enantiofacial adsorption can occur in the adsorbed state. Thus, the requirements for enantioselective reaction are not achieved, and this negative result makes a useful contribution to our understanding of these reactions.

Because both enantiomeric excess and the fraction of cinchonidine in the open-3 conformation decrease with increasing dielectric constant of the solvent, it follows that favorable enantioselectivity is observed under conditions where the open-3 conformation in solution is also favored. This correlation is strengthened by the recent report that enantiomeric excess in the hydrogenation of ketopantolactone over cinchonidine-modified Pt follows that described above for pyruvate ester [38]. These correlations must guide proposed mechanisms of enantioselective action.

Mechanism of enantioselective action and the origin of enhanced rate

The preceding section establishes our knowledge of cinchona alkaloid adsorption on platinum and the conformational characteristics of cinchona alkaloids in solution. Pyruvate ester undergoes slow hydrogenation to the racemic product at an alkaloid-free Pt surface, the reaction being of zero-order in ester and first-order in hydrogen which indicates strong adsorption of ester and relatively weak adsorption of hydrogen. These kinetic parameters are retained in the presence of alkaloid, with small deviations depending on the solvent used [53] but, more importantly, with rates greatly enhanced and the reaction enantioselective. Molecules of pyruvate ester may adsorb by either of their enantiofaces, as shown in Fig. 6.5. Hydrogenation of enantioface A gives *(R)*-lactate as the product, whereas enantioface B gives the *S*-product. Enantioselectivity may arise by the operation of either thermodynamic or kinetic factors. If competitive adsorption of ester in configurations A and B occurs at enantioselective sites cognate to adsorbed alkaloid with free energy changes $\Delta G^0_{ads,A}$ and $\Delta G^0_{ads,B}$ respectively, then, provided that adsorption equilibrium is achieved, the relative surface coverages θ_A/θ_B will be a direct function of $\exp(-\delta\Delta G^0_{ads}/RT)$, where $\delta\Delta G^0_{ads}$ is the difference in the free energies of adsorption of the ester in configurations A and B. Because of the exponential form of this expression, a chiral environment that distinguishes only modestly between A and B in terms of free energies of adsorption will provide a high value of the ratio θ_A/θ_B. If no kinetic factor operates, i.e. if $k_2^A = k_2^B$, then the enantiomeric excess is thermodynamically determined, $[(R)\text{-lactate}]/[(S)\text{-lactate}] = \theta_A/\theta_B$, and enantioselectivity is attributable to *selective enantioface adsorption.* There is, of course, the possibility that molecules adsorbed by the less stable enantioface may exhibit a higher inherent activity ($k_2^A \neq k_2^B$), in which case a kinetic factor also influences enantiomeric excess. Reference [54] describes a classic example of kinetic control in homogeneous catalysis in which the major enantiomer product is formed from the minor enantiofacial form of the reactant. Mathematical procedures for considering kinetic factors as applied to metal-catalyzed enantioselective re-

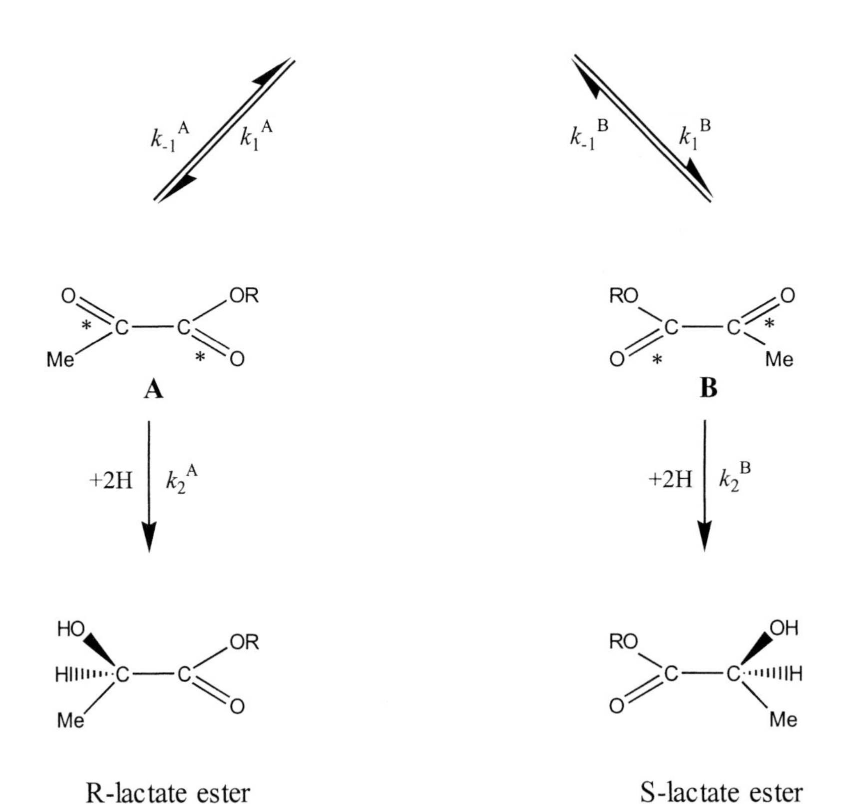

Figure 6.5. Reaction scheme showing pyruvate ester adsorbed by its two enantiofaces and the consequences of hydrogenation.

actions have been described [55–57]. For the present case in which the modifier and reactant are strongly adsorbed and achieve substantial surface coverages, semi-quantitative modelling of the thermodynamic factor can provide an estimate of the relative surface coverage of the two enantiofacial forms of the adsorbed reactant, and thereby definitely indicate the sense of the enantioselectivity that would be achieved if there were no kinetic factor operating. If the sense of the enantioselectivity is thus correctly predicted, the procedure is valuable, and the contribution of kinetic factors may be discernible from variations of enantiomeric excess with experimental variables. If the observed sense of enantioselectivity is the reverse of that predicted, then there is a *prima facie* case for the predominance of kinetic effects.

Modelling of selective enantioface adsorption can only be partial at the present time. Although the interactions of pyruvate ester molecules with cinchona alkaloids in their various low-energy conformations can be determined with some accuracy, the interactions with the metal surface cannot. Consequently, the available methodology in-

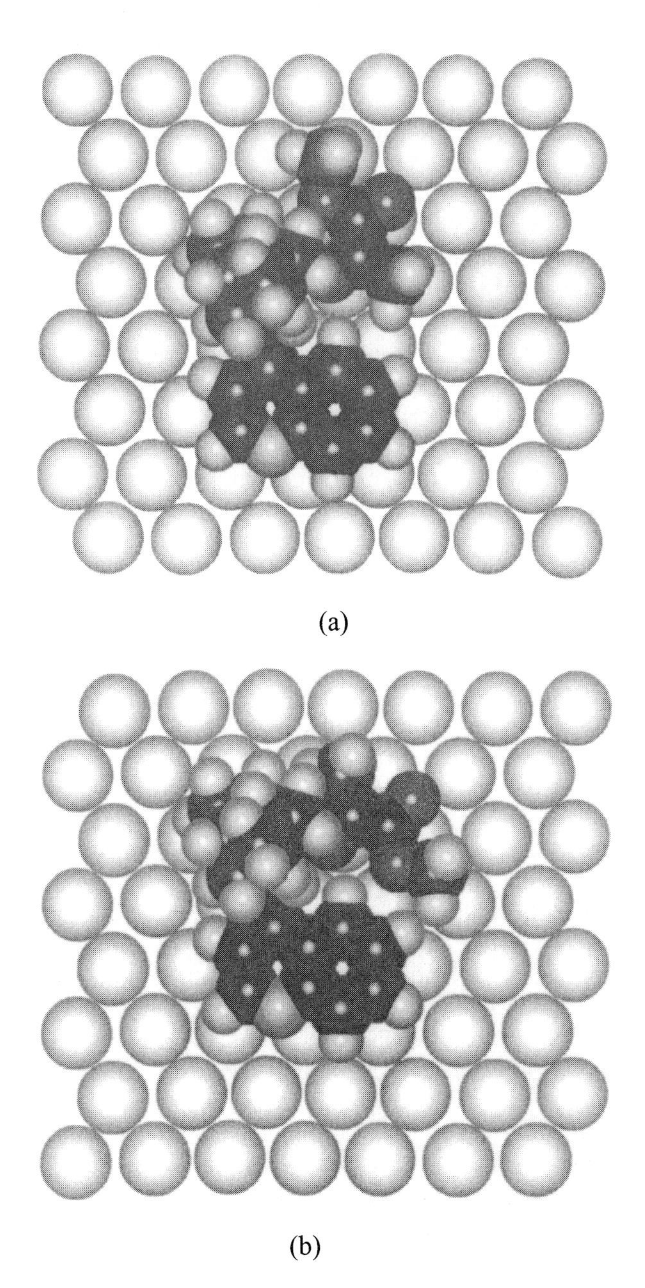

Figure 6.6. Methyl pyruvate adsorbed by each of its enantiofaces at the enantioselective site adjacent to cinchonidine adsorbed in the open-3 conformation. Upon hydrogenation, (a) gives (*R*)-lactate and (b) gives (*S*)-lactate.

volves: (i) determining the (repulsive) interactions as pyruvate molecules dock with alkaloid molecules in 1:1 arrangements, and (ii) assuming that relatively the same energy situations exist in the adsorbed state. Following this procedure, the docking of pyruvate ester molecules with the open-3 conformation of cinchonidine, for example, leads to the two geometrical situations shown in Fig. 6.6 [10, 42, 52]. The interaction energies are measured at the unique configurations such that, were the molecules adsorbed at a Pt(111) surface, the aromatic rings of the alkaloid and the carbonyl groups of the reactant would be located directly over surface Pt atoms. Under these conditions, the cinchonidine:pyruvate ester interactions are less severe for enantioface A than for enantioface B. Therefore, the free energy of adsorption of A is expected to exceed that of B, and selective enantioface adsorption will occur with $\theta_A > \theta_B$ leading to enantioselectivity in favor of *(R)*-lactate. This concurs with observation (Table 6.1). No conditions for selective enantioface adsorption of pyruvate are evident for 1:1 interactions with alkaloid in the closed conformations.

Analogous calculations for interactions between pyruvate ester and cinchonine identify less severe interactions for enantioface B than for enantioface A, leading to a predicted enantiomeric excess in favor of (*S*)-lactate ester, again in agreement with experiment. These arguments apply to substituted cinchonidines so that enantioselectivity in favor of *R*-product is expected for quinine, dihydrocinchonidine, dihydroquinine, methoxycinchonidine etc., concurring with results shown in Table 6.1. Likewise, *S*-product enantioselectivity is interpreted for the corresponding cinchonine analogs.

Since enantioselectivity and rate enhancement occur together in this system, the model just developed must also be capable of interpreting the improved rate. Figure 6.6 shows that the α-keto group undergoing reduction is in close proximity to the quinuclidine-N atom. It should be noted that the rate of racemic pyruvate hydrogenation is increased by the presence of N-bases [58, 59] and that generally the acceleration depends on the pK_a of the base. However, the rate enhancement caused by the cinchona alkaloids exceeds that attributable simply to their basicity. Interpretations of the enhanced rate therefore seek to rationalize the role of the quinuclidine-N in the adsorbed 1:1 diastereomeric complexes shown in Fig. 6.6. Wells and coworkers offer the following interpretation [9, 59]; others prefer the formalism of ligand acceleration of reaction rates [29]. D-tracer studies confirm that pyruvate ester hydrogenation occurs by the direct addition across the carbon-oxygen double bond [17]. This is conceived as occurring in two sequential H-atom addition steps, the first of which is reversible and the second rate-determining as required by the observed first-order in hydrogen [60]. Molecular mechanics calculations reveal that the half-hydrogenated state formed by addition of a H-atom to the O-atom of the carbonyl group leads to a minimum energy state in which that H-atom is involved in hydrogen bonding with the quinuclidine N-atom [42, 59]. Such H-bonding would stabilize the half-hydrogenated state, increasing its lifetime and, hence, its concentration, thereby increasing the observed reaction rate. This mechanism for rate acceleration applies to both diastereomeric complexes depicted in Fig. 6.6. Thus, the rates of formation of both enantiomeric products should be accelerated by action of the adsorbed alkaloid, as is indeed observed.

Some miscellaneous matters

The adsorption model provides a highly self-consistent interpretation of a wide range of observations based on kinetic, thermodynamic and theoretical considerations. Very high enantiomeric excess can be achieved, but values are normally decreased by the presence of sites on the catalyst surface where racemic reaction occurs. If the Pt particles contain just enough surface to accommodate one alkaloid molecule and one reactant molecule, then racemic reaction should be eliminated. Whether this was achieved in the case of the stabilized colloids referred to above [20] is an interesting speculation. Alternatively, racemic sites might be selectively poisoned, but experiments designed to achieve this objective have failed so far.

Table 6.1 shows that the e.e. values towards the *S*-product induced by cinchonine (or its analogs) are always slightly less than those towards the *R*-product provided by cinchonidine (or its analogs). It would be interesting to know whether this is attributable to a slightly lower population of the open-3 state for cinchonine in solution compared to cinchonidine.

The quinoline moiety provides the critical anchor onto the Pt surface for the alkaloid modifier. If reaction temperature is increased much above ambient temperature, enantioselectivity is considerably impaired, and surface science experiments show a change in orientation at about the same temperature [17, 44]. Partial hydrogenation which reduces the aromatic portion of the anchor to that of a pyridine ring system reduces enantioselectivity [24], and synthetic alternatives to the natural alkaloids which act as modifiers also require an aromatic anchor more extensive than the pyridine ring system [10]. This is consistent with the observation that pyridine itself is normally adsorbed at the Pt surface in a tilted, not parallel, configuration [61].

6.3.2 The Chemical Shielding Model

According to this model, the transfer of chiral information leading to enantioselective hydrogenation occurs in solution and not at the catalyst surface. The main proponents of this approach have been Margitfalvi and coworkers [62–66]. The concept of chemical shielding is well-known in organic chemistry [67], and the shielding of α-ketoester groups by large aromatic substituents resulting in enantiodifferentiation represents a particular example.

Changes in circular dichroism spectra and in NMR spectra on mixing cinchona alkaloid and pyruvate ester have been observed [62, 68]. Spectral changes, particularly those relating to shifts of the $-C_9H$ proton resonance in the alkaloid, in mixtures of various composition and in different solvents, have been attributed to conformational changes in the alkaloid, and thus to the formulation of chemically shielded complexes. The complex proposed to participate in enantioselective hydrogenation involves a closed conformation of cinchonidine and is illustrated schematically in Fig. 6.7. Calculated geometries show that the modifier provides an 'umbrella', concave with respect to the reactant, so that, as the complex approaches the surface, only one enantioface of the pyruvate molecule is exposed for adsorption. In this way, selective enantioface adsorption is achieved by the modifier in solution acting as host in a 'su-

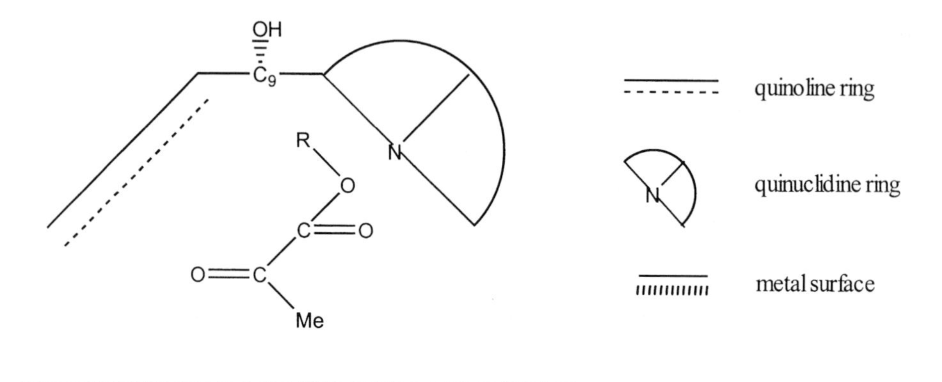

Figure 6.7. Representation of the chemically shielded complex involving cinchonidine and pyruvate ester; the chiral carbon atom C_9 is defined in Fig. 6.1.

pramolecular interaction' [62]. Any complexes involving an open conformation of the alkaloid would not provide effective shielding and would give a racemic product upon hydrogenation.

The chemically shielded complex is considered to be in adsorption-desorption equilibrium with the surface. Moreover, the reactant-host interaction causes activation of the α-keto group which, in turn, leads to a rate enhancement.

Further experimental work is required to determine whether the adsorption model and the chemical shielding model are incompatible alternatives, or whether they each represent processes that occur in solution to a greater or lesser extent. At least two apparently unambiguous tests can be envisaged. First of all, the performance as modifiers of cinchona alkaloids that exhibit only open or only closed conformations should be compared. Although (O-*p*-chlorobenzoyl)dihydroquinine is known to exist entirely in the closed conformation in the solid state and predominantly so in solution [49], it has not been tested as a modifier for enantioselective hydrogenation. On the other hand, α-isocinchonine, α-isoquinidine, and γ-isoquinidine, in which the O-atom at C_9 forms a bridge to C_{10}, are locked into an open conformation by their very structure. Bartok and coworkers have shown that these three alkaloids perform well as modifiers, affording respective e.e. values of 88%(*S*), 85%(*S*) and 22%(*S*) for ethyl pyruvate hydrogenation in acetic acid solution at a pressure of 1 bar and at 293 K [69 a]. Secondly, we may ask whether enantioselective hydrogenation can be conducted as a gas-phase reaction over a solid (modified) catalyst – conditions under which the chemical shielding model does not apply, since no solvent is present. A preliminary result suggests that enantioselectivity can be achieved in this way [69 b]. However, the catalyst used had some microporosity, and the work should be repeated with a non-porous catalyst to obtain results that pertain to conditions under which capillary condensation (and, hence, potential reaction in solution) does not occur.

Clearly, further investigations that try to distinguish between the adsorption model and the chemical shielding model are necessary so that the origin(s) of enantioselectivity in this formally simple reaction can be better described.

6.4 Enantioselective Hydrogenation of Activated Ketones Over Palladium

Silence in the literature implies that palladium has minimal activity for hydrogenation of the activated ketones found in Fig. 6.2. It does, however, show modest activity for pyruvate ester hydrogenation. That having been said, the Pd-catalyzed reaction strongly differs from the Pt reaction. The first report by Blaser and coworkers indicated that enantioselectivity over cinchonidine-modified Pd/C was low and favored (*S*)-lactate [70] – the reverse of the situation over Pt. Closer examination showed that reaction over conventionally supported Pd differed very significantly from the corresponding Pt-catalyzed reaction [12, 71]. Thus, (i) rate enhancement was absent over Pd (indeed, the rate decreased in the presence of alkaloid); (ii) enantioselective reaction was one-half order in hydrogen (first-order over Pt); (iii) reactions over Pd were only achieved in certain solvents (less solvent sensitivity with Pt); and (iv) Pd was enantioselective only if reduced at low temperatures (enantioselectivity sometimes improved by using higher reduction temperatures over Pt). Most importantly, reaction with deuterium yielded the product exchanged at the methyl group adjacent to the *α*-keto group, i.e. the product was $CX_3CX(OX)COOCH_3$ (where X=H or D) [71], whereas over Pt the participation of deuterium was limited to addition, thereby giving $CH_3CX(OX)COOCH_3$ [17]. These differences are consistent with pyruvate undergoing dissociative adsorption by H-atom loss from the methyl group followed by re-hydrogenation to give the enol in a rate-determining step (Fig. 6.8).

The process then becomes a carbon-carbon double bond hydrogenation, a process for which Pd is well-known to have high activity. Fast conversion of the enol to lactate, by conversion of intermediate C to (*R*)-lactate (Fig. 6.8), would provide an enantioselectivity in the product that is determined by any selective enantioface adsorption of pyruvate ester on Pd. If the characteristics of adsorption on Pd follow that on Pt, then enantioselectivity in favor of (*R*)-lactate would be expected with cinchonidine and the reverse with cinchonine. However, the observed reversed enantioselectivity can be simply interpreted if it is supposed that the latent intramolecular repulsion between the >C=C< and >C=O functions in adsorbed species C can be released by rotation around the carbon-carbon single bond to give adsorbed species C′ which yields (*S*)-lactate upon hydrogenation. This mechanism then provides an interpretation of the major experimental observations. Firstly, there is no increased rate, because the intermediate formed from pyruvate involved in the rate-determining step (B) does not contain an –OH group that can participate in hydrogen bonding with the quinuclidine-N atom. Secondly, the half-order in hydrogen is consistent with the rate-determining step involving the addition of one hydrogen atom (assuming that dissociative hydrogen adsorption under these conditions obeys the Langmuir equation [60]). Thirdly, solvent specificity may indicate that enolization occurs more effectively in one solvent than in another. Finally, the better performances of catalysts reduced at low temperature suggest that some $Pd^{\delta+}$ character at the surface may facilitate dissociative adsorption of the reactant and stabilization of the electron delocalisation present in intermediate B. The low enantioselectivities normally observed (e.g. 15% [71]) are attributable to the greater contribution of racemic (unmodified) sites to product formation that necessari-

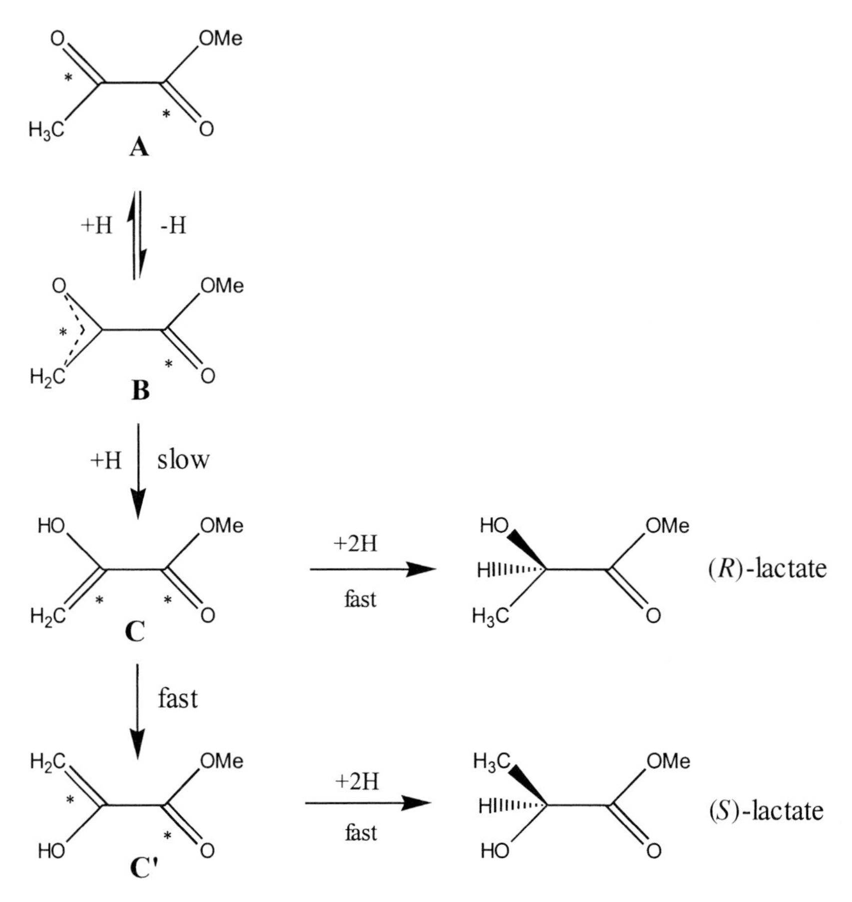

Figure 6.8. Mechanism for the enantioselective hydrogenation of pyruvate ester over palladium.

ly follows the absence of rate enhancement at enantioselective (modified) sites. They are also ascribed to the sensitivity imparted to the mechanism by the possible conversion of intermediates C and C′. Indeed, reactions giving low enantioselectivities in favor of the *R*-product over cinchonidine-modified Pd/alumina have recently been reported [72], and swings in the sense of the enantioselectivity have been obtained by varying the ester group. More D-tracer experiments are required to confirm or deny that common modes of adsorption and hydrogenation are capable of producing these enantioselective changes. Pending that information, the principle of minimum hypothesis requires that these complex variations in product composition are attributable to variable degrees of conversion of C to C′.

It would appear that the key to achieving the enantioselective hydrogenation of other activated ketones over palladium requires the selection of reactant and conditions such that the process is diverted into a carbon-carbon double bond hydrogenation.

6.5 Enantioselective Hydrogenation of Substituted Alkenes Over Palladium

Apart from Lipkin and Stewart's report of the (Pt-catalyzed) enantioselective hydrogenation of *β*-substituted cinnamates mentioned in the Introduction [2], the first study of the enantioselective hydrogenation of a substituted alkene was that of an *α,β*-unsaturated acid. In 1985 Perez and coworkers reported the hydrogenation of (*E*)-*α*-phenylcinnamic acid to 2,3-diphenylpropionic acid over cinchonidine-modified Pd/C; the enantiomeric excess was 30% in favor of the *S*-product [73]. Nitta and coworkers subsequently showed that the enantioselectivity in this reaction was strongly influenced by catalyst composition and mode of preparation [74–78]. Thus, at a 5% loading of metal, some Pd/carbons were enantioselective catalysts, whereas others were not. Moreover, the effectiveness of other supports varied in the sequence titania > zirconia > alumina > silica. Enantiomeric excess increased dramatically with the dielectric constant of the solvent. Thus for the series: ethyl acetate, tetrahydrofuran, *t*-butanol, *s*-propanol, ethanol, methanol, dimethylformamide, and dimethylformamide/water (9/1) for which the dielectric constant increases from 6 to 46, the enantiomeric excess provided by 5% Pd/alumina increased linearly from 2 to 30%. Over a 5% Pd/titania, a similar variation gave a best performance of 61% in dimethylformamide/water. Performance increased substantially with increasing concentration of cinchonidine and rose slightly with decreasing hydrogen pressure, decreasing reactant concentration, and decreasing temperature. Thus, an optimum performance of e.e. = 72%(*S*) at 100% conversion (Fig. 6.9) was achieved in the hydrogenation of 20 mmol dm^{-3} reactant in DMF/H_2O (9:1) at 283 K and a hydrogen pressure of 1 bar over 60 mg 5% Pd/titania modified by cinchonidine (6.0 mmol dm^{-3}) [77]. Although the support was rutile, in non-optimized experiments, the anatase form of titania functioned equally well as a support. This performance remains the best yet recorded for the enantioselective hydrogenation of a carbon-carbon double bond.

Reaction rates were *reduced* by nearly an order of magnitude in the presence of even small quantities of cinchonidine (0.02 mmol dm^{-3}), but the quantity of alkaloid necessary to achieve optimum enantioselectivity was very much greater than that required to modify Pt catalysts for *α*-ketoester hydrogenation. These characteristics also apply to the enantioselective saturation of the carbon-carbon double bond in alkenoic acids.

Hydrogenation of alkenoic acids having the double bond in the main chain (Fig. 6.9) over cinchonidine-modified Pd yields the corresponding alkanoic acid rich in the *S*-enantiomer, whereas modification with cinchonine favors the formation of the *R*-enantiomer [71, 79]. Thus, the sense of the enantioselectivity follows that observed in *α*-phenylcinnamic acid hydrogenation. However, the *opposite* sense of the enantioselectivity has been observed in the hydrogenation of 2-ethylpropenoic acid in which the carbon-carbon double bond is terminal [80]. A survey of the hydrogenation of (*E*)-2-methyl-2-butenoic acid (tiglic acid), (*Z*)-2-methyl-2-butenoic acid (angelic acid), (*Z*)-2-methyl-2-pentenoic acid, and of (*Z*)-2-ethyl-2-hexenoic acid, each in tetrahydrofuran solution over 1% Pd/silica at 293 K and a pressure of 10 to 50 bar, gave e.e. values ranging from 12 to 27% [71]. However, this catalyst system was not optimized with

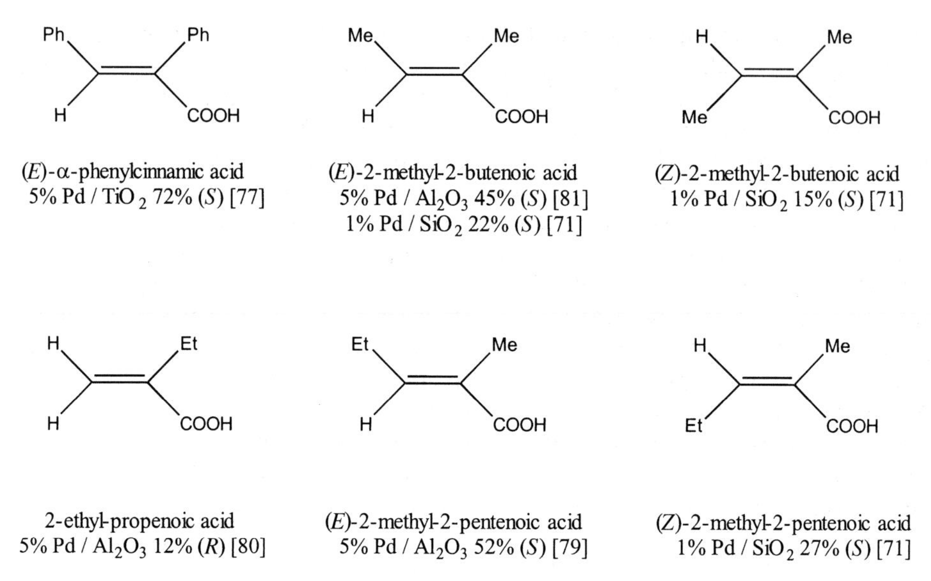

Figure 6.9. Some unsaturated acids which are hydrogenated enantioselectively over cinchonidine-modified Pd and values of the enantiomeric excess observed.

respect to loading or support. Thus, the hydrogenation of (*E*)-2-methyl-2-pentenoic acid over a 5% Pd/alumina modified by cinchonidine showed a superior performance, affording e.e. values as high as 52% (*S*) [79]. Curiously, the characteristics of this reaction are, in two important respects, the reverse of those outlined above for *α*-phenylcinnamic acid hydrogenation. Although the rate decreases in the presence of alkaloid and the enantioselectivity increases with the amount of alkaloid present in the reactor, enantiomeric excess *increases* with increasing hydrogen pressure, reaching a plateau at 70 bar, and *decreases* with increasing dielectric constant of the solvent. E.e. values measured at 60 bar fell continuously from 50 to 25% on varying the solvent in the series: cyclohexane, hexane, toluene, benzene, tetrahydrofuran, *i*-propanol, acetic acid, ethanol. There was strong evidence in this study that enantioselectivity was favored by experimental conditions that maximized the amount of hydrogen dissolved in the reaction medium.

It is intriguing that the best enantioselectivities observed in alkenoic acid hydrogenation are lower than those observed in *α*-phenylcinnamic acid hydrogenation. Isomerization possibly accompanies hydrogenation of the alkenoic acids. Geometrical (*E*- to *Z*-) isomerization does not occur – angelic acid is not formed during tiglic acid hydrogenation and *vice versa* [71]. However, isomerization can be rapid by double-bond migration in these systems [80]. For example, during the hydrogenation of 2-ethylpropenoic acid to 2-methylbutyric acid, isomerization to tiglic acid occurred in all solvents studied, and in ethanol a small yield of angelic acid was also formed. Moreover, this isomerization also occurred under a nitrogen atmosphere (i.e. in the absence of molecular hydrogen), thereby suggesting dissociative adsorption and the participation of *π*-allylic intermediates in the isomerization. Although conversion of tiglic acid (or angelic acid) to 2-ethylpropenoic acid would be thermodynamically disadvantageous, any process which allowed a low equilibrium concentration of 2-ethylpropenoic acid to be established would diminish enantioselectivity. Such an equilibrium might be manifested if the reaction was studied by the D-tracer technique. Such an isomerization is not possible in *α*-phenylcinnamic acid hydrogenation. Hence, the postulate that higher enantioselectivity accompanies hydrogenations that are uncomplicated by potential simultaneous isomerizations remains an intriguing possibility.

Esters of alkenoic acids are hydrogenated to the racemic product without a trace of enantiomeric preference, and quaternisation of the quinuclidine-N eliminates enantioselectivity [71, 81] (care must be taken to ensure that quaternary salts are not hydrogenolysed to the parent alkaloid over Pd [78, 81]). Thus, enantioselectivity in acid hydrogenations depends critically on the interaction of the carboxylic acid function of the reactant with the quinuclidine-N atom of the modifier. In considering mechanisms, we again draw on our knowledge of the conformations of the cinchona alkaloids in solution (as revealed by NMR spectroscopy and discussed in detail above) and assume that conformational preference is retained upon adsorption at the Pd surface.

The monomer-modifier interaction model

The sense of the observed enantioselectivity is interpreted by simply supposing that a precursor state is formed by an 1:1 H-bonding between acid in the fluid phase and the quinuclidine-N of the adsorbed modifier in the open-3 conformation. Rotation of the reactant molecule around this H-bond gives an energy profile containing two minima separated by an energy barrier of about 5 kcal mol^{-1} [71]. The reactant in either of these two low-energy precursor states may subsequently adsorb at the Pd surface by the carbon-carbon double bond, thus providing selective enantioface adsorption. Where the modifier is cinchonidine, the adsorbed enantiofacial variant corresponding to the precursor state of lower energy is that which would give the *S*-product upon hydrogenation, whereas the precursor state of higher energy would yield the *R*-product upon hydrogenation. Where the modifier is cinchonine, the preferred precursor state and preferred adsorbed enantiofacial variant are those that give the *R*-product on hydrogenation. The sense of the observed enantioselectivity is thereby interpreted. No enhanced rate is expected, because the carbon-carbon double bond undergoing hydrogenation is remote from the quinuclidine-N. Thus, as far as activity is concerned, the alkaloid reduces the reaction rate to the extent that it occupies a fraction of the active surface.

The dimer-modifier interaction model

Recent work suggests that the monomer model may represent an oversimplification of the true situation in solution and that other processes may occur alongside or may dominate over those just described. Firstly, in solution, monomers of these acids are in equilibrium with dimers, the extent of dimerization being greater where acid/solvent interactions are less strong, i.e. in apolar media. It is in such solvents that the best enantioselectivities are achieved, at least for the alkenoic acids. Spectroscopic evidence of dimerization of these alkenoic acids in apolar solvents has recently been reported [82], and just as acid dimers interact with simple amines such as triethylamine [83], so they may likewise interact with cinchona alkaloids at the quinuclidine-N atom. Secondly, there is evidence that not only the quinuclidine-N atom but also the C_9-OH group of the alkaloid interacts with α-phenylcinnamic acid during enantioselective hydrogenation [78], and conversion of the C_9-OH group in cinchonidine to C_9-OR (where R = methyl, allyl, or benzyl) destroys enantioselectivity in tiglic acid hydrogenation [81]. Furthermore, the alkaloid 9-deoxycinchonidine, in which C_9 is not a chiral carbon atom (being bonded to two H-atoms), is completely ineffective as a modifier in toluene. These two observations demonstrate that the quinuclidine-N atom, although a necessary feature in a successful modifier, is not a sufficient feature in itself. To test the postulate that cinchonidine (or cinchonine) can interact with an acid dimer in the enantiodiscriminating step in apolar media, Baiker and coworkers hydrogenated tiglic acid in toluene over dihydrocinchonidine-modified 5% Pd/alumina in the presence of the very strong base DBU (1,8-diazabicyclo[5.4.0]undec-7-ene) [81] which complexed with the dimer as $DBU(acid)_2$. Variation of the [DBU]/[tiglic acid] ratio in a series of reactions showed that normal enantioselectivities were obtained when free acid dimers remained in solution, but enantioselectivity fell dramatically when a majority of the dimer was complexed with the base. These observations, together with theoretical calculations, led to the conclusion that the acid dimer interacts with adsorbed cinchonidine (in the open-3 conformation) as shown in Fig. 6.10. Three hydrogen bonds are involved in the crucial diastereomeric complex: one maintains the dimer, one is formed between the carboxyl-H of an acid molecule and the quinuclidine-N, and one is formed between C_9OH and the carbonyl-O of the other acid molecule in the dimer.

For the diastereomeric complex involving cinchonidine, hydrogenation in the normal way would give the product in the *S*-configuration, whereas the complex involving cinchonine would preferentially yield the *R*-product. These expectations concur with experiment.

Further work is required to establish whether the monomer-modifier interaction model is the preferred mechanism in strongly polar solvents where dimer concentrations are low.

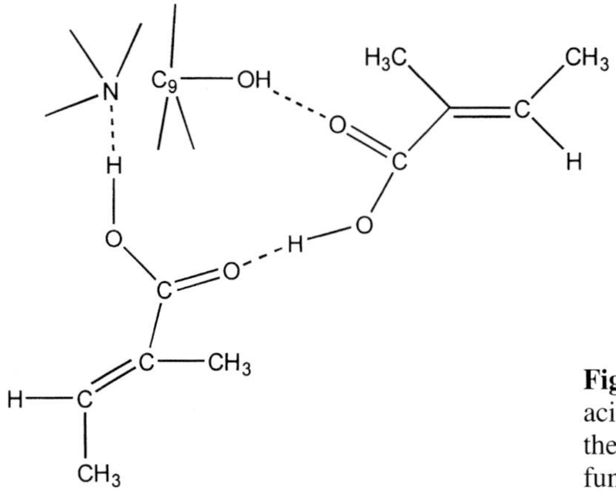

Figure 6.10. Representation of the tiglic acid dimer bonded to the $\equiv C_9$-OH and the protonated quinuclidine N-atom functions of cinchonidine.

6.6 Enantioselective Hydrogenation Involving Carbon-Nitrogen Unsaturation

Enantioselective hydrogenation of the carbon-nitrogen double bond has received comparatively little attention. There is an early report from 1941 that Pt modified by chiral acids is active for the enantioselective hydrogenation of the oxime Ph(C=NOH)Me giving an enantiomeric excess of 18% [7, 84]. Moreover, in their study of Pd/silk fibroin, Akabori and coworkers reported *inter alia* the hydrogenation of α-benzildioxime, Ph(C=NOH)(C=NOH)Ph, and of ethyl α-acetoxyimino-3-phenyl-propionate, $PhCH_2(C=NOAc)COOEt$, in which e.e. values of 15 and 30% were reported, respectively [3, 7, 13]. However, the combined hydrogenation and hydrogenolysis of pyruvic acid oxime to alanine over palladium (Scheme 6.2) is the only reaction to have been studied in detail.

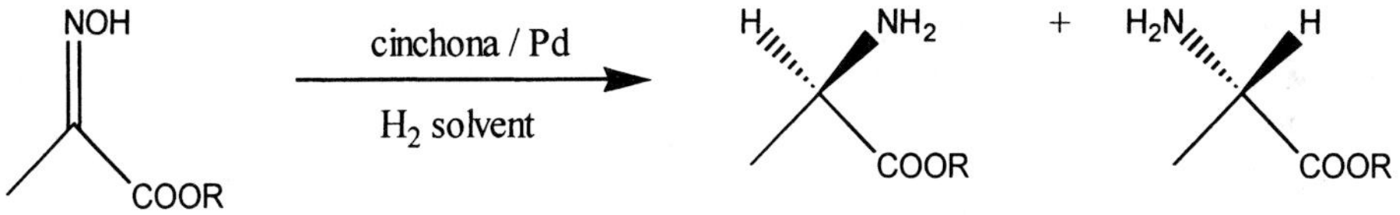

NOH, COOR — cinchona / Pd, H_2 solvent → H, NH_2, COOR + H_2N, H, COOR

Scheme 6.2

In 1971 Yoshida and Harada reported this reaction being catalyzed by Pd/C modified by the alkaloid ephedrine, whereby the modest yield of alanine showed an enantiomeric excess of 10% [85]. A more recent investigation of this reaction has shown the reaction to be complex [86]. Firstly, the enantiodirecting influence of the modifier depends on the amount used. Small quantities of cinchonidine or cinchonine in ethanolic solution (modifier: oxime molar ratios in the range 10^{-4} to 10^{-2}) induce enantioselectivities in favor of (*R*)- and (*S*)-alanine, respectively, with e.e. values being low at 1 to 5%. Secondly, higher molar ratios in the range from 10^{-1} to 1 cause a progressive increase in enantiomeric excess to values as high as 14%. With equal quantities of modifier and oxime, both cinchonidine and cinchonine direct the reaction in the same stereochemical direction to create preferential formation of (*S*)-alanine. Under the same conditions, use of (1*R*,2*S*)-ephedrine in place of cinchona alkaloid as modifier increased the enantiomeric excess to 26% (*S*), which is the highest value yet reported. Once again (as for the hydrogenation of alkenoic acids over Pd), the reaction rate is substantially decreased in the presence of the alkaloid, and enantioselectivity is completely lost when the reactant is changed from the acid oxime to its ethyl ester. A model was proposed, based on concepts drawn from the crystal structures of the 1:1 cinchonidinium pyruvate oxime salt and the 1:1 ephedrinium pyruvate oxime salt. These salts exhibit complex hydrogen-bonded chain structures and, correspondingly, the proposed reaction mechanism involved surface complexes comprising U-shaped hydrogen-bonded complexes containing the sequence ephedrine-oxime-oxime-ephedrine in which chemisorption is achieved by π-bonding of the aromatic ring of each ephedrine and of the >C=N- group of each oxime to the Pd surface. Enantioselectivity is predicted to favor (*S*)-alanine, as was observed. These hydrogen-bonded structures are so large that in molecular terms, enantioselectivity would be expected to depend on palladium particle morphology. This may provide opportunities for further optimization of the system.

The hydrogenation of substituted N-heterocycles, such as 2-methylpyridine or methyl pyridine-2-carboxylate, to the corresponding piperidines occurs as a six H-atom addition process. Accomplishing such reactions in an enantioselective fashion is a largely unfulfilled challenge. The only report in the literature is that of Wood and coworkers who claim to have achieved the enantioselective hydrogenation of 2-methylpyridine over tartaric acid-modified Ni with an enantiomeric excess of 5% [87]. Certainly, cinchona-modified Pt and Pd catalysts fail to induce enantioselectivity in these reactions, probably because of the characteristic (mentioned above [61]) that pyridine adsorbs at an angle to the metal surface. However, it has recently been observed that the situation is transformed when a second N-atom is present in the aromatic ring, as in substituted pyrazines. The hydrogenation of methyl pyrazine-2-carboxylate occurs rapidly over Pd/C and, when modified by a cinchona alkaloid, the product methyl piperazine-2-carboxylate shows an enantiomeric excess (Table 6.3) [88]. Cinchonidine favors the formation of the *S*-product and cinchonine formation of the *R*-product. However, quinine and quinidine are less effective than the parent alkaloids. In this process, (i) dihydro-, tetrahydro-, and hexahydro-products are formed simultaneously; (ii) the enantioselective six H-atom addition reaction, which is initially fast, is poisoned by tetrahydro-product that accumulates during reaction, and (iii) the enantiomeric excess in the initial hexahydro-product is progressively degraded by racemic hydrogenation of the tetrahydro-product. Ephedrine is also an effective modifier (Table 6.3). The four stereochemical variants direct reaction in such a way that (1*S*,2*R*)- and (1*S*,2*S*)-ephedrine facilitate (*R*)-piperazine ester formation, whereas (1*R*,2*R*)- and (1*R*,2*S*)-ephedrine favor formation of the (*S*)-piperazine ester.

In all of these reactions, the product enantiomer in excess was the one having a configuration opposite to that of the carbon atom adjacent to the aromatic moiety of the modifier. The significance of this relationship was revealed by NMR studies of solutions of (1*R*,2*S*)-ephedrine and methyl pyrazine-2-carboxylate in perdeuterio-toluene.

Table 6.3. Methyl pyrazine-2-carboxylate hydrogenation to methyl piperazine-2-carboxylate (MPC) over alkaloid-modified Pd/C in ethanol at 293 K and a pressure of 60 bar. Yields and values of the enantiomeric excess are for the ring-saturated product.

Modifier	Conversion/%	Yield of MPC/%	e.e./%
cinchonidine	60	36	24(*S*)
quinine	56	22	3(*S*)
cinchonine	45	18	14(*R*)
quinidine	32	33	3(*R*)
(1*S*,2*S*)-ephedrine [a]	96	10	12(*R*)
(1*S*,2*R*)-ephedrine	97	7	10(*R*)
(1*R*,2*R*)-ephedrine	98	17	11(*S*)
(1*R*,2*S*)-ephedrine	99	8	12(*S*)

[a] Ephedrine = PhCH(OH)CH(Me)NHMe

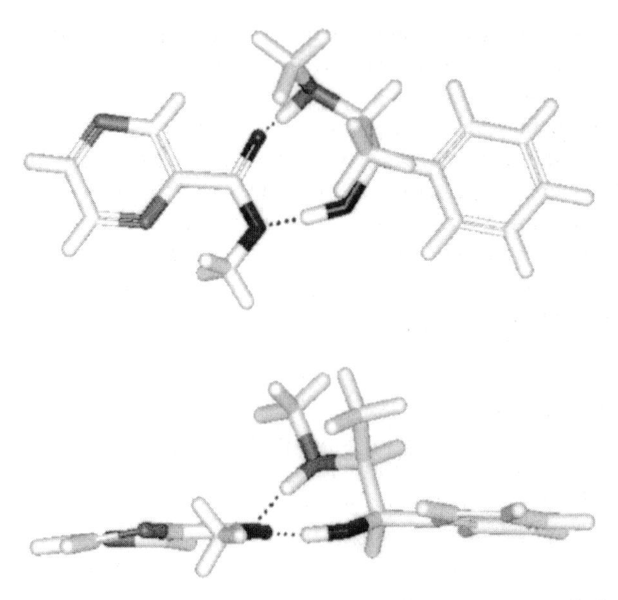

Figure 6.11. Hydrogen bonded complex between methylpyrazine-2-carboxylate (left) and (1*R*,2*S*)-ephedrine (right).

From the observed frequency shifts for the four types of CH-proton in the side chain of ephedrine and the absence of any change in the frequencies of the aromatic protons, it was concluded that there is a specific association of the pyrazine ester group with the ephedrine side chain via a bis(H-bonded) interaction, as shown in Fig. 6.11.

The adsorption of the complex shown in Fig. 6.11, or adsorption of the ester adjacent to Pd-adsorbed (1*R*,2*S*)-ephedrine utilizing these hydrogen bonding interactions, would constitute selective enantioface adsorption, and the transfer of six adsorbed H-atoms from the Pd surface to the adsorbed pyrazine from below the plane of the ring would produce (*S*)-methylpiperazine-2-carboxylate. The observed sense of the enantioselectivity is thereby interpreted. This mechanism applies *mutatis mutandis* to reactions modified by the other ephedrines so that the presence of an α-carbon atom having an *S*-configuration confers enantioselectivity to the benefit of the *R*-product. The β-carbon atom is only important insofar as it acts as a spacer between the α-carbon atom and the basic N-atom. Its configuration has no significance in determining the steric course of reaction. Potential modifiers in which that spacer is absent, for example in (*R*)-1-(1-naphthyl)ethylamine, are ineffective.

The cinchona alkaloids also contain the -C_α(OH)-C_β-N sequence except that the N-atom is now tertiary. The mechanism appears to follow the monomer model for the hydrogenation of alkenoic acids described above. Assuming that the quinuclidine-N atom is protonated in the active catalyst in ethanolic solution, a hydrogen-bonded precursor state may be formed between reactant and modifier,

$$\equiv N{-}H \cdots O{=}C(\mathrm{Py})(\mathrm{OMe})$$

where rotation around the hydrogen bond precedes adsorption of the ester by either enantioface. Molecular modelling shows that adsorption by one enantioface involves less steric interaction with the modifier than adsorption by the other, and hence preferential enantioface adsorption may occur. For cinchonine as the modifier, the preferred enantioface is that which would yield (*S*)-methylpiperazine-2-carboxylate as the product, whereas for cinchonine the reverse is the case.

Thus, the requirements for the enantioselective hydrogenation of this N-heterocycle are that the reactant should adsorb with the aromatic ring approximately parallel to the metal surface, and that the modifier should contain a -C_{α}(OH)-C_{β}-NH- unit bonded to an aromatic moiety also capable of adsorbing parallel to the metal surface.

6.7 Enantioselectivity Induced by Other Families of Alkaloids

Clearly, our present capability to induce enantioselectivity in hydrogenations catalyzed by conventionally supported metal catalysts relies heavily on the use of cinchona alkaloids as modifiers. Diversification of the system may be achieved in two ways. Simpler synthetic compounds which mimic the cinchona alkaloids may be sought, and indeed this has been successfully achieved [89–91] (see Chapter 7). Alternatively, other families of alkaloids can be examined as modifiers in order to discover how new chiral environments may be created. Three of the families that have been investigated are the strychnos, vinca and morphine alkaloids of which pertinent members are brucine, dihydrovinpocetine and codeine, respectively. Their structures are shown in Fig. 6.12, and their performances as modifiers in the enantioselective pyruvate ester hydrogenation are shown in Table 6.4.

Like the cinchonas, they each contain an aromatic moiety which provides for strong adsorption at a metal surface, and a tertiary basic N-atom for hydrogen bond-

Table 6.4. Typical e.e. values obtained in pyruvate ester hydrogenation at a pressure of 10 bar and at 293 K, catalyzed by Pt modified by brucine, dihydrovinpocetine, codeine, oxycodone, and naloxone

Modifier	Catalyst	Reactant [a]	e.e./%	Ref.
brucine	Pt/silica	MePy [c]	20(*S*)	19, 92
dihydrovinpocetine [b]	Pt/alumina	EtPy [d]	30(*S*)	67, 93
codeine	Pt/silica	MePy [c]	5(*S*)	92, 94
oxycodone	Pt/silica	MePy [c]	15(*R*)	19, 94
naloxone	Pt/silica	MePy [c]	5(*R*)	94

[a] MePy = methyl pyruvate; EtPy = ethyl pyruvate.
[b] cis-epimer.
[c] Solvent = ethanol.
[d] Solvent = 200:1 methanol:acetic acid.

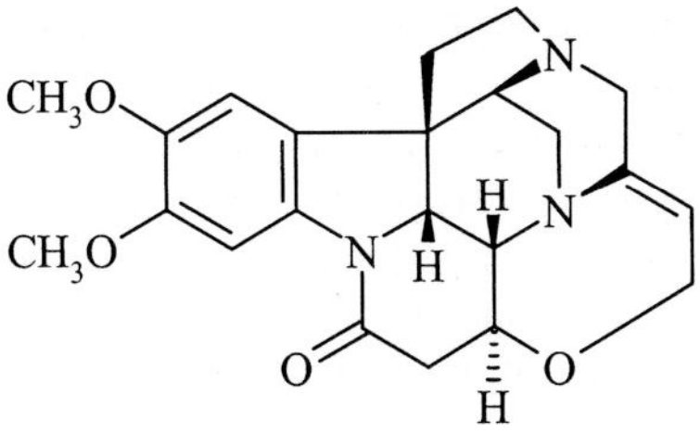

a 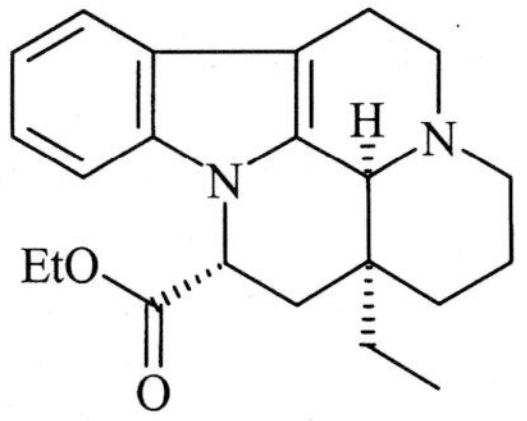 b 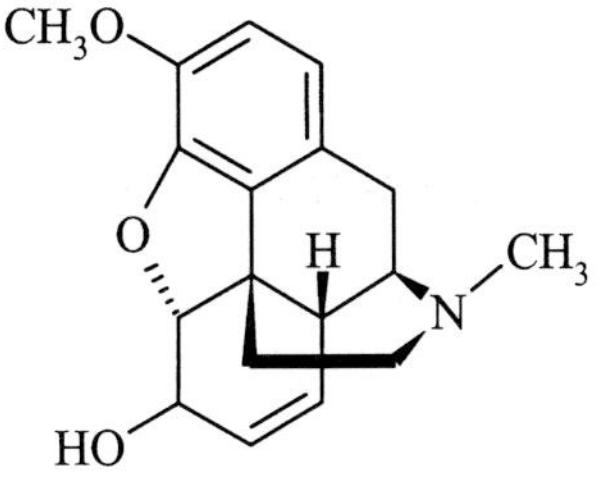 c

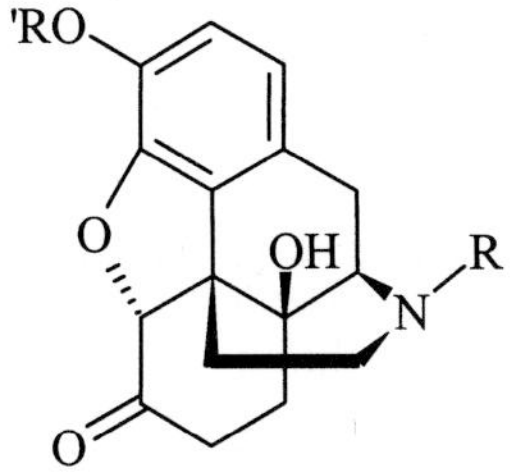

d $R = 'R = CH_3$

e $R = CH_2CH{=}CH_2$
$'R = H$

Figure 6.12. Structures of (a) brucine, (b) dihydrovinpocetine, (c) codeine, (d) oxycodone and (e) naloxone.

ing interactions with reactants or half-hydrogenated states. However, they differ from the cinchonas in being comparatively rigid molecules.

Brucine, codeine and oxycodone are T-shaped molecules, and their adsorption may be restricted to the edges of terraces at the Pt surface [94] so that the aromatic moiety may be oriented parallel to the terrace plane. When adsorbed in this way, brucine and the Pt atom surface together form a 'cavity' into which reactant molecules may diffuse and react [19]. The rings containing the ether O-atom and the basic-N atom constitute the roof of the cavity and the platinum surface the floor. Molecular interactions between the ester-type O-atom of methyl pyruvate adsorbed in its two enantiofacial forms and the ether-type O-atom of the brucine structure differ, thereby providing selective enantioface adsorption of the reactant and preferential formation of *S*-product. Moreover, a five-fold rate enhancement is observed, and the basic N-atom in the brucine structure is appropriately placed to provide such a rate enhancement by the mechanism described earlier for reaction modified by cinchona alkaloids. This mechanism is consistent with enantioselectivity and rate enhancement being absent when brucine quaternized with methyl iodide is used as a modifier. This mechanism is supported by the observation that butane-2,3-dione is not enantioselectively hydrogenated over brucine-modified Pt. This molecule may be regarded as methyl pyruvate from which the ester-O has been 'removed', and hence, selective enantioface adsorption does not occur. Molecules of dione do, however, enter the cavity and react there, as is seen from the observed ten-fold rate enhancement.

Dihydrovinpocetine (more correctly named dihydro-apovincaminic acid ethyl ester) exists as two epimers, and the effectiveness of each as a modifier of Pt for ethyl pyruvate hydrogenation has been described [67]. The *cis*-epimer, in which the ester group occupies the equatorial position and can more readily interact with the metal surface, gives an e.e. = 30%; the other epimer under the same conditions gives an e.e. = 14%. A slight rate enhancement is observed (factor = 1.5). Dihydrovinpocetine is less strongly adsorbed by its indole moiety than dihydrocinchonidine by its quinoline group, as is shown from the behavior of catalysts modified by graded mixtures of these alkaloids. Interestingly, palladium black modified by dihydrovinpocetine is enantioselective for carbon-carbon double bond hydrogenation in isophorone, the more effective epimer affording an e.e. of 40% and the other of 10%.

Codeine, dihydrocodeine, and the closely related alkaloid thebaine each provide enantioselectivity in pyruvate ester hydrogenation towards the *S*-product [94]. The effect is so weak (Table 6.4) that it would not merit close study except for the remarkable fact that enantioselectivity is not affected by quaternization of the basic N-atom. This is, therefore, the first system in which enantioselectivity has been generated without the participation of a basic N-atom. Moreover, modelling supports the view that selective enantiofacial adsorption occurs on chemisorption of pyruvate ester adjacent to the O-decorated edge of chemisorbed codeine [94]. Reactions in the presence of codeine also show a six-fold rate enhancement, which is not lost when the alkaloid is quaternized with methyl iodide. However, the process by which this occurs is not known.

Butane-2,3-dione undergoes rate-enhanced hydrogenation in the presence of codeine, but the product is racemic, thereby indicating that (as with brucine) the simpler molecule does not undergo selective enantioface adsorption. This enantioselectivity in pyruvate ester reduction indicates that searches for novel modifiers of Pt and Pd should not be restricted to molecules containing basic nitrogen functions, and thus other chiral pools (such as the sugars) may provide materials worthy of study as modifiers.

Finally, the morphine family of alkaloids springs an interesting surprise in that oxycodone and naloxone (Fig. 6.12) behave differently from codeine. They give a reversal in the sense of the observed enantioselectivity (Table 6.4), and the enhanced rate and enantiomeric excess are both destroyed by quaternization of the alkaloids with methyl iodide. These alkaloids contain the sequence –N-C-C-OH, which is also present in the cinchonas, and in many ways they behave as cinchonidine. The proposed mechanism [19, 94] is that these molecules follow codeine in being adsorbed at edge and step sites, that the reactant utilizes its two carbonyl groups to chelate via hydrogen bonding to the oxygen and (protonated) nitrogen atoms of the –N-C-C-OH unit as a precursor state, and that subsequent adsorption on the adjacent terrace (i.e. the terrace *below* that on which the alkaloid is adsorbed) leads to hydrogenation with the observed enantiomeric preference. This mechanism predicts that butane-2,3-dione should also be enantioselectively hydrogenated in the same sense, which is indeed the case. Such a specific mechanism places very particular demands on the morphology of the catalyst. Enantiomeric excess should be very sensitive to catalyst preparation parameters, although this has yet to be investigated.

6.8 Forward Look

The ability of natural alkaloids to induce enantioselectivity in hydrogenations catalyzed by conventionally supported platinum and palladium has brought a new and important dimension to heterogeneous catalysis. High e.e. values have been obtained only with the cinchona alkaloids. This may be attributable in part to their conformational flexibility, since flexible synthetic mimics also perform well whereas relatively rigid alkaloids less effectively. However, the possession of conformational flexibility on the part of the modifier means that the origins and roles of that flexibility have to be correctly understood. Indeed, the roles may differ for different reactions, or they may differ for the same reaction under different experimental conditions, in ways not yet explored.

Hydrogen bonding is a most important feature of the various modifier-reactant interactions described above, and this emphasizes a similarity between this type of catalysis and that facilitated by enzymes.

A comparative view of the chemistry described above suggests that assessments of novel modifier systems (perhaps involving new chiral pools or considerations of other effective natural processes) are currently as important or more important than increasingly close attention to the cinchona system.

Platinum and palladium dominate the enantioselective catalysis scene as far as alkaloid modifiers are concerned. Of the other platinum group metals, Ir follows Pt in cinchona-modified pyruvate ester hydrogenation [95], Rh seldom features perhaps because it can hydrogenolyse alkaloids adsorbed at its surface [41], and Ru may be too easily oxidized by adventitious oxygen in these reactions [40].

Investigators have so far used conventional heterogeneous catalysts, whereas some proposed mechanisms appear to require the active phase to possess specific surface geometries. There is certainly a challenge here with respect to the development of catalysts having novel metal particle morphologies. There is a natural thermodynamic tendency for metal particles to assume morphologies in which their surfaces expose closely packed low-index planes. A requirement for stepped surfaces, for example, might be difficult to achieve in a catalyst from which an extended period of enantioselective activity is required. On the other hand, the strong adsorption characteristic of these alkaloids may herald their ability to effect the reconstruction of low-energy metal surfaces, thereby meeting their requirements as effective modifiers. Studies in this area would be valuable.

Finally, we should ask how removed we are from being able to create inherently chiral active phases of metal which do not need a chiral modifier in order to induce enantioselectivity. In a review in 1998, one of the present authors stated that such a development was "not even on the horizon" as far as the preparation of supported metal catalysts is concerned. Notably, in the following year, Attard and coworkers described the preparation of chiral Pt electrodes which exposed either the $\{643\}^R$ face or the $\{643\}^S$ face and which showed enantiomeric selectivity in the adsorption and electrooxidation of D- and L-glucose [96, 97]. This behavior was ascribed to the inherent left- or right-handedness of the kink sites at the surface. Pt$\{531\}^R$, which also contains only kink sites but at a higher density, gave a stronger discrimination.

Pt{332} and Pt{211}, which are stepped without having kinked surfaces, showed no discrimination. This use of electrochemistry to explore molecular recognition at metal surfaces, the application of this technique to evaluate chiral modifier adsorption at the geometrically complex surfaces of supported metals, and the translation of this chemistry into the mainstream of enantioselective heterogeneous catalysis is a major challenge for the future.

References

[1] H. Erlenmeyer, *Helv. Chim. Acta* (1930) **13** 731.
[2] D. Lipkin and T.D. Stewart, *J. Am.Chem. Soc.* (1939) **61** 3295, 3297.
[3] S. Akabori, S. Sakurai, Y. Izumi and Y. Fujii, *Nature,* (1956) **178** 323.
[4] Y. Orito, S. Imai and S.Niwa, Collected Papers of the 43rd Catalyst Forum, Japan, (1978) p. 30.
[5] Y. Orito, S. Imai and S. Niwa, *Nippon Kagaku Kaishi* (1979) p. 1118.
[6] Y. Orito, S. Imai and S. Niwa, *Nippon Kagaku Kaishi* (1980) p. 670.
[7] H.-U. Blaser and M. Müller, *Studs. Surf. Sci. Catal.,* (1991) **59** 73.
[8] H.-U. Blaser, *Tetrahedron Asymm.,* (1991) **2** 843.
[9] G. Webb and P.B. Wells, *Catal. Today,* (1992) **12** 319.
[10] A. Baiker, *J. Mol. Catal. A: Chemical,* (1997) **115** 473.
[11] H.-U. Blaser, H.P. Jalett, M. Muller and M. Studer, *Catal. Today* (1997) **37** 441
[12] P.B. Wells and A.G. Wilkinson, *Topics in Catal.,* (1998) **5** 39.
[13] A. Baiker and H.-U. Blaser, in *Handbook of Heterogeneous Catalysis*, G. Ertl, H. Knözinger and J. Weitkamp (eds.), WILEY-VCH Weinheim, (1997) **5** 2422.
[14] I.M. Sutherland, A. Ibbotson, R.B. Moyes and P.B. Wells, *J. Catal.,* (1990) **125** 77.
[15] H.-U. Blaser, H.P. Jalett, D.M. Monti, A. Baiker and J.T. Wehrli, *Stud. Surf. Sci. Catal.,* (1991) **67** 147.
[16] P.A. Meheux and P.B. Wells, unpublished work
[17] P.A. Meheux, A. Ibbotson and P.B. Wells, *J. Catal.,* (1991) **128** 387.
[18] H.-U. Blaser, H.P. Jalett and J. Wiehl, *J. Mol. Catal.,* (1991) **68** 215.
[19] P.B. Wells, K.E. Simons, J.A. Slipszenko, S.P. Griffiths and D.F. Ewing, *J. Mol. Catal. A: Chemical,* (1999) **146** 159.
[20] X. Zuo, H. Liu and M. Liu, *Tetrahredron Lett.,* (1998) **39** 1941.
[21] J.T. Wehrli, A. Baiker, D.M. Monti, H.-U. Blaser and H.P. Jalett, *J. Mol. Catal.,* (1989) **57** 245.
[22] U.K. Singh, R.N. Landau, Y. Sun, C. LeBlond, D.G. Blackmond, S.K. Tanielyan and R.L. Augustine, *J. Catal.,* (1995) **154** 91.
[23] J. Wang, Y. Sun, C. LeBlond, R.N. Landau and D.G. Blackmond, *J. Catal.,*(1996) **161** 759.
[24] J.L. Margitfalvi, P. Marti, A. Baiker, L. Botz and O. Sticher, *Catal. Lett.,* (1990) **6** 281.
[25] J.T.Wehrli, A.Baiker, D.M. Monti and H.-U. Blaser, *J. Mol. Catal.,* (1990) **61** 207.
[26] T.J. Hall, J.E. Halder, G.J. Hutchings, R.L. Jenkins, P. Johnston, P. McMorn, P.B. Wells and R.P.K. Wells, *Topics in Catalysis,* (2000) **11** 351.
[27] S.D. Jackson, M.B.T. Keegan, G.D. McLellan, P.A. Meheux, R.B. Moyes, G. Webb, P.B. Wells, R. Whyman and J. Willis, in *Preparation and Characterisation of Catalysts V,* G. Poncelet et al. (eds.), Elsevier, Amsterdam, (1991) p. 135.
[28] B. Torok, K. Felfoldi, G. Szakonyi, K. Balazsik and M. Bartok, *Catal. Lett.,* (1998) **52** 81.
[29] M. Garland and H.-U. Blaser, *J. Am.Chem. Soc.,* (1990) **112** 7048.
[30] G. Bond, K.E. Simons, A. Ibbotson, P.B. Wells and D.A. Whan, *Catal. Today,* (1992) **12** 421.
[31] W.A.H. Vermeer, A. Fulford, P. Johnston and P.B. Wells, *J. Chem. Soc. Chem. Commun.,* (1993) 1053.
[32] J.A. Slipszenko, S.P. Griffiths, P. Johnston, K.E. Simons, W.A.H. Vermeer and P.B. Wells, *J. Catal.,* (1998) **179** 267.

[33] B. Torok, K. Felfoldi, K. Balazsik and M. Bartok, *J. Chem. Soc. Chem. Commun.*, (1999) 1725.
[34] M. Studer, S. Burkhardt and H.-U. Blaser, *J. Chem. Soc. Chem. Commun.*, (1999) 1727.
[35] T. Mallat, M. Bodmer and A. Baiker, *Catal. Lett.*, (1997) **44** 95.
[36] M. Schurch, O. Schwalm, T. Mallat, J. Weber and A. Baiker, *J. Catal.*, (1997) **169** 275.
[37] M. Schurch, N. Kunzle, T. Mallat and A. Baiker, *J. Catal.*, (1998) **176** 569.
[38] T. Burgi and A. Baiker, *J. Am. Chem. Soc.*, (1998) **120** 12920.
[39] W.A.H. Vermeer, Ph.D. thesis, University of Hull, 1995.
[40] S.P. Griffiths, P. Johnston and P.B. Wells, *J. Appl. Catal.*, (2000) **191** 193.
[41] G. Bond and P.B. Wells, *J. Catal.*, (1994) **150** 329.
[42] K.E. Simons, P.A. Meheux, S.P. Griffiths, I.M. Sutherland, P. Johnston, P.B. Wells, A.F. Carley, M.K. Rajumon, M.W. Roberts and A. Ibbotson, *Recl. Trav. Chim. Pays-Bas*, (1994) **113** 465.
[43] A.F. Carley, M.K. Rajumon, M.W. Roberts and P.B. Wells, *J. Chem. Soc. Faraday Trans.*, (1995) **91** 2167.
[44] T. Evans, A.P. Woodhead, A. Gutiérrez-Sosa, G. Thornton, T.J. Hall, A.A. Davis, N.A. Young, P.B. Wells, R.J. Oldman, O. Plashkevych, O. H. Vahtras, H. Agren and V. Carravetta, *Surf. Sci.*, (1999) **436** L691.
[45] S.R. Watson, Ph.D. thesis, University of Hull, 1995.
[46] R.L. Augustine, S.K. Tanielyan and L.K. Doyle, *Tetrahedron Asymmetry*, (1993) **4** 1803.
[47] H.-U. Blaser, H.P. Jalett, D.M. Monti, A. Baiker and J.T. Wehrli, in *Symposium on Structure-Activity Relationships in Heterogeneous Catalysis* American Chemical Society, Boston, (1990) p. 79.
[48] G.D.H. Dijkstra, R.M. Kellogg and H. Wynberg, *Recl. Trav. Chim. Pays-Bas*, (1989) **108** 195.
[49] G.D.H. Dijkstra, R.M. Kellogg and H. Wynberg, J.S. Svendsen and I. Marko and K.B. Sharpless, *J. Am. Chem. Soc.*, (1989) **111** 8070.
[50] G.D.H. Dijkstra, R.M. Kellogg, H. Wynberg, *J. Org. Chem.*, (1990) **55** 6121.
[51] F.I. Carrol, P. Abraham, K. Gaetano, S.W. Mascarella, R.A. Wohl, J. Lind and K. Petzoldt, *J. Chem. Soc. Perkin. Trans.1*, (1991) **1** 3107.
[52] K.E. Simons, PhD thesis, University of Hull 1994.
[53] A. Gamez, J.U. Kohler and J.S. Bradley, *Catal. Lett.*, (1998) **55** 73.
[54] C.R. Landis and J. Halpern, *J. Am. Chem. Soc.* (1987) **109** 1746.
[55] M. Boudart and G. Djega-Mariadassau, *Catal. Rev. Sci. Eng.*, (1994) **29** 7.
[56] J. Wang, C. LeBlond, C.F. Orella, Y. Sun, J. S. Bradley and D.G. Blackmond, in *Heterogeneous Catalysis and Fine Chemicals IV*, H.-U. Blaser, A. Baiker, and R. Prins (eds.), Elsevier, Amsterdam (1997) p. 183.
[57] Y. Sun, R.N. Landau, J. Wang, C. LeBlond and D.G. Blackmond, *J. Am. Chem. Soc.*, (1996) **118** 1438.
[58] H.-U. Blaser, H.P. Jalett, D.M. Monti, J.F. Reber and J.T. Wehrli, *Stud. Surf. Sci. Catal.*, (1988) **41** 153.
[59] G. Bond, P.A. Meheux, A. Ibbotson and P.B. Wells, *Catal. Today*, (1991) **10** 371.
[60] K.J. Laidler, in *Chemical Kinetics* 2nd edition, McGraw Hill, New York, (1965) Chapter 6.
[61] A.L. Johnson, E.L. Muetterties, J. Stohr and F. Sette, *J. Phys. Chem.*, (1985) **89** 4071.
[62] J.L. Margitfalvi, E. Talas, E. Tfirst, C.V. Kumar and A. Gergely, *Appl. Catal.*, (2000) **191** 177.
[63] J.L. Margitfalvi, M. Hegedus and E. Tfirst, *Tetrahedron Asymmetry* (1996) **7** 571.
[64] J.L. Margitfalvi, M. Hegedus and E. Tfirst, *Stud. Surf. Sci. Catal.*, (1996) **101** 241.
[65] J.L. Margitfalvi and E. Tfirst, *J. Mol. Catal. A: Chemical*, (1999) **134** 81.
[66] J.L. Margitfalvi and M. Hegedus, *J. Mol. Catal. A: Chemical*, (1996) **107** 281.
[67] A. Tungler, T. Mathe, T. Tarnai, K. Fodor, G. Toth, J. Kajtar, I. Kolossvary, B. Herenyi and R.A. Sheldon, *Tetrahedron Asymmetry* (1995) **6** 2395.
[68] U. Maitre and P. Mathivasan, *Tetrahedron Asymmetry*, (1994) **5** 1171.
[69] (a) M. Bartok, K. Felfoldi, G. and T. Bartok, *Catal. Lett.*, (1999) **61** 1.; (b) P. Johnston and P.B. Wells, *J. Catal.*, (1995) **156** 180.
[70] H.-U. Blaser, H.P. Jalett, D.M. Monti, J.F. Reber and J.T. Wehrli, *Stud. Surf. Sci. Catal.*, (1988) **41** 153.
[71] T.J. Hall, P. Johnston, W.A.H. Vermeer, S.R.Watson and P.B. Wells, *Stud. Surf. Sci. Catal.*, (1996) **101** 221.

[72] P.J. Collier, T.J. Hall, J.A. Iggo, P. Johnston, J.A. Slipszenko, P.B. Wells and R. Whyman, *J. Chem. Soc. Chem. Commun.,* (1998) 1451.
[73] J.R.G. Perez, J. Malthete and J. Jaques, *Comp. Rend. Acad. Sci. Paris* , Serie II, (1985) t**300** 169.
[74] Y. Nitta, Y. Ueda and T. Imanaka, *Chem. Lett.,* (1994) 1095.
[75] Y. Nitta and K. Kobiro, *Chem. Lett.,* (1995) 165.
[76] Y. Nitta and K. Kobiro, *Chem. Lett.,* (1996) 897.
[77] Y. Nitta, K. Kobiro and Y. Okamoto, *Proc. 70th Ann. Meeting Chem. Soc. Japan,* (1996) **1** 573.
[78] Y. Nitta and A. Shibata, *Chem. Lett.,* (1998) 161.
[79] K. Borszeky, T. Mallat and A. Baiker, *Catal. Lett.,* (1996) **41** 199.
[80] K. Borszeky, T. Mallat and A. Baiker, *Catal. Lett.,* (1999) **59** 95.
[81] K. Borszeky, T. Burgi, Z. Zhaohui, T. Mallat and A. Baiker, *J. Catal.,* (1999) **187** 160.
[82] K. Borszeky, T. Mallat and A. Baiker, *Tetrahedron: Asymmetry* (1997) **8** 3745.
[83] G.M. Barrow and E.A. Yeger, *J. Am. Chem. Soc.,* (1954) **76** 5211.
[84] Y. Nakamura, *Bull. Chem. Soc. Japan,* (1941) **16** 367.
[85] T. Yoshida and K. Harada, *Bull. Chem. Soc. Japan,* (1971) **44** 1062.
[86] K. Borszeky, T. Mallat, R. Aeschiman, W.B. Schweizer and A. Baiker, *J. Catal.,* (1996) **161** 451.
[87] R.M. Laine, G. Hum, B.J. Wood and M. Dawson, *Stud. Surf. Sci. Catal.,* (1981) **7** 1478.
[88] J.R. Caroll, D.F. Ewing, S.R. Korn, M.S. Howarth and P.B. Wells, submitted for publication; J.R. Caroll, Ph.D. thesis, University of Hull, (2000).
[89] A. Pfaltz and T. Heinz, *Topics in Catalysis,* (1997) **4** 229.
[90] A. Baiker, *J. Mol. Catal. A: Chemical,* (1997) **115** 473.
[91] M. Schurch, T. Heinz, R. Aeschimann, T. Mallat, A. Pfaltz and A. Baiker, *J. Catal.,* (1998) **173** 187.
[92] S.P. Griffiths, P. Johnston, W.A.H. Vermeer and P.B. Wells, *J. Chem. Soc. Chem. Commun.* (1994) 2431.
[93] A. Tungler, T. Tarnai, T. Mathe, J. Petro and R.A. Sheldon, in *Chiral Reactions in Heterogeneous Catalysis,* G. Jannes and V. Dubois (eds.), Plenum, New York, (1995) p. 121.
[94] S.P. Griffiths, P.B. Wells, K.G. Griffin and P. Johnston in *Catalysis of Organic Reactions,* F.E. Herkes (ed.), Marcel Dekker, New York, (1998) p. 89.
[95] K.E. Simons, A. Ibbotson, P. Johnston, H. Plum and P.B. Wells, *J. Catal.,* (1994) **150** 321.
[96] G.A. Attard, A. Ahmadi, J. Feliu, A. Rodes, E. Herrero, S. Blais and G. Jerkiewicz, *J. Phys. Chem.,* (1999) **103** 1381.
[97] A. Ahmadi, G.A. Attard, J. Feliu and A. Rodes, *Langmuir,* (1999) **15** 2420.

7 Design of New Chiral Modifiers for Heterogeneous Enantioselective Hydrogenation: A Combined Experimental and Theoretical Approach

Alfons Baiker

7.1 Introduction

The concept of using a chiral auxiliary for introducing enantiodifferentiation into a solid catalyst has probably first been demonstrated by Erlenmeyer in 1922 [1]. He described a ZnO/fructose catalyst that promoted the addition of Br_2 to cinnamic acid, affording an enantiomeric excess (e.e.) of 50%. In 1932, Schwab [2] applied for the first time a chiral support to induce enantiodifferentiation in metal catalysts. He demonstrated that quartz as a support can induce significant enantiodifferentiation in the dehydrogenation of racemic 2-butanol over Cu, Ni, Pd and Pt. However, enantiomeric excesses did not exceed 10%, and this conceptual approach did not prove to be favorable in later work. More successful was the chiral modification of metal catalysts introduced by Nakamura [3] in 1940 and later by Izumi [4]. They used chiral acids to modify Pt and Ni catalysts. From these efforts the well-known *nickel-tartrate-NaBr system* [5–8] for the hydrogenation of β-ketoesters, β-diketones and methyl ketones emerged (see Chapter 8). The other highly efficient catalyst system based on the chiral modification of metal catalysts – the *platinum-cinchona alkaloid system* [8–10] for the hydrogenation of α-ketoesters, α-ketoacids and lactones – was first described by Orito [11] in 1979. Both systems provide enantiomeric excesses reaching 95% in the hydrogenation of specific reactants (substrates), which render them relevant for technical applications. However, the range of reactants which can be transformed with high enantioselectivity is still relatively narrow.

Further progress in the development of catalytic systems based on this concept requires suitable design strategies for new efficient modifiers. This prompted us some years ago to start a systematic study aimed at a better mechanistic understanding and an efficient strategy for the design of new modifiers. Considerable progress has been made recently in the design of new modifiers for enantioselective hydrogenation of α-ketoesters and derivatives with chirally modified platinum. This example is used here to illustrate the potential of a combined experimental and theoretical approach for designing new chiral modifiers. First we consider some general aspects of chiral modification of metals, then we turn to the *platinum–cinchona* system and illustrate how the mechanistic information gathered could be utilized in the search for new modifiers.

7.2 Chiral Modification of Metal Catalysts

Chiral modification of metal catalysts is achieved by a very small quantity (usually a sub-monolayer) of an adsorbed chiral auxiliary (modifier), which is either simply added to the reaction mixture (*in situ*), or deposited on the catalyst surface in a special pretreatment step before reaction (*ex situ*). The modifier adsorbs on the active metal surface and by interacting with the reactant(s), it can control the stereoselectivity of the catalytic process. Thus, knowledge concerning the adsorption (anchoring) of the modifier on the metal surface and its interaction with the reactant(s) is of prime importance for designing this class of catalysts.

7.3 Prerequisites for Rational Design of Chiral Modifiers

Until recently, only natural modifiers were applied for chiral modification of metal catalysts [8]. These modifiers were found empirically, i.e. by trial-and-error methods. A rational design of suitable modifiers requires some understanding of the reaction mechanism and of the crucial interactions occurring in the catalytic system. Interactions to be considered are summarized in Fig. 7.1. In principle, all these interactions must be considered in designing new modifiers. This requires considerable experimental and theoretical effort, which makes a rational design extremely demanding. The ultimate goal is to gain some understanding of the transition state which determines the stereochemical outcome of the reaction. However, knowledge of the transition state structure is generally not available. One of the rare examples where suitable basic information has been gathered over the past decade is the platinum-cinchona system for

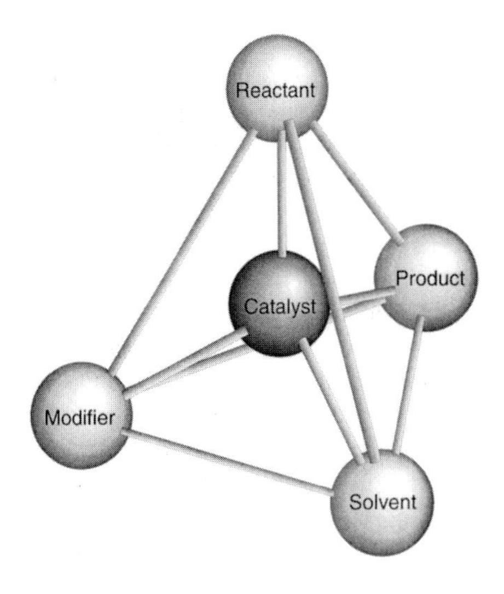

Figure 7.1. Interactions to be considered in optimization of an enantiodifferentiating catalytic system based on chirally modified metals.

the enantioselective hydrogenation of α-ketoesters. A feasible structure for the diastereomeric transition state, which was a key for the design of new modifiers, was suggested based on mechanistic information and molecular modelling. The design strategy as well as the emerging new modifiers will be described next.

7.4 A Case Study – Chiral Modification of Platinum by Cinchona Alkaloids

7.4.1 Experimental Findings

Crucial parameters of the catalytic systems

Since its discovery by Orito et al. [11] in 1979, the *platinum-cinchona alkaloid system* has been investigated by several groups, and considerable knowledge has been collected concerning the crucial parameters of this catalytic system and its functioning. The progress made until about 1996 is discussed in recent reviews [8-10]. Some characteristic features of this catalytic system are summarized in Scheme 7.1. The reaction is generally carried out at ambient temperature and a hydrogen pressure of 10–

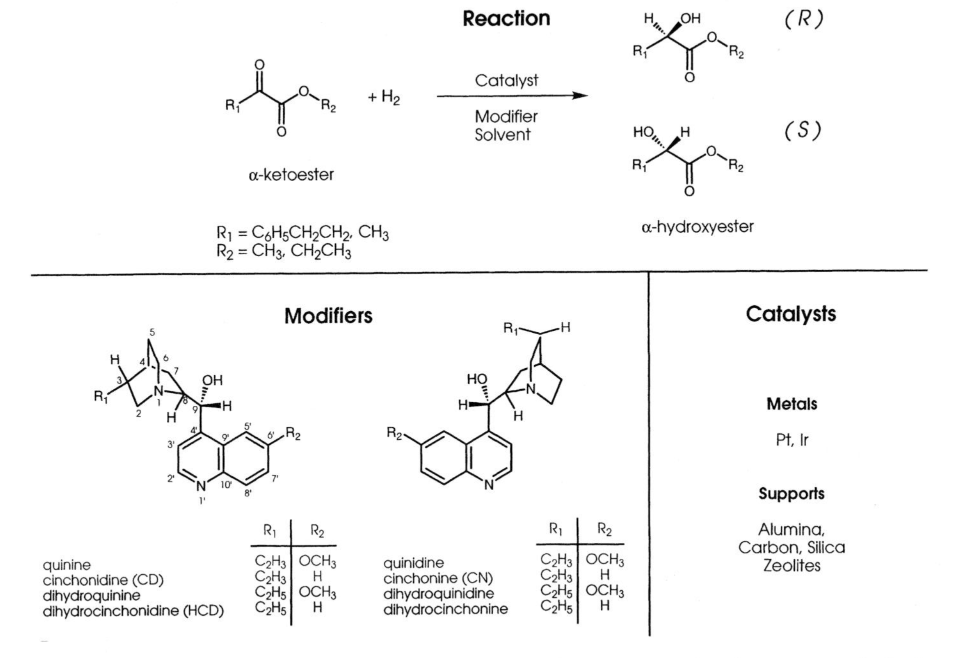

	R_1	R_2
quinine	C_2H_3	OCH_3
cinchonidine (CD)	C_2H_3	H
dihydroquinine	C_2H_5	OCH_3
dihydrocinchonidine (HCD)	C_2H_5	H

	R_1	R_2
quinidine	C_2H_3	OCH_3
cinchonine (CN)	C_2H_3	H
dihydroquinidine	C_2H_5	OCH_3
dihydrocinchonine	C_2H_5	H

Scheme 7.1. Characteristic features of platinum-cinchona system for the asymmetric hydrogenation of α-ketoesters.

70 bar. Most information has been gained by studying the asymmetric hydrogenation of methyl or ethyl pyruvate to the corresponding lactates. The most decisive parameters for the efficiency of the platinum-cinchona system are: (i) the structure and concentration of the modifier, (ii) the structural properties of the supported platinum, and (iii) the type of solvent used. Although all these properties have to be optimized to achieve high optical yields, the structure of the modifier is most crucial for enantiodifferentiation. Already very small quantities of modifier (modifier: $Pt_{surf} \ll 1$) are sufficient to induce both enantiodifferentiation and rate acceleration. The most suitable cinchona alkaloids (Scheme 7.1) used for chirally modifying platinum are cinchonidine (CD), 10,11-dihydrocinchonidine (HCD), and 10,11-dihydro-O-methylcinchonidine (OH at C-9 in CD substituted by OCH_3). Concerning the properties of the platinum catalyst, proper platinum dispersion [12], support material [13, 14] and pore size distribution [15] are important. Suitable supports [13] are alumina, silica, graphite and zeolites [14]. The solvent can influence the catalytic behavior due to different solubility of reactants (α-ketoester and hydrogen), and can change interactions between the modifier, α-ketoester and platinum surface. Apolar solvents with dielectric constants (d.c.) of between 2 and 10 were found to be suitable [16]. Alcohols and acids also afford high e.e.; however, these solvents interact with the reactant and modifier. Acetic acid (d.c. = 6.2) yields the highest e.e. (up to 95%) under optimized conditions [17].

The early systematic studies [15] of the influence of structure variation of cinchonidine, indicated three structural elements which are crucial for the functioning of the cinchona alkaloids as chiral modifiers: (i) an *anchoring part*, represented by the flat aromatic ring system (quinoline) which is assumed to be adsorbed on the platinum by multicenter π-bonding; (ii) a stereogenic center embracing C-9 and C-8; the latter is decisive for the sense of enantiodifferentiation; and (iii) a *basic nitrogen* (quinuclidine), which is directly involved in the interaction with the reactant α-ketoester, resulting in a 1:1 interaction.

Mechanistic information

Kinetics. Several kinetic studies of the enantioselective hydrogenation of ethyl pyruvate (EP) revealed the dependence of the observed reaction rate and enantiomeric excess (e.e.) on crucial reaction parameters (concentration of reactants and modifier, solvent, temperature) and indicated the role of external and intraparticle mass transfer in this complex reaction system. Some of these studies are summarized and discussed in a recent review [9]. Although a complete mechanistic model is beyond reach as yet, the kinetic results can be described reasonably well by using models [18–21] which are based on a two-cycle mechanism as presented in Fig. 7.2. The relative contributions of the modified "selective" metal sites and the bare "nonselective" metal sites are decisive for the enantiodifferentiation of the system.

The chiral modification of the platinum is supposed to lead to two enantiofacially distinct types of sites. The α-ketoester from the fluid phase adsorbs reversibly on these sites in its two enantiofacial forms, (R*) leading to the (R)-product and (S*) to the (S)-product upon hydrogenation. The modified sites have been suggested to interact with the adsorbed α-ketoester via hydrogen bonding between the quinuclidine N- and the O-atom of the α-carbonyl moiety. The rate acceleration and the enantiodiscri-

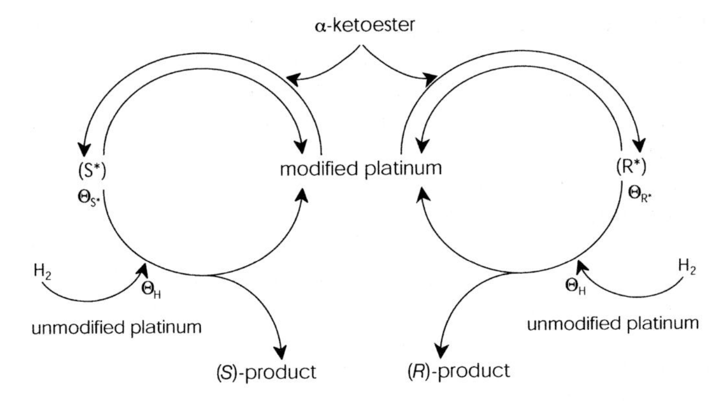

Figure 7.2. Two-cycle mechanism proposed for enantioselective hydrogenation of *α*-ketoesters on chirally modified platinum.

mination are considered to originate from preferential stabilization of one of the two diastereomeric intermediates (R^* and S^*) formed by the interaction of the *α*-ketoester and the adsorbed cinchona modifier. However, kinetic control of the enantioselectivity, in the case of different activation energies for the hydrogenation of the intermediates, (R^*) and (S^*), also cannot be ruled out under certain conditions. The question of whether the enantioselectivity is thermodynamically (stability of adsorbed R^* versus adsorbed S^*) or kinetically controlled (different activation energies of pro-(*R*)- and pro-(*S*)-routes) has not yet been definitively answered. The same holds for the question of whether the addition of the first or the second hydrogen atom is rate-determining [21]. In order to discriminate between these steps, one may study the dependence of the rate acceleration and enantioselectivity as a function of hydrogen pressure. Recently a model providing a quantitative rationalization of the effect of hydrogen concentration on enantioselectivity has been proposed [20]. It is based on the assumption of a mechanism involving reversible substrate adsorption and irreversible hydrogenation for both (*R*)- and (*S*)-pathways.

Structure of reactant – modifier complex. As pointed out above, systematic variation of the cinchona alkaloid structure proved to be an efficient tool to elucidate the reaction mechanism. Fig. 7.3 illustrates how structural changes of the cinchonidine affect the enantiomeric excess in the hydrogenation of ethyl pyruvate. Changing the absolute configuration at C-8 (*S*) and C-9 (*R*) of cinchonidine, i.e. substituting cinchonidine by the diastereomer, cinchonine, alters the chirality of the product from (*R*)- to (*S*)-lactate. The fact that the use of 9-deoxy-10,11-dihydrocinchonidine (DHCD) as modifier produces the *(R)*-ethyl lactate product indicates that the sense of enantiodifferentiation is mainly determined by the absolute configuration of C-8. Most important is the finding that the enantiodifferentiation is completely lost upon alkylation of the quinuclidine nitrogen atom, which indicates that this center plays a crucial role in the mechanism of enantioselection. Partial hydrogenation of the quinoline ring causes a drop in e.e. to below 50%. The selectivity is only marginally influenced by O-methylation, whereas replacing the OH by hydrogen (DHCD) or using the acylated derivative results in a decrease in e.e.

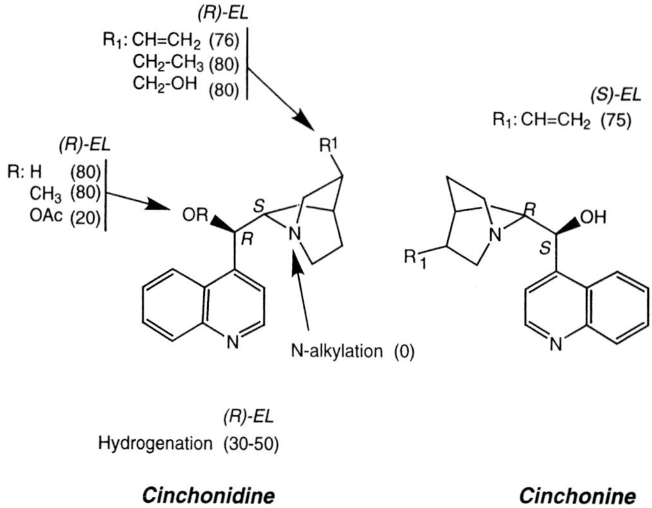

(R)-EL
R_1: $CH{=}CH_2$ (76)
CH_2-CH_3 (80)
CH_2-OH (80)

(S)-EL
R_1: $CH{=}CH_2$ (75)

(R)-EL
R: H (80)
CH_3 (80)
OAc (20)

N-alkylation (0)

(R)-EL
Hydrogenation (30-50)

Cinchonidine **Cinchonine**

Figure 7.3. Structural modification of cinchonidine (CD) and its effect on the enantio-differentiation in hydrogenation of ethyl pyruvate to ethyl lactate (EL). Numbers in parentheses represent enantiomeric excesses achieved with correspondingly modified CD. Note the change of the sense of enantiodifferentiation when the diastereomer cinchonine is used, which differs in the absolute configuration at C-9 (*S*) and C-8 (*R*). Hydrogenation (e.e. = 30–50%) refers to partial hydrogenation of quinoline ring.

by more than 20%. Interestingly, protonation of the quinuclidine nitrogen slightly increases e.e., indicating a favorable interaction of this center with the reactant ethyl pyruvate in the enantiodifferentiating modifier-reactant complex.

Conformation of modifier. The first systematic experimental studies on the conformation of cinchona alkaloids and derivatives by NMR techniques were reported by Dijkstra et al. [22, 23] Although they did not study the conformation of cinchonidine – the most frequently used modifier for platinum – they reported the conformational behavior of dihydrocinchonidine, which is expected to be similar to that of cinchonidine. The term "open" and "closed" conformers were coined to distinguish conformers in which the quinuclidine N points away or towards the heteroaromatic moiety (quinoline ring). Dihydrocinchonidine was suggested to adopt 90% open(3) conformation in the solvents investigated. A recent study [24] of the conformational behavior of cinchonidine in different solvents showed that the major conformers of cinchonidine coexisting at room temperature are closed(1), closed(2) and open(3) (Fig. 7.4), with the latter being most stable in apolar solvents. The earlier observed strong dependence of the enantioselectivity in ketoester hydrogenation on solvent polarity [16, 25] could be traced to the influence of the solvent polarity on the cinchonidine conformation [24]. Both the population of the open(3) conformer and the enantioselectivity achieved in ketopantolactone hydrogenation over cinchona-modified platinum in different solvents were found to show a similar dependence on solvent polarity, as illu-

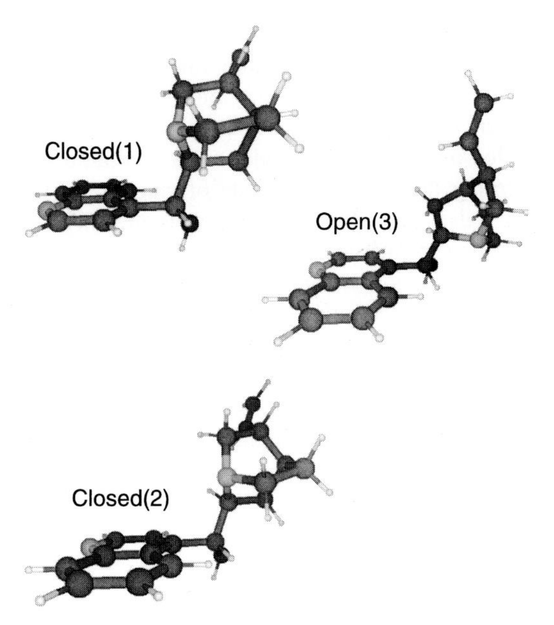

Figure 7.4. Ball-and-stick models of most stable conformers of cinchonidine, open(3), closed(1) and closed(2). The terms "open" and "closed" are used to indicate that the quinuclidine nitrogen atom respectively points away or towards the quinoline ring.

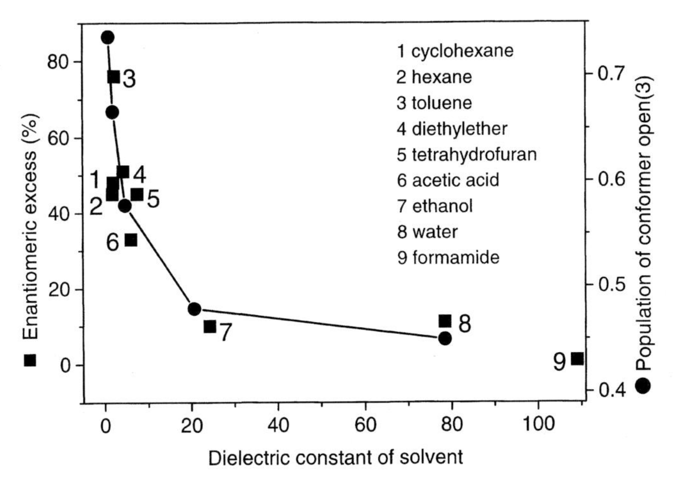

Figure 7.5. Dependence of population of conformer Open(3) on polarity (dielectric constant) of solvent and its influence on the enantiomeric excess (e.e.) in the hydrogenation of ketopantolactone [24].

strated in Fig. 7.5. Further support for the outstanding role of the open(3) conformer was recently provided by Bartok et al. [26], who showed that with the rigid cinchona alkaloids (no C8–C9 rotation), α-isocinchonine and α-isoquinidine, which exist in the open conformation, high e.e.'s are achievable.

Adsorption mode of modifier. The two most feasible adsorption modes of cinchonidine are adsorption via the aromatic bonding system of the quinoline or adsorption via the quinoline N lone pair. The former results in a flat adsorption of the quinoline ring parallel to the platinum surface, whereas the latter leads to an adsorption where the orientation of the quinoline ring is tilted or even perpendicular to the platinum surface. Although direct information concerning the adsorption mode of cinchonidine is still not available, considerable evidence exists that planar adsorption is prevalent, at least at lower surface coverages. This evidence emerges from H/D exchange experiments [27] and a comparative study of the behavior of 2-phenyl-9-deoxy-10,11-dihydrocinchonidine (PDHCD) [28], a cinchonidine derivative arylated at the quinoline ring, and 9-deoxy-10,11-dihydrocinchonidine (DHCD). Adsorption via the quinoline nitrogen is not possible with PDHCD due to steric hindrance caused by the bulky phenyl group at C-2, whereas the interaction of the π-bonding system of quinoline is possible. The fact that both modifiers show similar enantiodifferentiation corroborates the hypothesis of parallel adsorption via the quinoline π-bonding.

7.4.2 Theoretical Studies

Molecular simulation techniques have proven to be a valuable complement to experimental work for gaining information on the conformation of the modifier and the structure and stability of the diastereomeric 1:1 reactant – modifier complexes.

Conformation of modifier. The conformational behavior of cinchona alkaloids has been the focus of several theoretical studies using various computational methods ranging from empirical potentials to ab initio calculations. The most important degrees of freedom are the two torsional angles τ_1: $C_3-C_4-C_9-C_8$ and τ_2: $C_4-C_9-C_8-N$, which determine the relative orientation of the quinoline and quinuclidine moieties. In the two-dimensional conformational subspace, τ_1 and τ_2, six conformers exist on the potential energy surface [24]. The ball-and-stick models of the three most stable conformers, open(3), closed(1) and closed(2), are illustrated in Fig. 7.4. In apolar solvents the open(3) conformer was found to be most stable. *Ab initio* and density functional reaction-field calculations using cavity shapes determined by an isodensity surface were applied to simulate the dependence of the population of the open(3) conformer on the polarity (dielectric constant) of the solvent [24]. A good agreement with the population derived from NMR investigations was obtained for solvents that do not exhibit strong specific interactions (Fig. 7.5).

Structure and stability of reactant–modifier complexes. Theoretical studies aimed at rationalizing the structure and stability of the reactant–modifier complex have been undertaken using quantum chemistry techniques, at both *ab initio* and semiempirical levels, and molecular mechanics. The calculations provided feasible structures of the enantiodifferentiating complexes, which are supposed to resemble the corresponding transition complexes. It was assumed that the enantioselectivity is thermodynamically controlled, i.e. by the difference in the free energies of adsorption ΔG^0_{ads}, of the pro-*R*

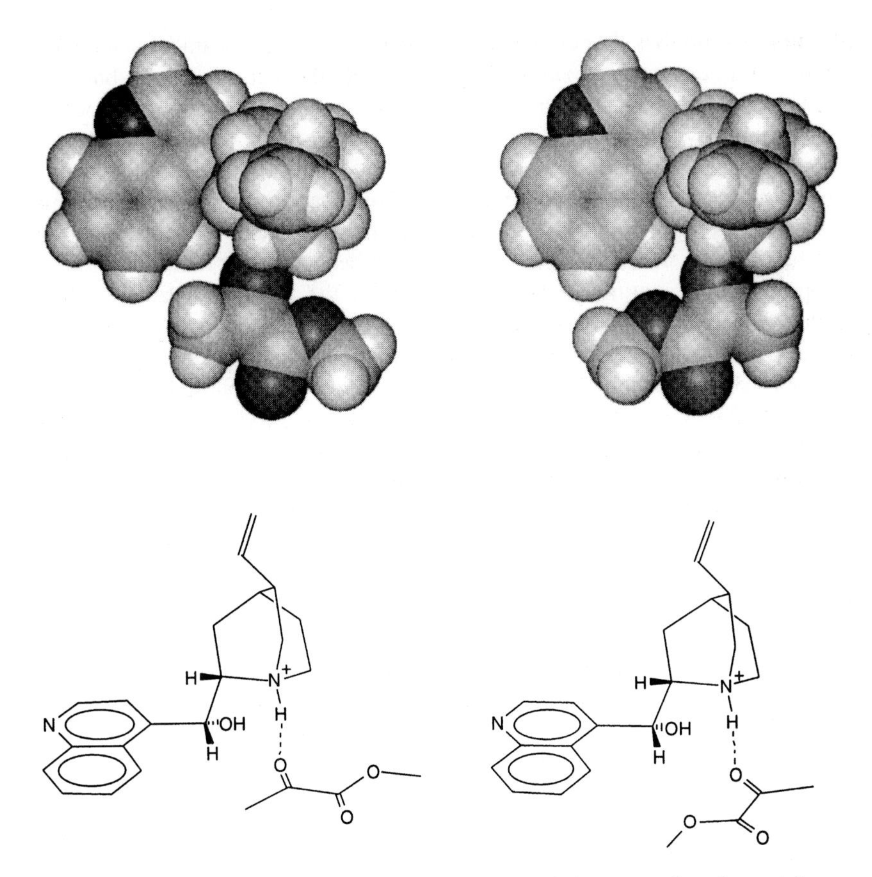

Figure 7.6. Proposed structure of enantiodifferentiating complex formed between protonated cinchonidine and methyl pyruvate which yield (*R*)-(+)-methyl lactate and (*S*)-(–)-methyl lactate, respectively, upon hydrogenation. Top: Platinum surface lies parallel to drawing plane. CD is adsorbed via π-bonding of the quinoline ring on the Pt surface. Adsorption of pro-(*R*)-methyl lactate complex is energetically and sterically favored.

(*R**) and pro-*S* (*S**) complexes (cf. Fig. 7.2). When adsorption equilibrium is established, the relative surface coverages of (*R**) and (*S**) can be expressed as $\Theta_{R*}/\Theta_{S*} = \exp(-\Delta\Delta G^0_{ads,}/RT)$. If no kinetic factor controls, i.e. if activation energies and preexponential factors of the subsequent hydrogen additions are the same, the ratio of the product enantiomers is determined by the ratio of the surface coverages $(R)/(S) = \Theta_{R*}/\Theta_S$. Thus under these conditions the sense of enantiodifferentiation should be predictable by stability considerations of the intermediate complexes (*R**) and (*S**). Figure 7.6 shows the optimized structure of the diastereomeric complexes formed on interaction of protonated cinchonidine (acetic acid as solvent) and methyl pyruvate, which upon hydrogenation would yield (*R*)-methyl lactate and (*S*)-methyl lactate, respectively. Note that in the transition complexes the α-ketoester is stabilized in its semihydrogenated form [29, 30]. The pro-(*R*)-methyl lactate complex was found to be energetically favored [30–33]. Experimental evidence [28] suggests parallel adsorption of cinchonidine in a conformation similar to open(3) on the Pt surface via π-bonding of the aromatic quinoline ring. This adsorption mode allows simultaneous interaction of the quinuclidine N with the α-carbonyl group of methyl pyruvate. In the complex affording (*R*)-methyl lactate, both carbonyl moieties of the reactant methyl pyruvate lie in a plane parallel to the Pt surface providing optimal adsorptive interaction. This adsorption mode is sterically hindered in the pro (*S*)-methyl lactate complex. The opposite behavior is found when the complexes formed between protonated cinchonine and methyl pyruvate are optimized [10]. In this case, the complex leading to (*S*)-methyl lactate upon hydrogenation is energetically favored and can be adsorbed without significant steric hindrance, while the one affording (*R*)-methyl lactate is sterically hindered. Thus calculations, show that – in agreement with the experimental observations – a change in the chirality of the stereogenic region (C-8, C-9) of the cinchona alkaloid results in a change of the chirality of the product lactate.

The origin of the hydrogen responsible for the stabilizing interaction between the cinchona alkaloid and the α-ketoester is assumed to change depending on the solvent [10]. The hydrogen can either come from the solvent (protonation of basic nitrogen, in acetic acid) or from dissociatively adsorbed hydrogen [10]. The enantiodiscrimination is considered to occur due to stabilization of one of the diastereomeric transition complexes.

The molecular modelling approach, taking into account the methyl pyruvate–cinchona alkaloid interaction and the steric constraints imposed by the adsorption on the platinum surface, leads to a reasonable explanation for the enantiodifferentiation. Although the predictions of the complexes formed between methyl pyruvate and the cinchona modifiers have been made for an ideal case (neglecting the quantum description of adsorptive interaction with platinum), this approach proved to be extremely useful in the search for new modifiers, which will be considered next.

7.4.3 Design of New Modifiers

Strategy

The theoretical calculations aimed at elucidating the structure and stability of the diastereomeric transition complexes leading to (*R*)- and (*S*)-products, respectively, proved to be a valuable guide for screening potential chiral modifiers. The search strategy, which included a systematic reduction of the cinchona alkaloid structure to the essential functional parts and rationalization of the structure of the corresponding complex formed between the new modifier and methyl pyruvate by means of molecular modelling, indicated that simple chiral 2-hydroxy-2-aryl ethylamines should be promising substitutes for cinchona alkaloid modifiers. An example of an efficient modifier emerging from this search is (*R*)-2-(1-pyrrolidinyl)-1-(1-naphthyl) ethanol (PNE) [34], which is used for illustrating the strategy. The optimized structures of the enantiodifferentiating complexes yielding (*R*)- and (*S*)-methyl lactate, respectively, are illustrated in Fig. 7.7. The similarity of these complexes to the corresponding complexes formed between methyl pyruvate and protonated CD (Fig. 7.6) is striking: both complexes are

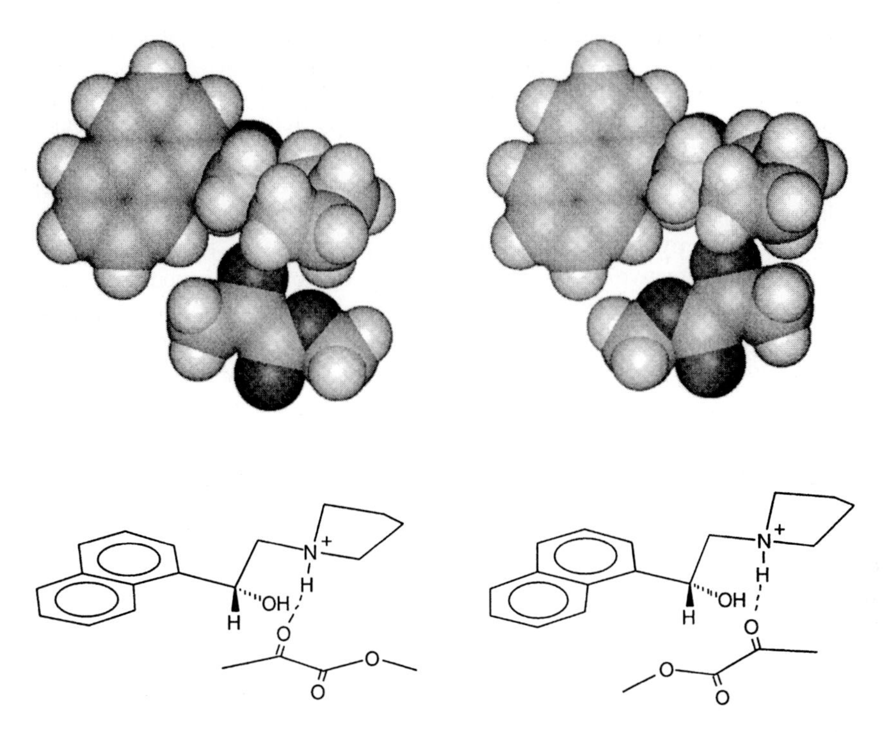

Figure 7.7. Proposed structures of complexes formed between protonated (*R*)-(2-(1-pyrrolidinyl)-1-(1-naphthyl)ethanol (PNE) and methyl pyruvate (MP), which yield upon hydrogenation (*R*)-(+)-methyl lactate and (*S*)-(–)-methyl lactate, respectively. Top: Platinum surface lies parallel to drawing plane. PNE is adsorbed via π-bonding of the naphthyl on the Pt surface. Adsorption of the pro-(*R*)-methyl lactate complex is energetically and sterically favored.

stabilised by a N-H-O interaction and parallel adsorption of the two carbonyl moieties is sterically hindered in the pro-(*S*)-complex. The pro-(*R*) complex was found to be energetically favored, indicating that the (*R*)-enantiomer product should prevail. Catalytic tests in ethyl pyruvate hydrogenation corroborated these findings; PNE proved to be a remarkably efficient modifier inducing enantiomeric excesses as high as 75% of (*R*)-ethyl lactate at low hydrogen pressures (1–10 bar). This is comparable to the results obtained with cinchonidine under these conditions. However, in contrast to CD, which is more effective at higher pressure (70–100 bar), the enantioselectivity of PNE decreases at higher hydrogen pressures due to partial hydrogenation of the naphthalene ring and concomitant loss of the π-bonding. Replacing the naphthyl by a quinolyl anchoring group had no marked effect on the e.e. under otherwise similar conditions, indicating that the quinolyl nitrogen does not play an important role, neither for adsorption nor for enantiodifferentiation.

New modifiers

The rationally guided design strategy described for PNE was used to scrutinize various 2-hydroxy-2-aryl ethylamines and other structurally related chiral compounds [35, 36]. Using the Sharpless asymmetric dihydroxylation as a key step, a series of enantiomerically pure 2-hydroxy-2-aryl ethylamines were synthesised from commercially available aromatic aldehydes or arylolefins and tested in the enantioselective hydrogenation of ethyl pyruvate [10, 34–36]. The importance of the extended aromatic ring system for effective anchoring is illustrated by comparing modifiers which only differ in the anchoring moiety [37] (Fig. 7.8). Replacing the naphthyl anchoring moiety by quinoline had no significant effect on the enantiodifferentiation, whereas benzene and pyridine derivatives of PNE are ineffective (e.e. ~0%). A striking increase in e.e. from 75 to 87% is achieved when the naphthyl anchoring moiety is substituted by an anthracenyl group resulting in 1-(9-anthracenyl)-2-(1-pyrrolidinyl)ethanol (APE, Fig. 7.8). This behavior is attributed to the different adsorption strength and steric constraints imposed by these anchoring groups. The different adsorption strength was not predictable by means of the model calculations, because the electronic interaction of the anchoring group with the platinum surface is still out of reach of accurate quantum chemical calculations. The geometrical constraints imposed on the transition complex seem to be insufficient when a phenyl or pyridyl ring are used for anchoring. Substituting the 9-anthracenyl group by a 9-triptycenyl moiety 1-(9-triptycenyl)-2-(1-pyrrolidinyl)ethanol, TPE, Fig. 7.8) also results in a complete loss of enantiodifferentiation [37], further demonstrating that the extended flat aromatic

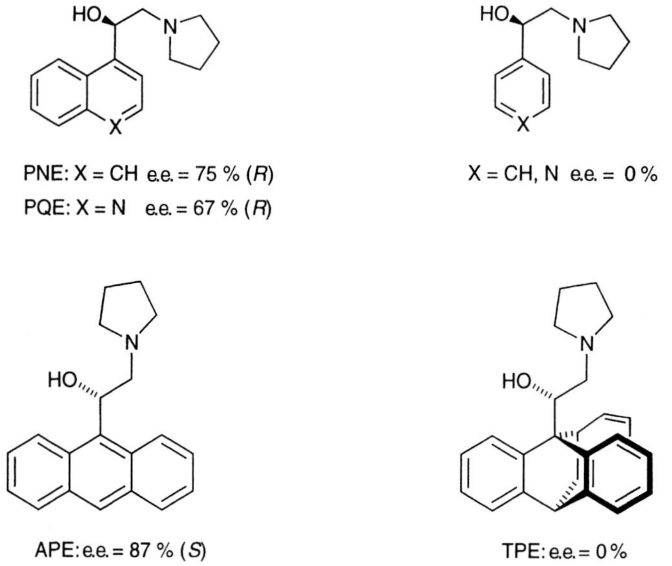

PNE: X = CH e.e. = 75 % (*R*)
PQE: X = N e.e. = 67 % (*R*)

X = CH, N e.e. = 0 %

APE: e.e. = 87 % (*S*)

TPE: e.e. = 0 %

Figure 7.8. Influence of anchoring group on catalytic performance of 1,2-aminoalcohol type modifiers in enantioselective hydrogenation of ethyl pyruvate.

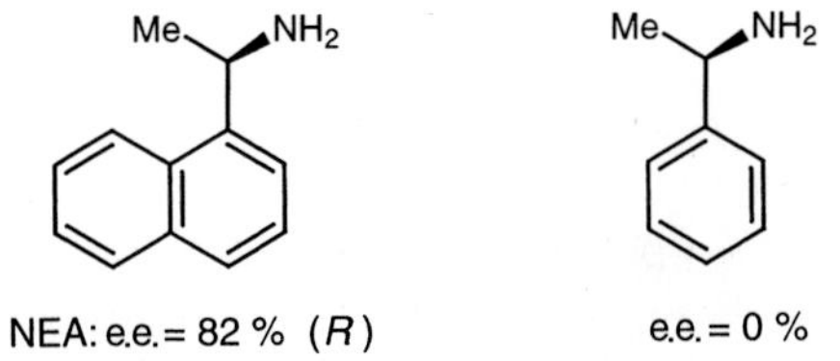
Me
NH_2
NEA: e.e. = 82 % (*R*)
Me
NH_2
e.e. = 0 %

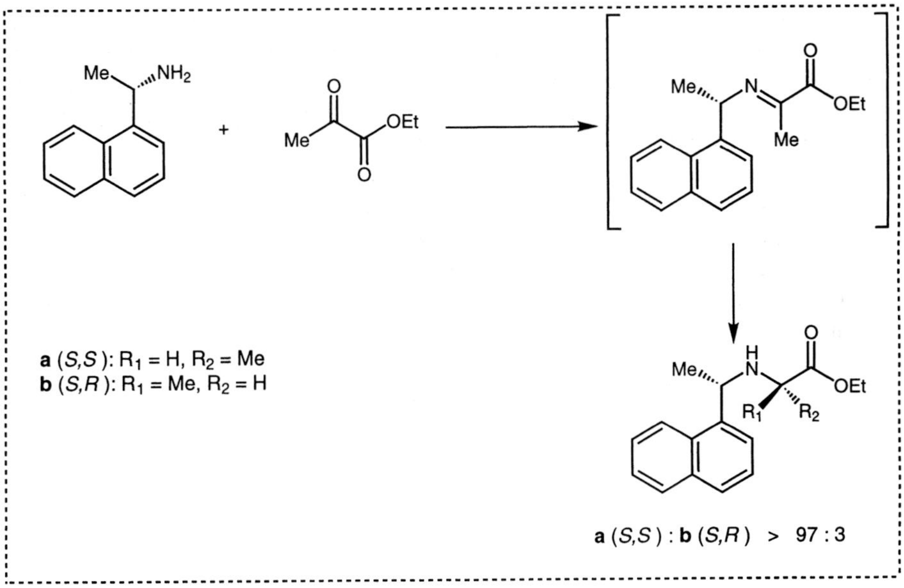
Me
NH_2
+
O
OEt
Me
O
Me
N
O
OEt
Me
a (*S,S*): R_1 = H, R_2 = Me
b (*S,R*): R_1 = Me, R_2 = H
Me
H
N
O
OEt
R_1
R_2
a (*S,S*) : **b** (*S,R*) > 97 : 3

Figure 7.9. *In situ* formation of modifier from 1-(1-naphthyl)ethylamine (NEA) precursor and ethyl pyruvate by condensation to the corresponding imine and subsequent reduction of the C=N bond.

ring system is a crucial structural element of an efficient modifier in α-ketoester hydrogenation.

An attractive alternative to the novel aminoalcohol type modifiers is the use of 1-(1-naphthyl)ethylamine (NEA, Fig. 7.9) and derivatives thereof as chiral modifiers [38–40]. Trace quantities of (*R*)- or (*S*)-1-(1-naphthyl)ethylamine induce up to 82% e.e. in the hydrogenation of ethyl pyruvate over Pt/alumina. Note that naphthylethylamine is only a precursor of the actual modifier, which is formed *in situ* by reductive alkylation of NEA with the reactant ethyl pyruvate. This transformation (Fig. 7.9), which proceeds via imine formation and subsequent reduction of the C=N bond, is highly diastereoselective (d.e. >95%). Interestingly, the absolute configuration of the stereogenic center in α-position to the ester group has no influence on the enantioselection of modifiers **a** and **b** (Fig. 7.9, bottom). Reductive alkylation of NEA with different aldehydes or ketones provides easy access to a variety of related modifiers [38–40]. The enantioselection achieved with the modifier derived from NEA could be ra-

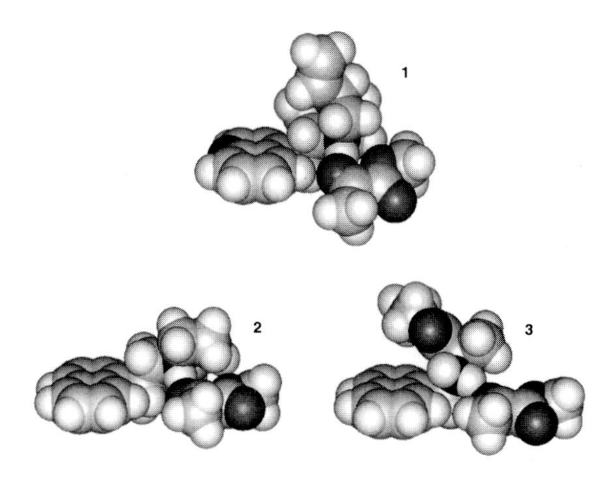

Figure 7.10. Proposed structures of diastereomeric transition complexes formed upon interaction of methyl pyruvate (MP) with different modifiers, all leading to (*R*)-methyl lactate. **1**: cinchonidine (CD)–MP, **2**: (*R*)-2-(1-pyrrolidinyl)-1-(1-naphthyl) ethanol (PNE)–MP, **3**: (2*R*,1′*R*)-N-[1′-(1-napthyl)ethyl]-2-amino propionic acid methylester–MP.

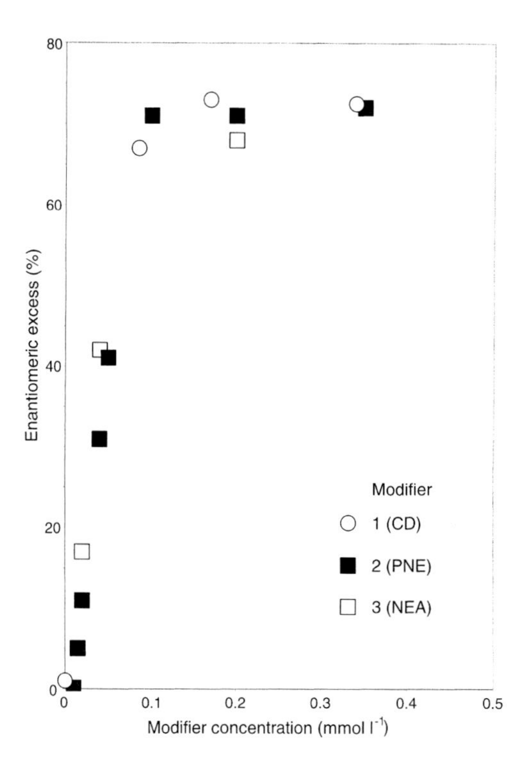

Figure 7.11. Dependence of enantiodifferentiation in ethyl pyruvate hydrogenation over supported platinum modified with **1**: cinchonidine (CD) [10], **2**: (R)-2-(1-pyrrolidinyl)-1-(1-naphthyl)ethanol, PNE [36], **3**: (2*R*,1′*R*)-N-[1′-(1-naphthyl) ethyl]-2-amino propionic acid ethylester (modifier formed in situ from NEA) [40]. Conditions: 100 mg catalyst, 10 ml (0.09 mol) ethyl pyruvate, 20 ml acetic acid, 25 °C.

tionalized with the same strategy of molecular modelling as demonstrated for the CD and PNE modifiers. Figure 7.10 compares the proposed structures of the enantiodifferentiating diastereomeric modifier – reactant complexes formed with different modifiers and methyl pyruvate. Rationalizing the optimized structures of modifier – reactant complexes and comparing them to that of the well-functioning cinchonidine-reactant pair provided a powerful tool for prescreening potential modifier-reactant pairs. Structural similarity of these complexes proved to be a strong indication for proper stereocontrol of the reaction system, as is proved by the experimental tests summarized in Fig. 7.11.

The strategy applied to search for new modifiers was also successfully applied to extend the scope of reactants, as illustrated elsewhere for the enantioselective hydrogenation of ketopantolactone [41], 2,2,2-trifluoroacetophenone [42], α-ketoamides [43], and pyrrolidine-2,3,5-triones [44].

7.5 Conclusions and Outlook

In the past decade considerable insight has been gained into the functioning of cinchona-modified platinum for the enantioselective hydrogenation of α-ketoesters. This knowledge paved the way for a rationally guided design of new efficient modifiers for this catalytic system. Computational methods, including *ab initio*, semiempirical and force field calculations were applied to rationalize the structure and stability of various reactant – modifier complexes. The structures of these complexes are considered to resemble the structures of corresponding enantiodifferentiating transition states and thereby provide a helpful criterion for judging the potential of a chiral compound to act as a suitable modifier. Steric constraints can be easily inspected and energetic considerations allowed to identify the more stable diastereomeric complex favoring one of the enantiomers. Limitations of this approach arise from the fact that binding energies and structures of large complexes adsorbed on metal surfaces are still beyond reach for quantum chemical calculations. Thus model predictions for such complex systems are only possible in conjunction with experimental verification of important boundary conditions (e.g. adsorption mode) imposed to the model.

The combined theoretical and experimental design strategy led to a series of new modifiers suitable for chiral modification of platinum. Two classes of suitable modifiers were found: 2-hydroxy-arylethylamine derivatives prepared from corresponding arylaldehydes or arylolefins using the Sharpless dihydroxylation as a key step, and modifiers prepared from a naphthylethylamine precursor by reductive alkylation with different carbonyl compounds. In some cases the latter class of modifiers can be prepared *in situ* on the catalyst surface by reductive alkylation of the naphthylethylamine with the reactant α-ketoester.

Acknowledgements

It is my pleasure to thank past and present coworkers for their valuable contibutions, their enthusiasm and perseverance, which have greatly stimulated our research. Valuable contributions originating from collaboration with the groups of A. Pfaltz (synthesis of modifiers) and J. Weber (modelling) are highly appreciated. Financial support by the Swiss National Science Foundation is gratefully acknowledged.

References

[1] E. Erlenmeyer, H. Erlenmeyer, *Biochem. Zeitschr.* **1922**, 233, 52.
[2] G.M. Schwab, L. Rudolph, *Naturwiss.* **1932**, 20, 363.
[3] Y. Nakamura, Bull. *Chem. Soc. Jpn.* **1941**, 16, 367.
[4] Y. Izumi, *Adv. Catal.* **1983**, 32, 215.
[5] A. Tai and T. Harada, in: Taylored Metal Catalysts, (Y. Iwasawa, Ed.) D. Reidel, Dordrecht, 1986, p. 265.
[6] E.I. Klabunovskii, *Russian Chem. Rev.* **1991**, 60, 980.
[7] G. Webb and P.B. Wells, *Catal. Today* **1992**, 12, 319.
[8] A. Baiker and H.U. Blaser, in: Handbook of Heterogeneous Catalysis, (G. Ertl, H. Knözinger and J. Weitkamp, Eds.), Vol. 5, WILEY-VCH, Weinheim, 1997, pp. 2422–2436.
[9] H. U. Blaser, H.P. Jalett, M. Müller, and M. Studer, *Catal. Today* **1997**, 37, 441.
[10] A. Baiker, *J. Mol. Catal. A: Chemical* **1997**, 115, 473.
[11] Y. Orito, S. Imai, S. Niwa and G-H. Nguyen, *J. Synth. Org. Chem. Jpn.* **1979**, 37, 173. Y. Orito, S. Imai and S. Niwa, *J. Chem. Soc. Jpn.* **1979**, 1118; *J. Chem. Soc. Jpn.* **1980**, 670; *J. Chem. Soc. Jpn.* **1982**, 137.
[12] J.T. Wehrli, A. Baiker, D.M. Monti, and H.U. Blaser, *J. Mol. Catal.* **1989**, 49, 195.
[13] H.U. Blaser, H.P. Jalett, D.M. Monti, J.F. Reber and J.T. Wehrli, *Stud. Surf. Sci. Catal.* **1988**, 41, 153.
[14] W. Reschetilowski, U. Böhmer, J. Wiehl, *Stud. Surf. Sci. Catal.* **1994**, 84, 2021.
[15] H.U. Blaser, H.,P. Jalett, D.M. Monti, A. Baiker, and J.T. Wehrli, *Stud. Surf. Sci. Catal.* **1991**, 67, 147.
[16] J.T. Wehrli, A. Baiker, D.M. Monti, H.U. Blaser, and H.P. Jalett, *J. Mol. Catal.* **1989**, 57, 245.
[17] H.P. Jalett, J. Wiehl, *J. Mol. Catal.* **1991**, 68, 215.
[18] T. J. Wehrli, PhD Thesis ETH-Zürich, No. 8833, 1989.
[19] M. Garland and H.U. Blaser, *J. Am. Chem Soc.* **1990**, 112,7048.
[20] J. Wang, C., C.F. LeBlond, Orella, Y. Sun, J.S. Bradley, D.G. Blackmond, *Stud. Surf. Sci.* **1997**, 108, 183.
[21] H.U. Blaser, P.P. Jalett, M. Garland, M. Studer, H. Thies and A. Wirth-Tijani, *J. Catal.* **1998**, 173, 282.
[22] G.D.H. Dijkstra, R.M. Kellogg, H. Wynberg, J.S. Svendsen, I. Marko, and B. Sharpless, *J. Am. Chem. Soc.* **1989**, 111, 8069.
[23] G.D.H. Dijkstra, R.M. Kellogg, and H. Wynberg, *J. Am. Chem. Soc.* **1990**, 55, 6212.
[24] T. Bürgi and A. Baiker, *J. Am. Chem. Soc.* **1998**, 120, 12920.
[25] B. Minder, T. Mallat, P. Skrabal, and A. Baiker, *Catal. Lett.* **1994**, 29, 115.
[26] M. Bartok, K. Felföldi, G. Szöllösi, T. Bartok, *Catal. Lett.* **1999**, 61,1.
[27] G. Bond, P.B. Wells, *J. Catal.* **1994**, 150, 329.
[28] T. Bürgi, Z. Zhou, N. Künzle, T. Mallat and A. Baiker, *J. Catal.* **1999**, 183, 405.
[29] G. Bond, P.A. Meheux, A. Ibbotson and P.B. Wells, *Catal. Today* **1991**, 10, 371.
[30] O. Schwalm, B. Minder, J. Weber, and A. Baiker, *Catal. Lett.* **1994**, 23, 268.
[31] O. Schwalm, J. Weber, B. Minder, and A. Baiker, I. *J. Quant. Chem.* **1994**, 52, 1191.
[32] O. Schwalm, J. Weber, B. Minder and A. Baiker, *J. Mol. Struct. (Theochem)* **1995**, 330, 353.
[33] K.E. Simons, P.A. Meheux, S.P. Griffiths, I.M. Sutherland, P. Johnston, P.B. Wells, A.F. Carley, M.K. Rajumon, M.W. Roberts and A. Ibbotson, *Recl. Trav. Chim. Pays-Bas* **1994**, 113, 465.
[34] G. Wang, T. Heinz, A. Pfaltz, B. Minder, T. Mallat , A. Baiker, *J. Chem. Soc. Chem. Commun.* **1994**, 2047.
[35] K.E. Simons, G. Wang, T. Heinz, A. Pfaltz, A. Baiker, *Tetrahedron: Asymmetry,* **1995**, 6, 505.
[36] B. Minder, T. Mallat, A. Baiker, G. Wang, T. Heinz A. Pfaltz, *J. Catal.* **1995**, 154, 371.
[37] M. Schürch, T. Heinz, R. Aeschimann, T. Mallat, A. Pfaltz, A. Baiker, *J. Catal.* **1998**, 173, 187.
[38] B. Minder, M. Schürch, T. Mallat and A. Baiker, *Catal. Lett.* **1995**, 31, 143.
[39] T. Heinz, G. Wang, A. Pfaltz, B. Minder, M. Schürch, T. Mallat, and A. Baiker, *J. Chem. Soc. Chem. Commun.* **1995**, 1421.

[40] B. Minder, M. Schürch, T. Mallat, A. Baiker, G. Wang, T. Heinz, A. Pfaltz, *J. Catal.* **1996**, 160, 261.
[41] M. Schürch, O. Schwalm, T. Mallat, J. Weber, A. Baiker, *J. Catal.* **1997**, 169, 275. M. Schürch, N. Künzle, T. Mallat, A. Baiker, *J. Catal.* **1998**, 176, 569.
[42] T. Mallat, M. Bodmer, A. Baiker, *Catal. Lett.* 1997, 44, 95. M. Bodmer, T. Mallat, A. Baiker, in: "Catalysis of Organic Reactions", F.E Herkes ed., Marcel Dekker, New York, 1998, p. 75.
[43] G.Z. Wang, T. Mallat, A. Baiker, *Tetrahedron Asymmetry* **1997**, 8, 2133.
[44] N. Künzle, A. Szabo, G.Z. Wang, M. Schürch, T. Mallat, A. Baiker, *J. Chem. Soc. Chem. Commun.* **1998**, 1377. A. Szabo, N. Künzle, T. Mallat, A. Baiker, *Tetrahedron Asymmetry* **1999**, 10, 61.

8 Modified Ni catalysts for Enantio-differentiating Hydrogenation

Akira Tai and Takashi Sugimura

8.1 Introduction

More than 35 years have passed since Izumi and Akabori introduced asymmetrically modified nickel (MNi). The research on this catalyst was continued by Izumi in Osaka University until 1984, based on organic and stereochemical approaches. When Izumi retired, the research was taken over by coworkers: Harada (Ryukoku University), Osawa (Tottori and later Toyama University), and the present authors (Himeji Institute of Technology). Recently, our cooperative studies enabled us to achieve almost perfectly enantiodifferentiating hydrogenations. The MNi catalysts have been studied in depth by numerous catalytic scientists from Europe, Japan and the USA, mostly based on inorganic and physicochemical approaches. Their results disclosed many features of MNi and provided useful information for our studies.

Izumi and coworkers obtained the first MNi by soaking a freshly prepared Raney nickel catalyst (RNi) in a boiling aqueous solution of monosodium L-glutamate. The hydrogenation of methyl acetoacetate (MAA) with this catalyst (L-Glu-MRNi) gave methyl 3-hydroxybutanoate (MHB) in 15% e.e. [1]. It was sheer luck that MAA was employed as a substrate. Even today MAA is one of the best substrates, and it has long been employed as a standard. The enantiodifferentiating ability (e.d.a.) as expressed in the product e.e. was quite significant in those days, and provided motivation for further intensive investigations.

The total process of the enantiodifferentiating hydrogenation with MNi is shown in Fig. 8.1. The preparation of MNi is simple, and no special reagents or expensive materials are required. In addition, the hydrogenation is smooth and the product is easily separated from the catalyst. However, many variables affect the performance of the system and mutually interact. Moreover, the MNi system must be studied at the underdeveloped frontier between surface chemistry and organic chemistry. To gain insight into the mechanism, the state of all molecules on the Ni surface should be simultaneously determined in reaction conditions. Even today, this is a too ambitious goal. To overcome such difficulties, one may simulate and predict the behavior of MNi with the aid of hypothetical models. These are based on the accumulation of various types of fragmentary information. For these reasons, it took more than 35 years to understand the general features of the MNi catalyst and to improve its e.d.a. from 15% to over 98%.

8.2 General Characteristics of MNi

The early work on MNi has already been reviewed in detail [2]. In this section, only major information is recapitulated, in addition to some recent results.

8.2.1 Variables Affecting the Enantiodifferentiating Ability (e.d.a.) of MNi

As shown in Fig. 8.1, there are three groups of variables that affect the e.d.a. of the MNi catalyst: modification variables, variables concerning the preparation of the Ni catalyst, and reaction variables including the choice of the substrate.

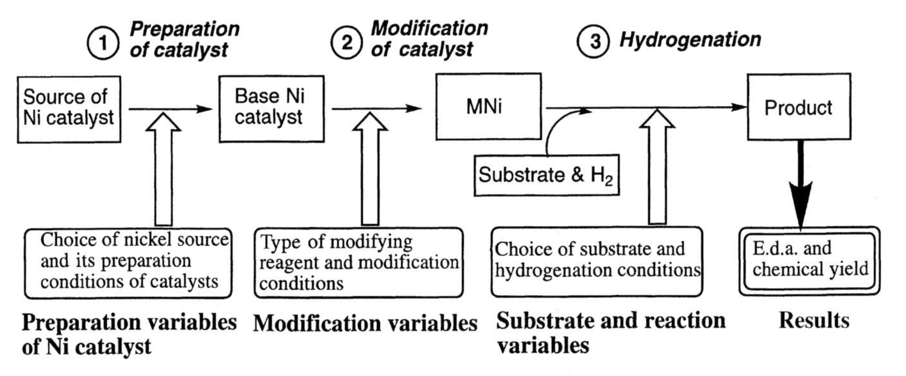

Figure 8.1. The total procedure for the enantiodifferentiating hydrogenation with MNi catalysts.

8.2.1.1 Modification Variables

Izumi's early work focused on finding suitable modification reagents and conditions, using freshly prepared RNi (W-1 type) as a base catalyst and MAA as a substrate [2a, 2c]. Screening of various water-soluble chiral compounds (amino acids and their derivatives, peptides, amino alcohols, hydroxy acids and their derivatives) indicated that tartaric acid (TA) was clearly superior. (–)-*Threo*-methyltartaric acid and (+)-1,2-dihydroxycyclohexane-1,2-dicarboxylic acid were also tested, and the former compound gave a slightly better e.e. than TA. However, this compound has not been widely utilized because it is difficult to prepare. Figure 8.2 shows the stereochemistry and the e.d.a. for some representative chiral modifiers.

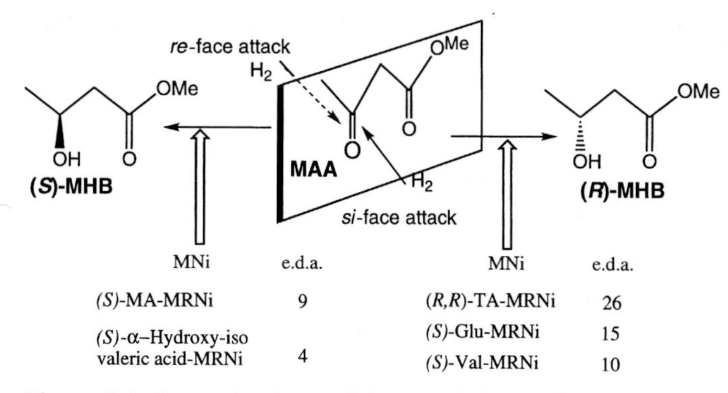

MNi	e.d.a.	MNi	e.d.a.
(*S*)-MA-MRNi	9	(*R,R*)-TA-MRNi	26
(*S*)-α–Hydroxy-iso valeric acid-MRNi	4	(*S*)-Glu-MRNi	15
		(*S*)-Val-MRNi	10

Figure 8.2. Stereochemistry of the enantioface-differentiating hydrogenation of MAA over various MRNi catalysts.

The pH and the temperature of the modifying solution strongly affect the e.d.a., even if this dependence was initially poorly understood. Only later, the effect of unstable aluminum contaminants on the freshly prepared RNi was recognized. Modification with a slightly acidic TA solution (pH adjusted to 3.5–5.5 with NaOH) at 100 °C gave over 44% e.d.a. in the MAA hydrogenation. Modification at lower temperature or outside this pH-range gave poorer results. As an optimum TA concentration, 1 to 5% was recommended. Better MRNi catalysts were obtained via modification under ambient than under N_2, Ar or H_2 atmosphere. These findings were confirmed by Bartók and Smith [3–4]. Adjustment of the pH with an alkaline hydroxide other than NaOH, such as LiOH, KOH, RbOH or NH_4OH, resulted in a poor e.d.a. [5]. It seems that the cation radius may affect the mode of adsorption of TA on the Ni surface.

The modification procedure as established for RNi entirely applies to other Ni catalyst types [6, 7]. While Webb and a coworker recommended a low TA concentration for modification of a Ni/SiO_2 catalyst (0.02–0.04%), the maximum observed e.d.a. was only 22%, which is too low for practical use [8].

8.2.1.2 Variables Concerning the Preparation of the Ni Catalyst

One may consider numerous metal catalysts as starting materials for chiral modification. While cinchona modification is well known for Pt and Pd (Chapters 6 and 7), TA modification of these metals is ineffective. Klabunovskii found that enantiodifferentiation is possible with various Raney-type catalysts [9]. In the reaction of ethyl acetoacetate, maximum e.d.a.'s were 7% for Ru, 8% for Co, and 20% for Cu, compared with 40% for Ni. Similar results were obtained with 4-hydroxy-2-butanone as a substrate [10]. Hence, Ni appears to be the best base catalyst for TA modification.

The active Ni catalysts can be prepared either by chemical reaction in aqueous solution, with 'wet' precipitation of metal black, or by hydrogenolysis or thermal decomposition of nickel compounds in the gas phase, with 'dry' deposition of metal black. Representative e.d.a.'s of TA-modified Ni catalysts are listed in Table 8.1. For materials obtained by wet preparation, only RNi gives a favorable e.d.a. after TA modification [11, 12]. Raney-type leaching at high temperature (RNiH) results in a

Table 8.1. Representative E.d.a.'s of TA-modified Ni black catalysts

	Preparation Ni-catalyst (carrier source/conditions)	Code	Modifying conditions				Reaction conditions				Ref.
			pH	T (°C)	MNi (g)	MAA (ml)	Solvent (ml)	T (°C)	P_{H_2} (atm)	E.d.a. (%)	
	Wet preparation										
1	Al/Ni-alloy (digested 20 °C; 20% aq. NaOH solution; aged 75 °C for 1 h)	(RNiL)	5.1	0	0.8	11.5	MPr[a] (23)	100	90	35	[13]
2		(RNiL)	5.1	0	0.6	17.5	Neat	70	1	26	[15]
3	Al/Ni-alloy (digested 80 °C; 20% aq. NaOH solution; aged 100 °C for 1 h)	(RNiL)	5.1	0	0.8	17.5	MPr[a] (23)	100	90	40	[13]
4		(RNiL)	5.1	100	0.8	11.5	MPr[a] (23)	100	90	44	[13]
5		(RNiL)	3.5	100	1.25	5.4	EtOH (6.2)	50	1	40	[4]
6	Si/Ni-alloy (digested 80 °C; 20% aq. NaOH solution; aged 75 °C for 1 h)	(Si-Ni)	5.0	100	0.6	17.5	Neat	60	90	44	[6]
7	Mg/Ni-alloy (digested 80 °C; 20% aq. NaOH solution; aged at 25 °C for 45 min)	(Mg-Ni)	5.0	100	0.6	17.5	Neat	60	90	6	[6]
8	$Ni(AcO)_2$ reduced by $NaBH_4$	(Ni/B)	5.1	83	1.0	10	EtAc (10)	60	10	3.0	[12]
9	$Ni(OH)_2$ reduced by NaH_2PO_2	(Ni/P)	5.1	83	1.0	10	EtAc (10)	60	10	1.5	[12]
	Dry preparation										
10	$Ni(O_2CH)_2$ (thermally decomposed 300 °C; 20 mmHg; 1 h)	(DNi)	5.0	100	0.6	11.5	THF (23)	100	90	55	[6]
11	$Ni(O_2CH)_2$ (thermally decomposed 300 °C; 3 h; followed by hydrogen flushing)	(DNi)	5.1	83	1.0	10	EtAc (10)	60	10	42	[6]
12	Greenish NiO (reduced under hydrogen	(HNi)	5.0	100	0.6	11.5	THF (23)	100	90	59	[11]
13	stream; 350 °C; 1 h)		4.1	85						65–82	
14		(HNi)	5.1	83	1.5	10	EtAc (10)	60	10	42	[12]
15	Ni powder prepared from $Ni(CO)_4$ (Metal Foil and Powder MFG Co.)	(PNi)	4.1	85	0.8	11.5	THF (23)	100	90	58–75	[11]

[a] MPr: Methyl propionate.

higher e.d.a. than leaching at lower temperature (RNiL) [13, 14]. At contrast, various Ni powders obtained by dry preparation eventually give a relatively good e.d.a. However, their hydrogenation activity is inferior to that of TA-modified RNi. As a result, long reaction times are required, with low chemical yields (55%) and formation of oligomeric by-products [6].

The catalysts obtained by wet preparation consist of well-dispersed fine metal particles with high hydrogenation activity. They contain substantial amounts of contaminants such as Al, Mg, Si, B and P. On the other hand, the catalysts obtained by dry preparation are almost free from contaminants and consist of relatively large crystalline metal particles. With XPS and XRD techniques, Nitta and a coworker carefully investigated the relationships between e.d.a., surface contaminants, and the mean crystal size (D_c) of MNi [12]. Although it became clear that the contaminants on the Ni surface directly lower the e.d.a. to some extent, they also decrease the D_c of the catalysts, which is even more detrimental for the e.d.a. For a $D_c \geq 20$ nm, the e.d.a. is always excellent. However, for a D_c smaller than 20 nm, the e.d.a. steeply decreases. Thus, the size of the ensembles of regularly arranged Ni atoms seems an important parameter for the e.d.a. of MNi.

Various types of supported Ni catalysts (Ni/Sup) are known, and have been studied as starting materials for chiral modification. Some preparation variables are the type of the support, the loading procedure, the reduction conditions, etc. Representative results are listed in Table 8.2.

In the case of silica-supported Ni, the precursor prepared by addition of Na_2CO_3 to a suspension of the support in aqueous $Ni(NO_3)_2$ provides a better catalyst than precursors prepared by hydroxide precipitation, by urea precipitation and desorption, or by impregnation. The carbonate precipitation method leads to a Ni/SiO_2 with a relatively large D_c and with a narrow crystal size distribution (CSD). The critical Ni/Si ratio for achieving a high e.d.a. and hydrogenation activity is around 1 [16]. Preferred reduction conditions for Ni/SiO_2 are 3 h at 400 °C or 1 h at 500 °C (Table 8.2, entries 3, 4). With modified Ni/SiO_2, the relationship between D_c and e.d.a. is similar to that for unsupported Ni [16]. Reduction of Ni/SiO_2 for 1 h at 300 °C results in a small D_c and a poor e.d.a., whereas reduction for 3 h at 400 °C leads to a larger D_c and high e.d.a.'s (Table 8.2, entries 1, 3). The pore size of the silica gel also affects the e.d.a.. According to Sachtler and coworkers, the presence of gas bubbles in the pores, and slow diffusion of the modifier may prevent modification of the inner part of the catalyst [19, 21]. As hydrogenation on such non-modified sites decreases the e.d.a., the use of a wide pore support was recommended (Table 8.2, entry 10). For materials with micro- and mesopores, the mesopores negatively affect the e.d.a., while the micropores have no effect [17]. Such observations are in line with Sachtler's viewpoint (Table 8.2, entries 5 to 8).

Alumina (Al_2O_3), kieselguhr (Kg), and titanium oxide (TiO_2) have also been employed as a support. The best e.d.a. values are obtained with the reduction of the Ni/Al_2O_3, Ni/Kg, and Ni/TiO_2 precursors at rather low temperature (300 °C) (Table 8.2, entries 13, 16, 19 *vs.* entries 15, 17, 18, 22). In these cases, there is no systematic relationship between D_c and e.d.a. [16, 22]. As Al_2O_3, TiO_2, and Kg have an appreciably lower surface area than SiO_2, a homogeneous Ni salt distribution is difficult to achieve at high Ni loading (Ni/Sup = 1/1). While part of the Ni salt is strongly bound

Table 8.2. Representative E.d.a. of TA-modified supported Ni catalysts

	Preparation Ni-catalyst (carrier source/reducing conditions [a])	Modifying conditions					Reaction conditions				Ref.
		Conc TA (wt%)	pH	T (°C)	MNi (g)	MAA (ml)	Solvent (ml)	T (°C)	P_{H_2}, (atm)	E.d.a (%)	
	Ni/SiO_2										
	Precursor (aq. $Ni(NO_3)_2$ + aq. Na_2CO_3) reduction with H_2										
	Nakarai Silica gel-1 (60–200 mesh)										
1	300 °C; 1 h; 1/1 (= Ni/Si)	1.6	5.1	83	1.5	10	EtAc (10)	60	10	23.9	[16]
2	300 °C; 3 h; 1/1	1.6	5.1	83	1.5	10	EtAc (10)	60	10	40.4	
3	400 °C; 3 h; 1/1	1.6	5.1	83	1.5	10	EtAc (10)	60	10	56.6	
4	500 °C; 1 h; 1/1	1.6	5.1	83	1.5	10	EtAc (10)	60	10	50.5	
	Wakogel G (30–50 mesh), microporous										
5	400 °C; 3 h; 1/1	1.6	5.1	83	1.5	10	EtAc (10)	60	10	54.6	[17]
	Wakogel Q63 (325 mesh), microporous										
6	400 °C; 3 h; 1/1	1.6	5.1	83	1.5	11.5	THF (23)	60	10	43.7	
	Wakogel C-100 (30–50 mesh), mesoporous										
7	400 °C; 3 h; 1/1	1.6	5.1	83	1.5	11.5	THF (23)	60	10	47.2	
	Wakogel C-300 (200–300 mesh), mesoporous										
8	400 °C; 3 h; 1/1	1.6	5.1	83	1.5	11.5	THF (23)	60	10	37.6	
	Precursor aq. $Ni(NO_3)_2$ evaporation until dry										
	Merck 7754										
9	450 °C; 0.05/1	0.5	5	100	0.04	12.5	neat	100	100	17–18	[18]
	Precursor $Ni(NO_3)_2$ + NH_4OH										
	Shell (wide pores, 2.5 mm beads)										
10	450 °C; 0.2/1	2	5	50	0.08	5	neat	55	1	32	[19]
	Precursor (5 M $Ni(NO_3)_2$ + aq. Urea)										
	Cab-O-Sil										
11	450 °C; 0.11/1	0.1	5.1	70	0.05	10	n-BuOH (40)	70	1	27	[8]
12		2.0	5.1	70	0.05	10	n-BuOH (40)	70	1	8	

Table 8.2 (continued)

Preparation Ni-catalyst (carrier source/reducing conditions[a])	Modifying conditions					Reaction conditions				Ref.
	Conc TA (wt%)	pH	T (°C)	MNi (g)	MAA (ml)	Solvent (ml)	T (°C)	P_{H_2}, (atm)	E.d.a (%)	
Ni/kieselguhr(Kg)										
Precursor (aq. $Ni(NO_3)_2$ + aq. Na_2CO_3)										
Yoneyama Chemic										
13 300 °C; 3 h; 1/1	1.6	5.1	83	1.5	10	EtAc (10)	60	10	41.0	[16]
14 400 °C; 3 h; 1/1	1.6	5.1	83	1.5	10	EtAc (10)	60	10	29.1	
Shimalite SP-17										
15 350 °C; 1 h; 1/1	1	4.1	85	0.6	10	THF (23)	120	90	54	[11]
Ni/Al_2O_3										
Precursor (aq. $Ni(NO_3)_2$ + aq. Na_2CO_3)										
JRC-ALO-5										
16 300 °C; 1 h; 1/1	1.6	5.1	83	1.5	10	EtAc (10)	60	10	49.9	[16]
17 500 °C; 1 h; 1/1	1.6	5.1	83	1.5	10	EtAc (10)	60	10	11.9	
γ-alumina (Merck 1095)										
18 400 °C; 15 h; 0.05/1	0.5	5	100	0.04	12.5	Neat	60	100	0–2	[18]
Wolem 200, acidic, basic or neutral										
19 300 °C; 3 h; 1/1	1	4.1	85	1.5	10	THF (23)	60	90	67–68	[16]
Precursor (Ni-acetylacetonate)										
α-Al_2O_3 pyrolysed at 370 °C										
20 500 °C; 3 h; 0.4/1	1	3.2	100	0.4	10	THF (20)	100	100	71	[20]
Ni/TiO_2										
Precursor (aq. $Ni(NO_3)_2$ + aq. Na_2CO_3) reduced by H_2										
JRC-TID-1										
21 250 °C; 1 h; 1/1	1.6	5.1	83	1.5	10	EtAc (10)	60	10	38.2	[16]
22 400 °C; 1 h; 1/1	1.6	5.1	83	1.5	10	EtAc (10)	60	10	3.3	

[a] Reducing conditions: T; time; Ni/carrier.

to the Al_2O_3 or TiO_2, the remaining portion of the Ni is only weakly interacting with the support and is easily reduced at low temperature, with formation of rather large Ni metal particles. Modification of the resulting catalyst gives a high e.d.a. High-temperature reduction results in reduction of the strongly adsorbed Ni salt, with formation of dispersed Ni with a small D_c and a high hydrogenation activity, while the Ni crystallites produced from weakly bound salt undergo sintering to large, almost inactive crystallites. Thus, while such materials have a large apparent D_c value, their CSD is in fact bimodal, and their e.d.a. is limited.

A novel approach was proposed by Osawa and coworkers. They prepared a solid preparation Ni/Sup catalyst (SPC) by thermal decomposition of a well-pulverized mixture of solid Ni acetylacetonate and Al_2O_3 in Ar at 370 °C. Reduction for 3 h in a H_2 stream at 500 °C afforded SPC type Ni/Al_2O_3 which gave 71% e.e. after TA modification (Table 8.2, entry 21) [20].

Asymmetrically modified bimetallic catalysts such as Ni-Pd [23], Ni-Rh [24], Ni-Cu [24], and Ni-Co [25] have also been examined, but after TA modification, the e.d.a. does not exceed that of the pure Ni compounds. An e.d.a. of 90% has been reported for TA-M(Ni-Pd)/Kg [26], but this result has not been reproduced by other groups [22].

Summarizing, hydrogenation activity and e.d.a. vary widely with the precursor preparation method. For instance, TA-MNi/Sup catalysts display the highest activity, while TA modification of the HNi catalyst, obtained by dry preparation, gives the best e.d.a.'s. As both activity and e.d.a. are satisfactory for Raney Ni, we decided to develop improved catalysts using RNi as a precursor (*cfr.* Section 8.3.1).

8.2.1.3 Substrate and Hydrogenation Parameters

Studying the reaction variables is essential to improve the e.d.a. of the MNi catalyst. Izumi examined more than 50 prochiral substrates and found that only β-ketoesters and β-diketones gave good e.e. values [2 a]. For the β-ketoesters, the alkoxy group did not significantly affect the e.e. [27], while replacing the methyl group at the acyl side (MAA) by an ethyl group (methyl propionylacetate) increased the e.d.a. from 33 to 66% [28]. Other β-ketoesters, however, were not studied at that time, because the $[\alpha]_D$ values for the optically pure β-hydroxyesters were not available (see Section 8.3.3).

While Izumi performed the hydrogenation of MAA without solvent, others used methanol, ethanol, or butanol, without major effects [29]. Later, semipolar aprotic solvents such as THF and ethyl acetate appeared to give much higher e.d.a. values than alcohols [26]. In our experience, the best solvents are methyl propionate and THF. Ethanol or butanol should not be used for MAA because of possible transesterification. In the hydrogenation of MAA, infiltration of water significantly decreases the e.d.a. [29], while small amounts of acids such as acetic acid raise the e.d.a. [2 a, 11, 26].

High-pressure liquid phase hydrogenation in an autoclave is not usually carried out at constant pressure. Pressure fluctuations, e.g. between 80 and 120 atm, or between 25 and 100 atm do not lead to detectable fluctuations of the e.d.a. [29]. However, changes in the e.d.a. have been reported in the range of 1 to 30 atm H_2 pressure [34]. In our experience, and in the work of Webb [30], high pressure hydrogenation (e.g. 90 atm) gives better e.e.'s than low or atmospheric pressure reactions. Gas phase hydrogenation of MAA over MNi is also possible, but the e.d.a. is much lower than in the liquid phase hydrogenation [18, 31].

The important effects of hydrogenation temperature will be discussed in detail in Section 8.3.3.

8.2.2 Kinetics of Hydrogenation over MNi

Various groups have carried out kinetic studies in order to understand the origin of the hydrogenating and enantiodifferentiating functions of MNi catalysts. We initially studied rates and e.d.a.'s for the liquid phase hydrogenation at atmospheric H_2 pressure [15]. Examples of Arrhenius plots are shown in Fig. 8.3. For all MRNi's, parallel lines are obtained with an apparent activation energy (E_A) of 44.2 kJ mol^{-1}, regardless of the e.d.a. value of each MRNi.

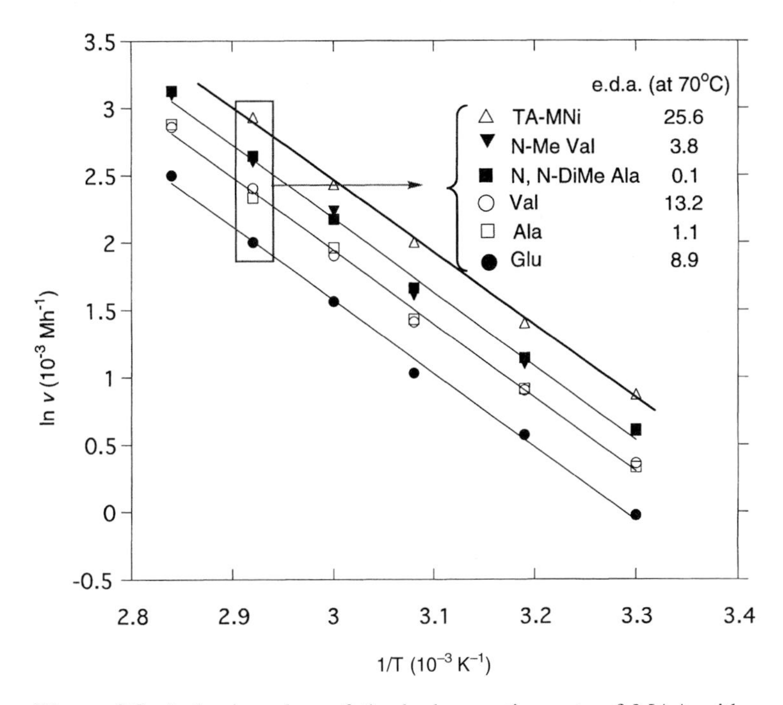

Figure 8.3. Arrhenius plots of the hydrogenation rate of MAA with various MRNi catalysts. Selected data from [15, 32].

Secondly, there is no evidence for a systematic relationship between the e.d.a. and the reaction temperature, e.g. ln([*S*-product]/[*R*-product]) $\propto$ (1/T) [32]. Thirdly, the relation between rate and the MAA concentration was investigated for TA-MNi and malic acid modified Ni (MA-MNi) [33]. From initial rates (v), the order with respect to the substrate was found to be 0.2–0.3 in both cases. Plots of $1/v$ *versus* 1/[MAA] were linear and overlapped each other. These results indicate that 1) the reaction has a

Langmuir-Hinshelwood mechanism, which is generally proposed for the liquid phase hydrogenation of ketones, and 2) the kinetic parameters for TA-MNi and MA-MNi are identical, regardless of the different e.d.a.'s. These facts lead to the proposal that 1) the rate-determining step of the hydrogenation is a surface reaction between adsorbed MAA and H_2, and 2) the nature of the modifier does not affect the rate-determining step. Hence the e.d.a. must be independently determined at a non rate-determining step.

The kinetics of the high-pressure hydrogenation with MNi/SiO_2 are quite similar to those for MRNi [46]. In a study by Keane, $TA-MNi/SiO_2$ and unmodified Ni/SiO_2 were compared in the liquid phase MAA hydrogenation at atmospheric H_2 pressure [35, 36]. Diffusion limitations were eliminated by high-speed stirring. E_A values for MNi and unmodified Ni are 46.5 and 52.0 kJ mol^{-1}, respectively. The decrease of E_A after modification with TA was explained by the association between MAA and TA on the catalyst surface. For gas phase MAA hydrogenation, several values for E_A can be found in literature, e.g. 44 for TA-MDNi [31], 46 for $TA-MNi/SiO_2$, and 57 kJ mol^{-1} for unmodified Ni/SiO_2 [2 d].

8.2.3 State of the Adsorbed TA

The state of the TA adsorbed on MNi directly influences the e.d.a.. While *α*-aminocarboxylic acids and *α*-hydroxycarboxylic acids have been adsorbed on Ni and studied with physicochemical techniques [38–40], direct detection of the TA adsorbed on Ni has not yet succeeded, and one has to rely on indirect information.

In a modifying solution prepared from TA and NaOH, TA can exist as free acid (TAH_2), monosodium salt (TAHNa), or disodium salt ($TANa_2$) depending on the pH. Since contaminants on the Ni surface might influence the TA speciation, HNi prepared from pure NiO was employed to study the effect of the pH of the modifying solution [11]. Figure 8.4 plots the e.d.a. *versus* pH, and the amount of adsorbed TA *versus* pH. The latter plot comprises two levels, namely a high value for a pH between 2 and 4.5, and a low value for pH's over 6, with a transition at pH 4.5–5.5. The surface concentration is low when both carboxylic groups of TA are neutralized with NaOH ($TANa_2$). At lower pH, the adsorbed species must be TAHNa and TAH_2. If nickel corrosion in this pH domain is taken into account, $(TAH)_2Ni$ and (TANa)(TAH)Ni may also be present on the surface. While the e.d.a. is extremely low at pH<2.5, it increases sharply to 50% at pH 3.5, and gradually rises to a plateau at pH 5.5. Such results clearly show that the TA must be present as $TANa_2$, TAHNa or $(TANa)_2Ni$ in conditions where a high e.d.a. is obtained. Note that the presence of Na^+ on the catalyst is indispensable: no other alkaline hydroxide allows to obtain a high e.d.a.

XPS studies on TA-modified DNi (pH 5.1) show that the C(1*s*) binding energy resembles that of $TANa_2$ or TANi, rather than that of TAH_2 [41, 42], in agreement with the prevalence of dianionic TA^{2-}. The Ni($2p_{3/2}$) spectrum of TA-MDNi shows the characteristic peak of pure Ni metal, with some peaks corresponding to NiO. The peak of pure Ni is preserved after exposure of the TA-MDNi catalyst to air, while unmodified DNi is oxidized upon air contact. This result proves that despite the corro-

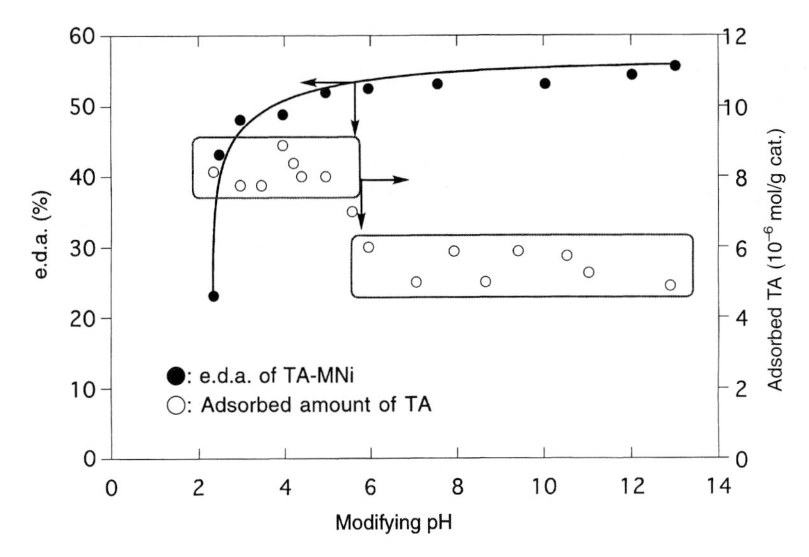

Figure 8.4. Effect of the modifying pH on the e.d.a. of TA-MHNi and the adsorbed amount of TA on TA-HNi. Selected data from Ref. [13].

sion of the Ni, the adsorption of TA^{2-}/Na^{+} protects the zerovalent metal. This protective effect is also shown by the resistance of MRNi to reaction with acetylacetone [37, 43].

The degree of coverage of the Ni surface with TA is an important parameter in order to visualize the TA-modified surface. From electrochemical measurements, Fish and Ollis determined the amount of adsorbed TA as 3.0×10^{-10} mol/cm^2 at pH 5 and 5.3×10^{-10} mol/cm^2 at pH 4 [44]. Based on a molecular model, adsorption of a complete monolayer was estimated to correspond to 8.8×10^{-10} mol/cm^2. Hence, fractional coverages are 0.34 at pH 5, and 0.6 at pH 4. Based on adsorption isotherms as in Fig. 8.4, and based on BET surface areas, we have estimated the optimum surface coverage of HNi at 2.7×10^{-10} mol/cm^2, or a fractional coverage of 0.3. For TA-MNi/SiO_2, Webb and Keane estimate that for various modifying conditions, a fractional surface coverage of 0.20 corresponds to an optimum e.d.a. [45].

Summarizing, a submonolayer coverage of the metallic Ni surface with TA is needed in order to allow adsorption of MAA.

8.3 Elucidation of the Functions of MRNi and Development of a Highly Efficient MRNi Catalyst Based on Hypothetical Models

The current situation with the MNi catalyst bears some similarities to the knowledge of enzymes in the early 1950's. In order to get to know the active site of enzymes, organic chemists made hypothetical models for each enzyme based on indirect information such as the enzyme's response to the action of various chemicals. Later, most enzyme models were found to be quite compatible with the real situation, as studied with the aid of spectroscopic and X-ray crystallographic measurements. Realizing the limitations of the physicochemical approach in the study of MNi, we started to build our own hypothetical models. Three models, the "catalyst region model", the "reaction process model", and the "stereochemical model" were proposed in order to answer three simple questions:

1) Which region on the MNi catalyst contributes to the enantiodifferentiation?
2) Which elementary reaction step is decisive for the enantiodifferentiation?
3) How does the adsorbed modifying reagent differentiate between the enantiofaces of the substrate?

These models not only simulate the functions of MNi well, but also have afforded important clues to improve the MNi catalyst.

8.3.1 Enantiodifferentiating and Non-Enantiodifferentiating Regions on MNi

8.3.1.1 The Catalyst Region Model

From early on, there was a general consensus that it was difficult to uniformly modify the active Ni surface. Hence the ratio of modified to unmodified Ni surface was believed to be an e.d.a. determining parameter. This parameter was later rephrased as the ratio of enantiodifferentiating (E) to non-enantiodifferentiating (N) surface, since it was discovered that the amount of modifying reagent did not directly determine the e.d.a. of the catalyst (Fig. 8.4). During our kinetic studies with TA-MRNi and MA-MRNi [33], we found that the rate of hydrogenation (v) decreased with an increasing conversion, while the e.d.a. increased with increasing conversion and reached a plateau (Fig. 8.5). Since the relationship $v \propto [\mathrm{MAA}]^{0.2-0.3}$ applies under the reaction conditions [15], the rate decrease is mainly caused by deactivation of part of the catalyst, rather than by changes in the MAA concentration. The coinciding rate plots for both catalysts indicate that this deactivation process does not depend on the type of modifying reagent, while the e.d.a. values are different for the two catalysts.

These results led to the following hypotheses: 1) The increase of the e.d.a. is related to the disappearance of the N-region which is initially present on the catalyst. 2) The e.d.a. of the E-region is the intrinsic e.d.a., and does not depend on the ratio of N- and E-regions. This intrinsic e.d.a. (i) is determined by the nature of the modifying agent and substrate. Bringing together these hypotheses, we formulated the catalyst region model as follows:

$$\text{e.d.a. (\% e.e.)} = \frac{i \times [\mathrm{E}]}{([\mathrm{E}] + [\mathrm{N}])} \qquad (1)$$

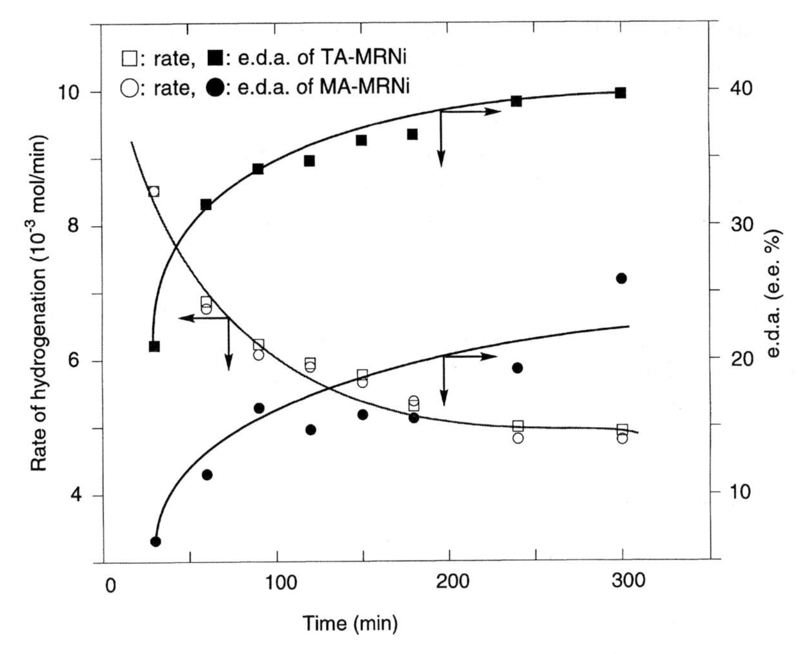

Figure 8.5. Time dependence of the rate and e.d.a. during the hydrogenation of MAA over TA-MRNi and MA-MRNi. Selected data from Ref. [33].

where [E] and [N] are the contributions of the E-region and N-region to the hydrogenation, respectively, and i is the intrinsic e.d.a. (% e.e.) for the modifier and substrate. Figure 8.6 shows an illustration of the MRNi catalyst region model.

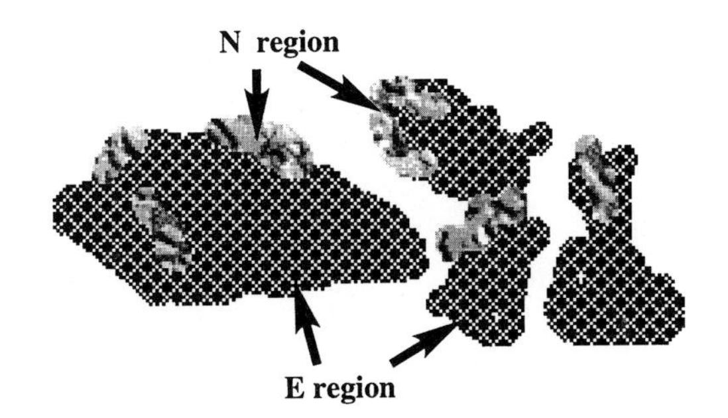

Figure 8.6. Representation of E- and N-regions in the MNi catalyst.

That the N-regions of a freshly prepared MRNi are liable to gradually loose their hydrogenation activity, suggests that the N-regions consist of well-dispersed fine metal particles, with a disordered Ni surface, and with substantial amounts of contaminants such as residual Al. In contrast, the E-regions are probably stable crystalline metal particles with an ordered surface. These ideas on the N- and E-regions are compatible with the observations on catalyst preparation (Section 8.2.1.2).

8.3.1.2 Enhancement of the e.d.a. of MRNi

The catalyst region model states that removal of the N-regions from the catalyst is essential to obtain a high e.d.a. Fortunately, the N-regions are expected to be sensitive to chemical or mechanical treatments.

Treatment with acids

It is well known that certain components in Raney alloys, e.g. NiAl or $NiAl_3$, are difficult to dissolve with NaOH. However, they can be dissolved by treatment with aqueous solutions of α-hydroxy acids. Selective elimination of Al, together with removal of disordered Ni, results in an improved precursor Ni catalyst, denoted as acid-treated RNi (RNiA). Suitable acids are glycolic acid (GA) or TA in a solution at low pH and at high temperature. Thus, treatment of a conventional RNiH catalyst (6% Al) with GA or TA at pH 3.5 and 100 °C yields {RNiA(GA)} or {RNiA(TA)} with an Al content lower than 3%. "Soft" TA modification (pH 5; 0 °C) does not eliminate Al, but nevertheless gives a MNi with a better e.d.a. than RNiH (Table 8.3, entries 1, 2, 4 and 5) [13]. The optimum modifying conditions (pH 3.5–5; 100 °C; see Section 8.2.1.1) thus not only deposit the right TA species, but also partly remove the N-regions.

The acid treatment even enhances the e.d.a. of HNi (Table 8.3, entries 6–8). Although HNi does not contain any Al, it is reasonable to assume that the surface of HNi contains both a crystalline and an amorphous component; the latter can effectively be removed by acid treatment [47].

Table 8.3. TA-MNi prepared from various Ni blacks.[a]

	Base Catalyst	Al Content (%)	Modifying conditions		E.d.a. of TA-MNi (%)
			pH	Temp (°C)	
1	RNiL	6	5.0	0	35
2	RNiH	5	5.0	0	40
3	RNiH	5	4.1	100	44
4	RNiA(GA)	3	5.0	0	72
5	RNiA(TA)	2	5.0	0	62
6	HNi	–	7.3	0	71
7	HNi		3.2	100	75
8	HNiA(TA)		7.3	0	81

[a] Reaction conditions: 11.5 ml MAA, 0.9 g MNi, 23 ml MP, 0.2 ml AcOH, initial H_2 pressure: 90 kg/cm^2, 100 °C.

Modification with NaBr

Although the acid treatment was effective, the maximum e.d.a. of 72% was still insufficient from a practical viewpoint. As an alternative approach, we tried to eliminate the activity of the N-region by partial poisoning. Using sulfur and nitrogen compounds resulted in failure. Although Keane has recently reported that thiophene is effective [35], this compound is too strong a catalyst poison to be utilized as a selective poisoning agent. While studying partial poisoning, Harada noticed an appreciable increment of e.d.a. when he accidentally used water contaminated with Cl^- and SO_4^{2-} in preparing the modifying solution. Subsequent screening of various sodium salts revealed that NaBr enhanced the e.d.a. of MRNi most (Table 8.4) [11, 48]. This increase was observed only when the TA and NaBr modification were carried out simultaneously with one solution (TA-NaBr-MRNi).

The NaBr concentration in the modifying solution strongly affects the e.d.a. as shown in Fig. 8.7. The amount of NaBr adsorbed on the catalyst [NaBr] increases with increasing NaBr concentration in the modifying solution, up to a saturation value $[NaBr]_{sat}$. Similarly, the e.d.a. increases until it reaches a maximum at $[NaBr]_{sat}$. In contrast, the amount of adsorbed TA steeply decreases as [NaBr] increases. A comparison of the e.d.a.'s for RNiH and RNiA indicates that NaBr modification enhances the e.d.a.'s to almost the same extent, but it does not change the difference in e.d.a. between these two catalysts.

The effect of NaBr has been explained in terms of partial poisoning of the N-region, or in terms of enhancement of the *i* factor in the E-region by interaction with adsorbed TA. However, the real mode of action is still open to discussion.

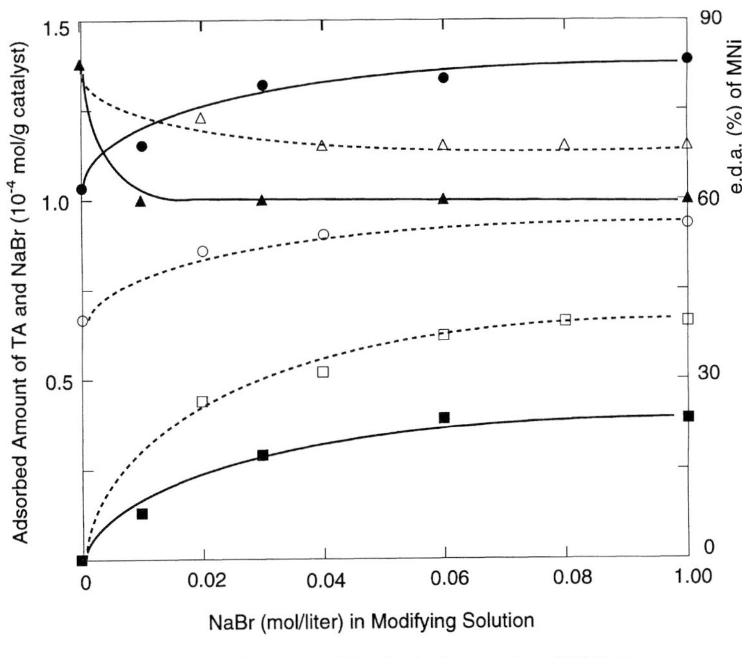

Figure 8.7. Effect of NaBr concentration of modifying solution on amount of adsorbed NaBr, adsorbed TA, and e.d.a. of the resulting MRNi. Selected data from Ref. [11, 13].

Table 8.4. Enantio-differentiating hydrogenation of MAA over TA-inorganic salt-MRNi.[a]

Inorganic salt	Amount of salt in modifying solution (100 ml containing 1 g TA)	E.d.a (%)
None	–	39
NaBr	10	83
NaF	3	61
NaCl	10	72
NaI	$5 \cdot 10^{-4}$	51
Na_2SO_4	10	56
LiBr	10	62

[a] Reaction conditions: 11.5 ml MAA, 0.9 g MNi, 23 ml MP, 0.2 ml AcOH, initial H_2 pressure: 90 kg/cm^2, 100 °C.

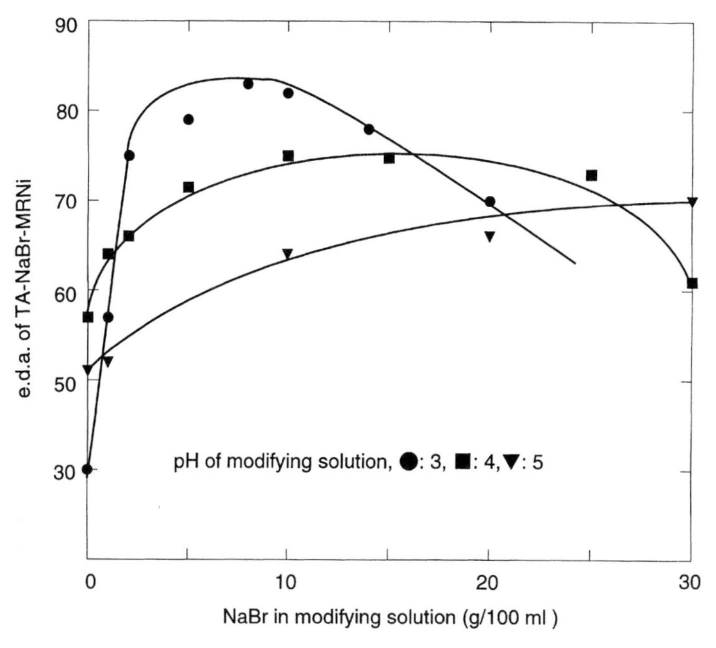

Figure 8.8. Effect of the amount of NaBr in modifying solution and modifying pH on e.d.a. of TA-NaBr-MNi. Modifying temperature 100 °C. Reprinted by permission from Ref. [11].

Some results for the one pot TA-NaBr-modification of RNi are shown in Fig. 8.8 [11]. An optimum e.d.a. (83%) is reached when RNi, prepared from 2 g Ni-Al alloy (Ni/Al = 42/58), is modified at 100 °C for 1 h with a 100 ml solution of 1 g TA and 10 g NaBr, adjusted to pH 3.2 with a 1 M NaOH solution (Fig. 8.8). The same e.d.a. value was obtained by TA-NaBr-modification of RNiA. By the nineties, TA-NaBr-MRNi was the best MNi catalyst. The TA-NaBr-modification is also effective for HNi [11] and for various Ni/Sup catalysts [45, 49, 50].

Mechanical treatments

As an additional approach, mechanical elimination of the N-region from RNi is promising, since the N-region is expected to contain fragile, disordered Ni domains with substantial amounts of residual Al. Ultrasonic irradiation was successfully applied to remove unfavorable parts from the RNi [51, 52]. A freshly prepared RNiH was suspended in water, and was subjected to ultrasonic irradiation (48 KHz, 600 W, 5 min), after which the turbid supernatant was removed. The same treatment was repeated until the supernatant became transparent. SEM photographs of the original RNi (a), ultrasound-irradiated RNi (RNiU) (b), and the supernatant materials (c) are shown in Fig. 8.9. In the original RNi, small particles (less than 1 μm) and large particles (10–20 μm) form an aggregate-like cluster. Ultrasonic irradiation crushes the cluster and reduces the amount of small particles. Thus, the resulting RNiU mostly consists of medium-sized particles (5–10 μm) with a smooth Ni surface. The solid in the supernatant is a mixture of small particles and rather large plates with a rough surface, and is estimated to contain ≈85% Al. The similar X-ray diffraction patterns of RNi and RNiU indicate that the ultrasonic irradiation does not change the crystallite size but merely homogenizes the particle size distribution, with removal of Al-enriched small particles.

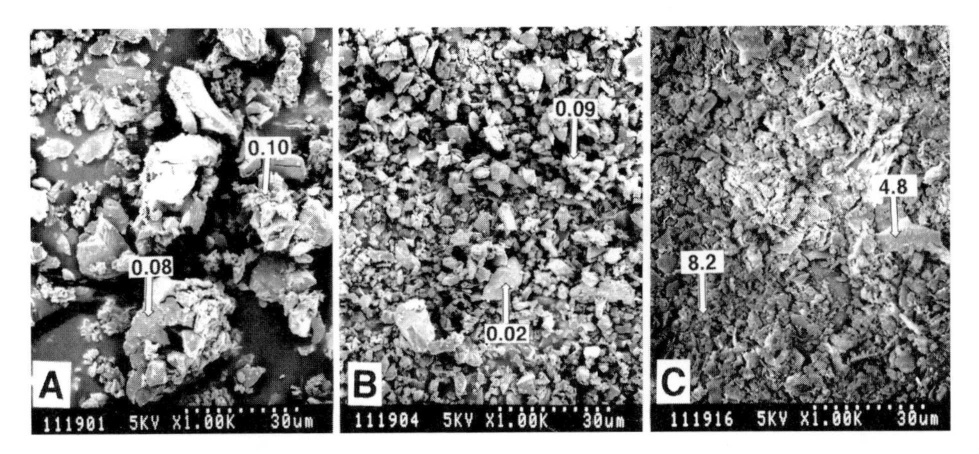

Figure 8.9. SEM photographs and EXD data of freshly prepared RNi (A), RNiU (B), and materials in supernatant (C); figures in the photograph are ratio of peak height (Al-K_a/Ni-K_a) at the point indicated by an arrow. Reprinted by permission from Ref. [51].

Under optimum modifying conditions, TA-NaBr-modified RNiU gives 86% e.d.a. in the hydrogenation of MAA, with an activity 3–8 times higher than that of conventional TA-NaBr-MRNi.

Parallel to our efforts on the RNi catalyst, Osawa and coworkers have recently prepared SPC-type Ni zeolites. These materials should contain pure Ni particles with a reasonable hydrogenation activity. Upon TA-NaBr-modification, they also reach an e.d.a. of 86% [53]. The progress of the e.d.a., as guided by the catalyst region model, is summarized in Table 8.5.

Table 8.5. Improvement of MNi monitored by the stereo-differentiating hydrogenation of MAA.[a]

Catalyst	E.d.a. (%)	Hydrogenation activity
TA-MRNi	52	high
TA-MRNiA(GA)	72	high
TA-MHNi	80	fair
TA-NaBr-MRNi	83	high
TA-NaBr-MRNi-U	86	very high
TA-NaBr-MNi/zeolite	86	high

[a] Reaction conditions: 10 g MAA, 0.9 g MNi, 20 ml THF, 0.1 ml AcOH, initial H_2 pressure: 100 kg/cm^2, 100 °C.

8.3.1.3 Catalytic Stability of MRNi

The durability of a catalyst and its recyclability for consecutive runs are important from a practical point of view. TA-NaBr-MRNi almost loses its e.d.a. after two runs. The hydrogenation activity also decreases, but not so drastically (Fig. 8.10). Thus, the deterioration of the catalyst seems attributable to the loss of modifier.

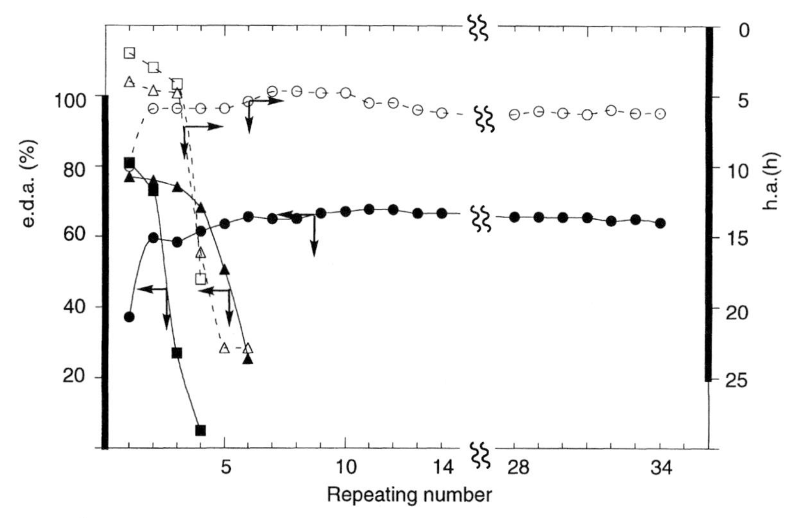

Figure 8.10. Durability of SR/MRNi's for repeated use. a) Hydrogenation activity determined by the time required to complete the hydrogenation of MAA (11.5 ml) over SR/MNi (0.8 g equivalent of Ni) in the presence of 0.2 ml of AcOH. Selected data from Ref. [54, 55].

In order to protect the modifying reagent, MNi was enveloped in a silicone rubber (SR) which vulcanizes at room temperature. This rubber has a high H_2 permeability and a moderate permeability for MAA. The results of the durability tests for TA-NaBr-MRNi embedded in an oxime type SR (OX-SR), and an acetic acid type SR (AC-SR) are shown in Fig. 8.10. The TA-NaBr-MRNi embedded in OX-SR more or less maintains its hydrogenation activity and a moderate e.d.a. (60–64%) during a long series of runs. Moreover, this catalyst has a long shelf life. After four months of storage in air, no changes are observed in its performance. In contrast, embedding in AC-SR was not beneficial, except for a good preservability [54, 55]. Activity and e.d.a., however, quickly vanished.

An explanation for such differences is found in the fact that during the curing process, OX-SR releases methyl ethyl ketone oxime, which is readily hydrogenated to 2-butylamine under reaction conditions. The constant supply of trace amounts of amine to the catalyst is assumed to maintain the activity of the catalyst. Starting from this assumption, the effect of amines on the reusability of TA-NaBr-MNi was examined. As shown in Fig. 8.11, treatment of the catalyst with trace amounts of amine effectively improved the reusability of the catalyst, without major effects on the e.d.a. of TA-NaBr-MRNi. In all cases, the catalyst maintained its original activity until about 10 recycles. Among the amines examined, the best results were obtained with heterocyclic amines. The productivity in the hydrogenation of MAA was improved more than ten times by acridine or pyridine treatment [56]. How these amines interact with the E-region of MNi is not yet clear.

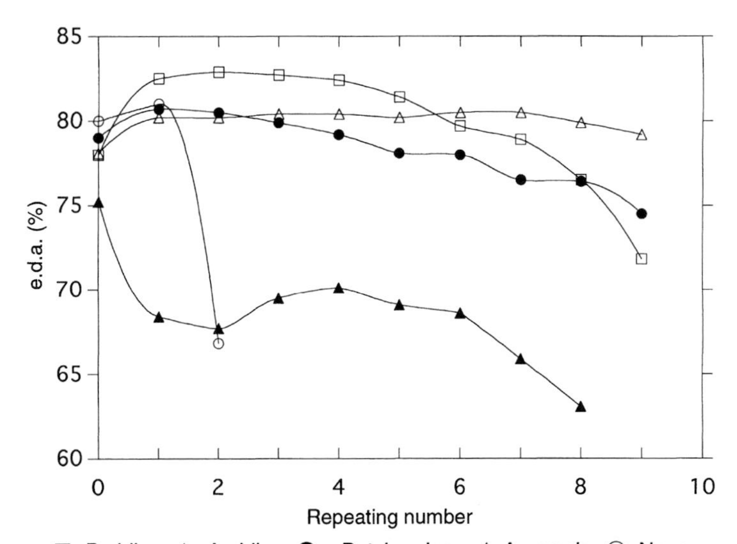

Figure 8.11. Change of e.d.a. of various amine-treated TA-NaBr-MRNi in repeated uses. Selected data from Ref. [56].

8.3.2 Enantiodifferentiation and Hydrogenation Steps in the Reaction Path (Reaction Process Model)

As discussed in Section 8.2.2, kinetics suggest that the e.d.a. is not determined in the rate-determining step; the enantiofaces of the substrate are probably differentiated in an adsorption step prior to the rate-determining step. To obtain direct evidence for an interaction between the substrate and the modifying reagent in the adsorption step, the hydrogenation of methyl 2-methyl-3-oxobutanoate (**1**) was attempted [57]. The configuration at the C-2 position of **1** is highly susceptible to racemization. However, as soon as **1** is hydrogenated to methyl 2-methyl-3-hydroxybutanoate (**2**), no further epimerization occurs at the C-2 position. Hence, the stereochemical changes at C-2 provide an internal probe for the detection of the stereodifferentiating step. The results of the hydrogenation of **1** with (*R*,*R*)-TA-MNi are shown in Fig. 8.12. Although racemic **1** is used as a starting product, the (2*S*)-isomers (2*S*,3*R*-**2** + 2*S*,3*S*-**2**) are formed in excess compared to the 2*R*-isomers (2*R*,3*R*-**2** + 2*R*,3*S*-**2**). This indicates that the interaction between **1** and (*R*,*R*)-TA leads to an excess of adsorbed (2*S*)-**1** on the catalyst surface. Since more (3*R*)-isomers (2*R*,3*R*-**2** + 2*S*,3*R*-**2**) are formed than (3*S*)-isomers (2*S*,3*S*-**2** + 2*R*,3*S*-**2**), the carbonyl group of **1** must be hydrogenated by predominant *si* face H_2-attack, in the same manner as in the MAA hydrogenation on (*R*,*R*)-TA-MNi. More generally, it was found that an excess of (3*R*)-isomer is accompanied by formation of the (2*S*) configuration. These results indicate that the configurations at the C-2 and C-3 positions of **2** are determined simultaneously, in an adsorption step that precedes the rate-determining hydrogenation step.

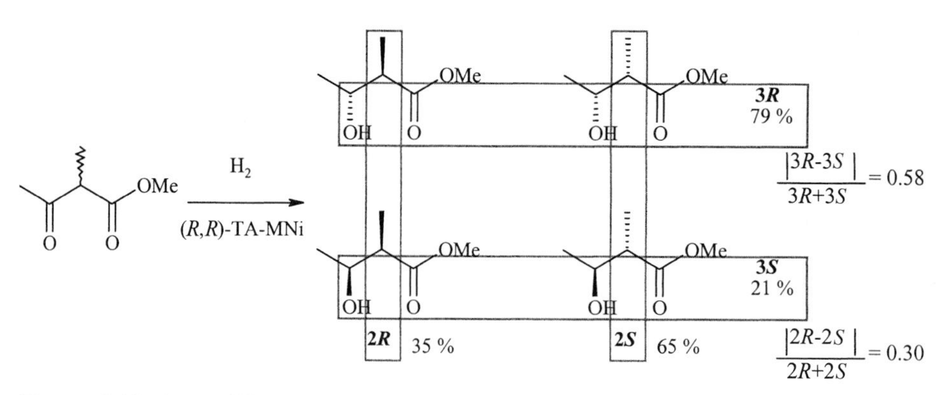

Figure 8.12. Stereodifferentiating hydrogenation of methyl 2-methyl-3-oxobutanoate. Selected data from Ref. [57].

Thus enantioface differentiation of the substrate occurs in an adsorption step, which is followed by a rate-determining hydrogenation step. This hypothesis, which is based on stereochemical and kinetic facts, is the focus of the "reaction process model", as illustrated in Fig. 8.13. An important implication is that the activated complex of the rate-determining step need not be considered in order to explain or improve the stereochemistry of the reaction.

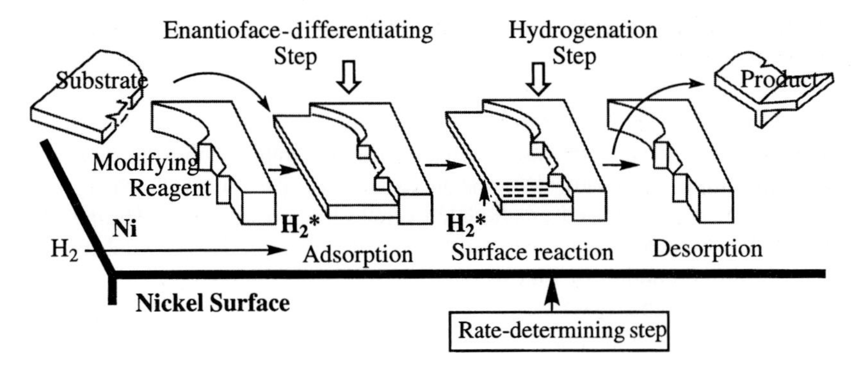

Figure 8.13. Schematic illustration of "reaction process model".

8.3.3 Interaction between Substrate and TA on MNi (Stereochemical Model)

As mentioned in Section 8.2.3, the adsorbed TA is present on the Ni surface as TA^{2-} with less than a monolayer coverage. Early studies on the hydrogenation of MAA with TA-MRNi indicated that the addition of extra TA to the reaction decreased the e.d.a., and that there is a linear relationship between the optical purity of the TA employed for modification and the e.d.a. of the resulting MNi [58]. Yasumori studied the thermal desorption of MAA from TA-MDNi and unmodified DNi, and proposed that hydrogen bonds are formed between MAA and the TA modifier [42].

In early studies, Sachtler [59] and Yasumori [42] intuitively proposed that a good e.d.a. in the MAA hydrogenation might be related to the facility of enol formation during the reaction. However, methyl *α,α*-dimethylacetoacetate, which cannot exist as an enol form, was later found to give almost the same e.e. as MAA. This means that the enolized form is not important for the enantioface differentiation of MAA [3]. For reactions of cyclic *β*-ketoesters, Bartók and coworkers found that the molecule largely takes part in the reaction in its enol form [60]. However, since the reported e.d.a.'s are very low, it seems that the contribution of the enol actually decreases the e.d.a. Therefore, our stereochemical model considers the keto form of MAA and related compounds as a starting point.

8.3.3.1 Stereochemical Model Based on the Interaction between TA and MAA through Two Hydrogen Bonds (2P Model)

This model was first proposed to explain the distribution of diastereomers in the hydrogenation of methyl 2-methyl-3-oxobutanoate (**1**) over TA-MNi (Section 8.3.2; Fig. 8.13) [57]. Later, it was generalized to understand the stereochemistry of the hydrogenation of MAA and its analogs, based on the information listed in Tables 8.6 and 8.7 [61]. In Table 8.6, e.d.a.'s are collected for reactions with MRNi catalysts prepared from TA or its analogs and NaBr. Of the modifiers examined, only TA gave more than 80% e.d.a. in the hydrogenation of MAA (Table 8.6, entry 1). When one of the hydroxyl groups of TA is replaced by hydrogen, a benzoyloxy or a methoxy group, an appreciable decrease in e.d.a. is observed (Table 8.6, entries 2, 3, 4). The e.d.a.'s for this group of modifying reagents were around 60%. When one of the carboxylate groups of TA is replaced by H, or by a methyl group, the e.d.a. is completely lost (Table 8.6, entries 5, 6), even if the compound possesses vicinal hydroxy groups. Apparently, the interactions of the two carboxyl groups with the Ni surface are essential to immobilize the modifying reagent on the surface and to create a uniform chiral environment on the catalyst. A drastic decrease in e.d.a. was also observed when both hydroxy groups of TA were acylated or alkylated (Table 8.6, entries 7, 8). The results clearly show that two hydroxy groups and two carboxyl groups in TA are vital for effective enantioface differentiation of MAA.

In the hydrogenation of the methyl esters of α-, β-, γ- and δ-keto acids, and 2-hexanone, an excellent e.d.a. was obtained only for the β-ketoester (MAA) (Table 8.7) [61]. Thus, the distance between the two carbonyl groups in the substrate is critical. If one considers the distance between the two hydroxy groups in TA, it is striking that only MAA can be bound with its two carbonyl groups *via* two hydrogen bonds to the two hydroxyl groups of adsorbed TA. Such a precise fit is not possible with the other substrates. In the case of 2-hexanone, which can form only one hydrogen bond with TA, the stereochemistry of the product was *S*, in contrast to the *R* configuration observed for the ketoesters. This issue will be discussed in Section 8.3.3.2.

The model depicted in Fig. 8.14 a is based on formation of two hydrogen bonds between the two hydroxyl groups of (*R*,*R*)-TA and the two carbonyl groups of MAA (2P model). Construction of a space-filling CPK model reveals that one of the hydroxyl groups of TA comes close to the catalyst surface (site 1), whereas the second hydroxyl group is somewhat remote from the surface (site 2). In Fig. 8.14 a, the carbonyl group of MAA to be hydrogenated is fixed on site 1, and comes within 0.1 nm from the surface with its *si*-face towards the catalyst. When MAA is adsorbed in this

Table 8.6. The e.d.a.'s of TA (and analogs)-modified RNi's determinated by the hydrogenation of MAA. [a]

		Modifying reagent [b]				Configuration of Products	E.d.a. (%)
		X_1	X_2	Y	Name of compound		
1	COOH	–OH	–OH	–COOH	(*R*,*R*)-TA	*R*	83
2		–OH	–H	–COOH	(*S*)-Malic acid (MA)	*S*	61
3	H–┼–X^1	–OH	–OCOPh	–COOH	(*R*,*R*)-O-Benzoyl-TA	*R*	65
4		–OH	–OMe	–COOH	(*R*,*R*)-O-Methyl-TA	*R*	63
5	X_2–┼–H	–OH	–OH	–Me	(2*S*,3*R*)-2,3-Dihydroxybutyric acid	*S*	1.2
6	Y	–OH	–OH	–H	(*S*)-Glyceric acid	–	0.0
7		–OCOPh	–OCOPh	–COOH	(*R*,*R*)-O,O′-Dibenzoyl-TA	*R*	8
8		–OMe	–OMe	–COOH	(*R*,*R*)-O,O′-Dimethyl-TA	*R*	0.2

[a] Reaction conditions: 11.5 ml MAA, 0.9 g MNi, 23 ml MPr, 0.2 ml AcOH, initial H_2 pressure: 90 kg/cm^2, 100 °C.

[b] Modifying condition: in a solution of 1 g TA, 10 g NaBr and 100 ml H_2O adjusted to pH 3.2 with aq. NaOH, 0.9 g RNi was soaked for 1 h at 100 °C.

Table 8.7. Enantio-differentiating hydrogenation of prochiral ketones with TA-NaBr-MRNi.[a]

	Substrate	Catalyst and solvent[b]	Configuration of product	E.d.a. (%)
1		(*R,R*)-TA-NaBr-MRNi MPr/AcOH	*R*	2
2		(*R,R*)-TA-NaBr-MRNi MPr/AcOH	*R*	83
3		(*R,R*)-TA-NaBr-MRNi MPr/AcOH	*R*	38
4		(*R,R*)-TA-NaBr-MRNi MPr/AcOH	*R*	0
5		(*R,R*)-TA-NaBr-MRNi THF/AcOH	*S*	28

[a] Reaction conditions: 0.09–0.10 mol substrate, 23 ml solvent, 0.2 ml AcOH, 0.0 g MNi, initial H_2 pressure: 90 kg/cm^2, 100 °C.
[b] MPr = Methyl propionate.

fashion, it is ready to be hydrogenated by the supply of an active hydrogen atom from the Ni surface, leading to (*R*)-MHB (hydrogen attack from *si* face; Fig. 8.14b). When (*S,S*)-TA-MNi is employed as a catalyst, (*S*)-MHB is obtained (hydrogen attack from *re* face; Fig. 8.14c).

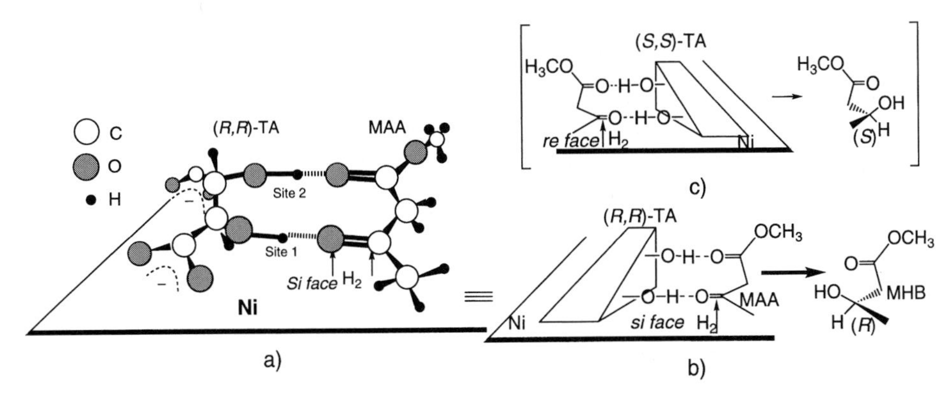

Figure 8.14. Schematic illustration of the interaction between MAA and TA adsorbed on the Ni surface through two hydrogen bonds "2P model". Selected data from Ref. [61].

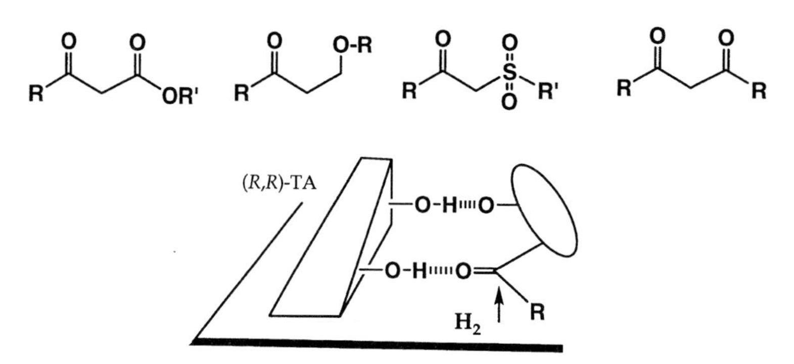

Figure 8.15. *β*-Functionalized prochiral ketones to which the "2P model" can be applied.

The 2P stereochemical model rationally explains the stereochemistry of the hydrogenation of MAA over TA-MNi. It also predicts that the TA-MNi system should be effective for the enantioselective hydrogenation of prochiral ketones with the structures shown in Fig. 8.15. As shown in Table 8.8, hydrogenation of these *β*-ketoesters and ketones, with an electronic structure similar to that of MAA, gives excellent e.d.a. values [10, 62, 63]. Consequently, these results suggest that the 2P model is valid for enantiodifferentiation of these molecules on the E-region of the catalyst. We have also developed practical methods to obtain optically pure *β*-hydroxy acids [63, 64], *β*-hydroxy sulfones [62], and 1,3-butanediol [10] from the corresponding reaction products listed in Table 8.8.

The double hydrogenation of acetylacetone (AA) or its analogs to 2,4-pentanediols (PD) is particularly suitable to test the 2P model for the TA-MNi system (Fig. 8.16) [65]. In this reaction, three stereodifferentiation mechanisms, i.e. enantioface, diastereoface, and enantiomer differentiation, contribute to increase the e.e. of the final product. In a first and fast step, the enantioface-differentiating reaction gives 4-hydroxy-2-pentanone (HP) in 74% e.e. When racemic HP is hydrogenated in the same conditions, diastereoface differentiation becomes important too. The *R*-enantiomer of HP is converted to (*R,R*)-PD in 80% diastereomer excess (d.e.), whereas the *S*-enantiomer of HP gives almost equal amounts of the (*R,S*) and (*S,S*) PD diastereomers (10% d.e.). In the former case, the diastereoface- and enantioface-differentiation work in the same direction, whereas in the latter case, they counteract each other. The overall product distribution obtained in the hydrogenation of AA is in excellent agreement with the distribution calculated based on the e.e. and d.e.'s of the individual reactions, as shown in Fig. 8.16.

In the second step of the AA hydrogenation, kinetic resolution of the two HP enantiomers is important, particularly if the reaction is stopped rather early. Indeed, the hydrogenation of racemic HP over (*R,R*)-TA-NaBr-MRNi indicated that the consumption of (*R*)-HP was much faster than that of (*S*)-HP. As a result, formation of (*R,R*)-PD is even more favored. Considering the material balance of the reaction in Fig. 8.16, the productivity for (*R,R*)-PD is mostly governed by the e.d.a. of the first step, while the e.e. of the (*R,R*)-PD can further be enhanced by kinetic enantiomer resolution, if the reaction is discontinued at an appropriate conversion. The major product of the reac-

Table 8.8. Enantiodifferentiating hydrogenation of prochiral ketones suitable for 2P model.[a]

	Substrate R =	Catalyst	Solvent[b]	Product E.e. (%)	Product Configuration
	$RCOCH_2COOMe \rightarrow RCH(OH)CH_2COOMe$				
1	CH_3–	(*R*,*R*)-TA-MRNi	neat	33	*R*
2		(*R*,*R*)-TA-NaBr-MRNi	MPr/AcOH	83	*R*
3		(*R*,*R*)-TA-NaBr-MRNiU	MPr/AcOH	86	*R*
4	CH_3CH_2–	(*R*,*R*)-TA-MRNi	Neat	66	*R*
5		(*R*,*R*)-TA-NaBr-MRNi	MPr /AcOH	87	*R*
6		(*R*,*R*)-TA-NaBr-MRNiU	MPr /AcOH	92	*R*
7	$CH_3(CH_2)_6$–	(*R*,*R*)-TA-NaBr-MRNi	MPr /AcOH	83	*R*
8		(*R*,*R*)-TA-NaBr-MRNiU	MPr /AcOH	89	*R*
9	$CH_3(CH_2)_{10}$–	(*R*,*R*)-TA-NaBr-MRNi	MPr /AcOH	86	*R*
10		(*R*,*R*)-TA-NaBr-MRNiU	MPr /AcOH	94	*R*
11	AcO-$(CH_2)_7$–	(*R*,*R*)-TA-NaBr-MRNiU	MPr /AcOH	85	*R*
	$CH_3COCH_2CH_2OR \rightarrow CH_3CH(OH)CH_2CH_2OR$				
12	H–	(*R*,*R*)-TA-MRNi	THF	27	*R*
13		(*R*,*R*)-TA-NaBr-MRNi	THF	70	*R*
14	CH_3–	(*R*,*R*)-TA-NaBr-MRNi	THF	68	–
	$RCOCH_2SO_2CH_3 \rightarrow RCH(OH)CH_2SO_2CH_3$				
15	CH_3CH_2–	(*R*,*R*)-TA-NaBr-MRNi	THF	71	–
16	$CH_3(CH_2)_4$–	(*R*,*R*)-TA-NaBr-MRNi	THF	68	–
17	$CH_3(CH_2)_7$–	(*R*,*R*)-TA-NaBr-MRNi	THF	67	*R*

[a] Reaction conditions: 0.10–0.05 mol substrate, 23 ml solvent, 0.2 ml AcOH, 0.9 g MNi, initial H_2 pressure: 90 kg/cm^2, 100 °C.
[b] MPr: Methyl propionate.

tion, (*R*,*R*)-PD, is readily separated from the *meso* form (*R**,*S**)-PD by recrystallization from ether at –20 °C. As the isolated crystals contain (*R*,*R*)-PD of 100% e.e., the isolated (*R*,*R*)-PD yield can be used as a measure for the stereo-differentiating ability of the catalyst. Results for the hydrogenation of AA and its homologues are listed in Table 8.9 [51, 53, 66, 67]. Particularly TA-NaBr-MRNiU is an excellent catalyst to obtain optically pure 1,3-diols with high chemical yield.

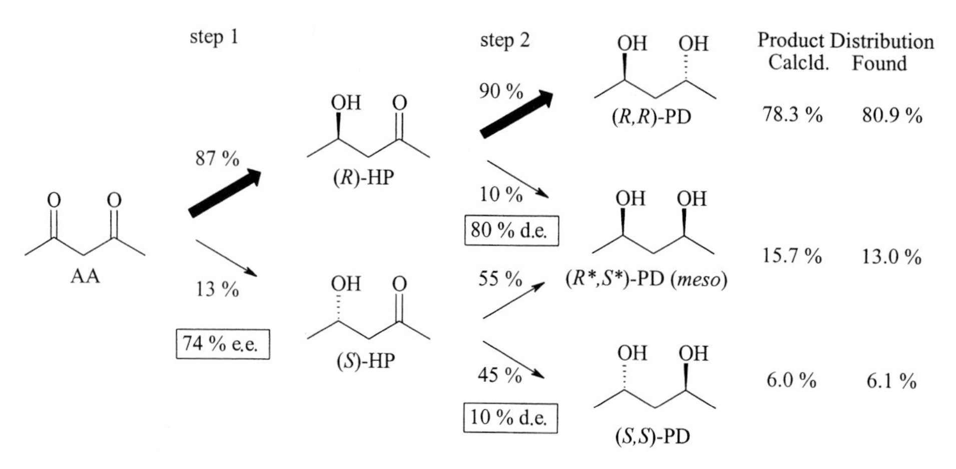

Figure 8.16. Stereodifferentiating hydrogenation of acetylacetone over TA-NaBr-MRNi. Selected data from Ref. [65].

Table 8.9. Stereodifferentiating hydrogenation of 1,3-diketones over MNi catalyst.

(1) —H_2→ (2) —H_2→ (3) (R, S) + (4) (R, R)

	Substrate (1) R =	Catalyst	Product composition				E.e. of 4 (%)	Yield (%) of optically pure 4 based on 1
			1	2	3	4		
1	$-CH_3$	(*R*,*R*)-TA-NaBr-MRNi	0	0	13	87	86	41 (*R*,*R*)
2			0	20	10	70	90	21 (*R*,*R*)
3		(*R*,*R*)-TA-NaBr-MRNiU	0	7	7	86	91	60 (*R*,*R*)
4	$-CH_2CH_3$	(*R*,*R*)-TA-NaBr-MRNi	0	0	20	80	–	25 (*R*,*R*)
5	$-CH(CH_3)_2$	(*R*,*R*)-TA-NaBr-MRNi	1	17	16	66	85	32 (*S*,*S*)
6		(*R*,*R*)-TA-NaBr-MRNiU	0	6	22	72	90	59 (*S*,*S*)
7	$-(CH_2)_2CH_3$	(*R*,*R*)-TA-NaBr-MRNi	0	0	15	85	–	13 (*R*,*R*)
8	$-(CH_2)_5CH_3$	(*R*,*R*)-TA-NaBr-MRNi	0	0	20	80	–	20 (*R*,*R*)
9	–Ph	(*R*,*R*)-TA-NaBr-MRNi	0	0	23	77	–	20 (*S*,*S*)

8.3.3.2 Stereochemical Model Based on the Interaction between TA and Methyl Alkyl Ketones through One Hydrogen Bond and a Steric Repulsion (1P Model)

Hydrogenation of a series of prochiral ketones (Table 8.7) reveals that (*R,R*)-TA-MNi can also differentiate to some extent between the enantiofaces of simple ketones (Table 8.7, entry 5). In the case of 2-hexanone, the adsorption can only be governed by the relative size of the methyl and butyl groups connected to the carbonyl group. The hydroxy groups of TA are expected to form a highly polar region on the catalyst. The carbonyl group of 2-hexanone may well interact with one of the hydroxy groups of TA (site 1) in the same manner as MAA. Meanwhile, the other hydroxy group (site 2) will repel the large hydrophobic butyl group to the other side. Consequently, the small methyl group is accommodated in the inside cavity. This interaction model, comprising one hydrogen bond and a steric repulsion (1P model) is depicted in Fig. 8.17a. The model is consistent with the stereochemistry of the reaction, since a catalyst modified with (*R,R*)-TA leads to an excess of the (*S*)-isomer via *re*-face hydrogen attack [61].

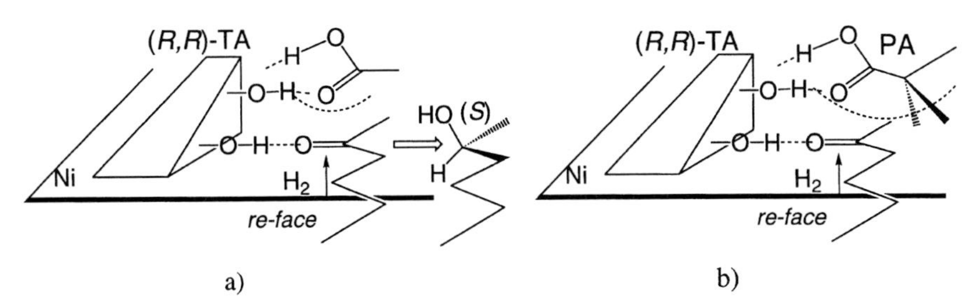

Figure 8.17. Schematic illustration of the interaction between 2-alkanone and TA adsorbed on the Ni surface through a hydrogen bond and steric repulsion; "1P model". Selected data from Ref. [61].

Initially, 2-alkanones were hydrogenated in the optimum conditions for MAA hydrogenation. The reaction mixture then also contains a small amount of acetic acid as an additive, which later turned out to be essential for the enantioselective 2-alkanone hydrogenation. A survey of the carboxylic acid additives was carried out in the 2-octanone hydrogenation [68]. As shown in Table 8.10, carboxylic acids with branching at the *α*-position were very effective in increasing the e.d.a. of the system. The best result is obtained with pivalic acid (PA) in a concentration more than twice that of the substrate [69]. The effect of PA can also well be rationalized by the 1P model, taking into account an association between PA and TA (Fig. 8.17b). PA associates with the hydroxy group of TA at site 2, forming a bulky, polar fence toward the long-chained alkyl group of the adsorbed substrate. As a result, the discrimination between methyl and alkyl groups becomes far more pronounced than in the presence of acetic acid [70].

While temperature has little effect on the e.d.a. of the MAA hydrogenation, its effect on the e.d.a. of the 2-octanone hydrogenation is significant (Fig. 8.18) [71, 72]. This is

Table 8.10. Effects of the additives in the reaction system on e.d.a. in the hydrogenation of 2-octanone.[a]

	Additive	Amount added (mol/20 ml solvent)	E.d.a. (%)
1	None	–	2
2	Acetic acid	0.05	33
3	Propionic acid	0.16	43
4	Isobutyric acid (iBuA)	0.16	57
5	Pivalic acid (PA)	0.11	60
6	1-Adamantanecarboxylic acid	0.11	53
7	(*S*)-2-Ethylhexanoic acid	0.11	50
8	(*R*)-2-Ethylhexanoic acid	0.11	51
9	TA	0.0001	1
10	3-Hydroxybutanoic acid	0.0001	3
11	t-Butanol	0.11	3
12	Triethyl amine	0.11	9

[a] Reaction condition: 0.8 g (*R,R*)-TA-NaBr-MRNi, 8.2 g substrate, 20 ml THF and the additives, initial H_2 pressure: 90 kg/cm^2, 100 °C.

not unexpected: while a single hydrogen bond may make the interaction loose with an increase in temperature, multiple hydrogen bonds are expected to be less sensitive to temperature. Thus, the different temperature dependences of the MAA and 2-octanone hydrogenation clearly reflect the distinction between the 2P and 1P model.

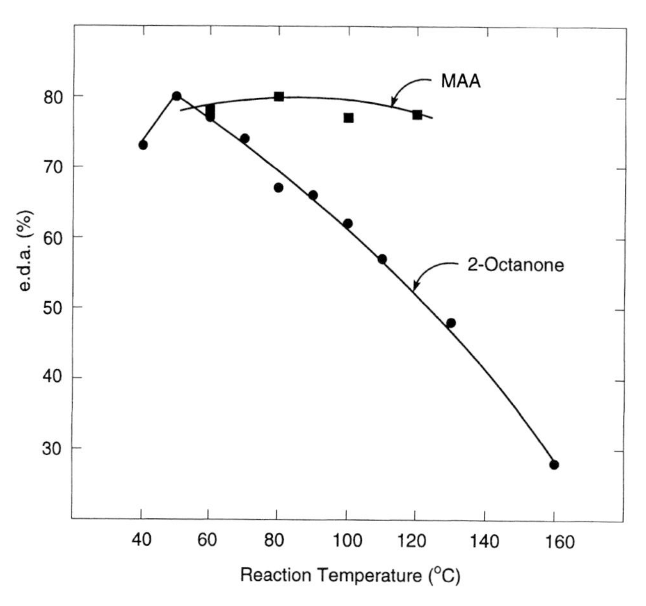

Figure 8.18. Relation between e.d.a. and the reaction temperature for the hydrogenation of MAA and 2-octanone. Selected data from Ref. [69, 70].

The e.d.a.'s for 2-octanone hydrogenation with various MNi's are listed in Table 8.11. The best catalyst for MAA hydrogenation always brings about the best e.d.a. in the alkanone reaction. This proves that MAA and 2-alkanone hydrogenation occur at the same E-region of the catalyst [71].

Table 8.11. Hydrogenation of 2-octanone over various kinds of MRNi. [a]

	Catalyst	E.d.a. (%)	Configuration
1	(*R,R*)-TA-MRNi	38	*S*
2	(*R,R*)-TA-NaBr-MRNi	51	*S*
3	(*S*)-MA-MRNi	10	*R*
4	(*R,R*)-O-Methyl-TA-MRNi	8	*S*
5	(*R,R*)-O,O′-Dimethyl-TA-MRNi	2	*S*

[a] Reaction conditions: 0.8 g MNi, 8.2 g 2-octanone, 10.9 g PA, 20 ml THF, initial H_2 pressure: 90 kg/cm^2, 100 °C.

Results for various 2-alkanones are collected in Table 8.12. Almost the same e.d.a.'s are obtained for straight chain 2-alkanones as for 2-octanone (Table 8.12, entries 1–4). As expected from the 1P model, branching in α of the carbonyl group increases the e.d.a. up to 85% (Table 8.12, entry 5). For 2-butanone, the e.d.a. is 60% under these conditions (Table 8.12, entry 1), and it can be raised to 72% by further tuning of the reaction conditions [72]. This is the first example of an effective differentiation between methyl and ethyl groups in an enantiodifferentiating reaction, with the exception of enzymatic reactions. We have also developed simple methods for the further optical enrichment of the produced 2-alkanols [69].

Table 8.12. Enantiodifferentiating hydrogenation of 2-alkanone over (*R,R*)-TA-NaBr-MRNi. [a]

	Substrate $RCOCH_3$ (R-=)	E.d.a. (%) Hydrogenation temperature	
		100 °C	60 °C
1	CH_3CH_2–	49	63
2	$CH_3(CH_2)_3$–	66	80
3	$CH_3(CH_2)_5$–	66	80
4	$CH_3(CH_2)_{10}$–	65	75
5	$CH_3CH(CH_3)$–	63	85
6	$CH_3CH(CH_3)CH_2$–	–	75
7	$CH_3CH(CH_3)CH_2CH_2$–	–	77
8	$CH_3CH(CH_3)CH_2CH_2CH_2$–	–	66

[a] Reaction conditions: 1.6 g MNi, 10 g substrate, 15 g PA, 20 ml THF, initial H_2 pressure: 90 kg/cm^2.

For 3-alkanone hydrogenation on TA-MNi, the same stereochemistry is observed as for 2-alkanones (Table 8.13) [73, 74]. The results are listed in Table 8.13. Reactions are slower with 3-alkanones than with β-ketoesters or 2-alkanones, and generally require a high reaction temperature (100–120 °C). As to the base catalyst, commercially available fine Ni powder (20 nm mean particle size) activated at 300 °C under H_2 gave the best results upon TA-NaBr modification (TA-NaBr-MFNiP). RNi was not favorable for this reaction, probably because of instability in the reaction conditions. Addition of carboxylic acids is also essential for this hydrogenation. Highly bulky acids such as 1-adamantanecarboxylic acid (AdA) or 1-methyl-1-cyclohexanecarboxylic acid (McA) give better results than medium-sized acids such as PA or isobutyric acid (iBuA). While PA or iBuA can self-associate to form dimers or polymers, the alkyl groups in AdA or McA are too large to allow such associations. Thus, PA at site 2 of TA is expected to be an almost polymeric cluster, which forms a mobile, large steric fence. At contrast, association of the monomeric acids AdA and McA with TA produces a rigid but rather slimmer fence, so that even an ethyl group can be accommodated inside the cavity. Consequently, the 1P stereochemical model can also explain the enantioface differentiation of 3-alkanones.

Table 8.13. Enantiodifferentiating hydrogenation of various 3-alkanones over TA-NaBr-MFNiP.[a]

	Substrate $R\text{-}COCH_2CH_3$	Carboxylic acid	E.d.a. (%)	Configuration of product
1	$R=CH_3(CH_2)_6-$	PA	23	*S*
2		McA	44	*S*
3	$R=CH_3(CH_2)_4-$	none	6	*S*
4		AcOH	22	*S*
5		iBuA	29	*S*
6		PA	30	*S*
7		McA	40	*S*
8		AdA	42	*S*
9	$R=CH_3(CH_2)_3-$	PA	17	*S*
10		McA	41	*S*
11	$R=CH_3(CH_2)_2-$	PA	2	*S*
12		McA	25	*S*
13	$R=(CH_3)_2CHCH_2-$	PA	15	*S*
14		McA	32	*S*

[a] Reaction conditions: 0.8 g MNi, 32 mmol substrate, 77 mmol carboxylic acid, 10 ml THF, initial H_2 pressure: 90 kg/cm^2, 100 °C.

8.3.3.3 Extended Stereochemical Model: Merging the 2P and 1P Models

Study of the 2P and 1P models teaches that both enantiodifferentiating reactions take place at the same region and site on the catalyst. Their relative contribution depends on the type of substrate. In order to prove this idea and to unify the two models, the enantiodifferentiating hydrogenation of a series of ketoesters and 2-octanone was reexamined with TA-NaBr-MRNiU, in the absence and presence of PA. The results are shown in Fig. 8.19. In the absence of PA, the reaction of the β-ketoester leads to the highest (*R*)-excess. The e.d.a. gradually decreases in going over the γ-ketoester to the δ-ketoester. Eventually, with the ε-ketoester, a slight excess of the (*S*)-configuration is obtained. Addition of a large amount of PA brings characteristic changes to the e.d.a. for all substrates. Lower (*R*)-excesses are obtained with PA for the β- and γ-ketoesters, while appreciable (*S*)-excesses are formed from the δ-ketoester, the ε-ketoester as well as from 2-octanone. The e.d.a. for the δ-ketoester and ε-ketoester has the same temperature dependence as that for 2-octanone. From these observations, it can be deduced that the relative contributions of 2P and 1P in the absence of PA are: 2P >> 1P for β-ketoester, 2P > 1P for γ-ketoester, 1P=2P for δ-ketoester, and 1P > 2P for ε-ketoester. Addition of PA evidently decreases the contribution of the 2P mode in favor of the 1P mode. As an example, the relative contributions of 2P and 1P in the hydrogenation of γ- and δ-ketoester in the absence and presence of PA are schematically illustrated in Figs. 8.20 a and 8.20 b, respectively.

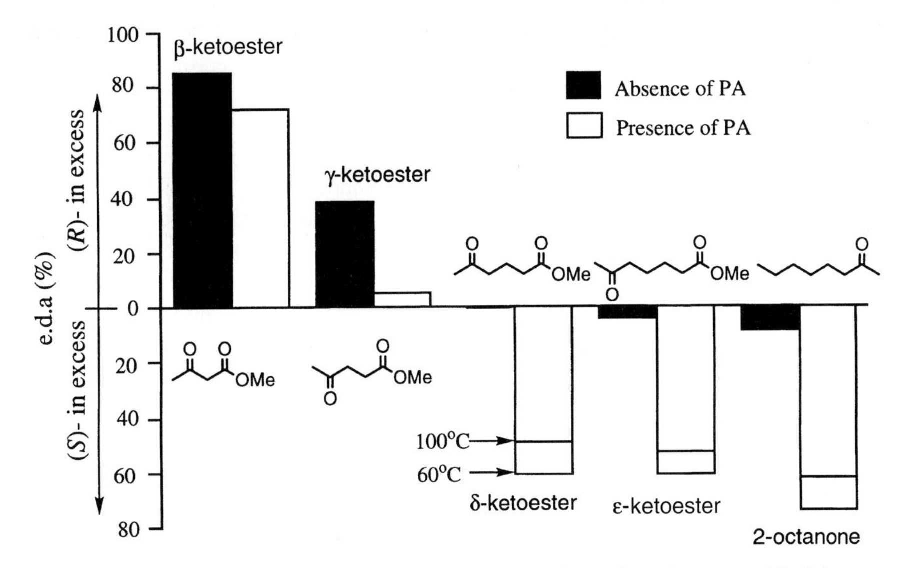

Figure 8.19. Enantioface-differentiating hydrogenation of various prochiral ketones over TA-NaBr-MRNiU in the absence and presence of PA in the system. Selected data from Ref. [75].

The results mentioned above brought about a new hypothetical model named "extended stereochemical model". This model states that the e.d.a. of the hydrogenation of ketoesters is determined by the relative contributions of 1P and 2P, which depend on the degree of structural matching between the substrate and TA to create the 2P adsorbed state. In view of the extended stereochemical model, the intrinsic e.d.a. *i*, as proposed in equation (1), may be replaced by the intrinsic e.d.a. resulting from the 1P and 2P contributions. The intrinsic e.d.a. is eventually determined by the interaction between the modifying reagent and substrate [75].

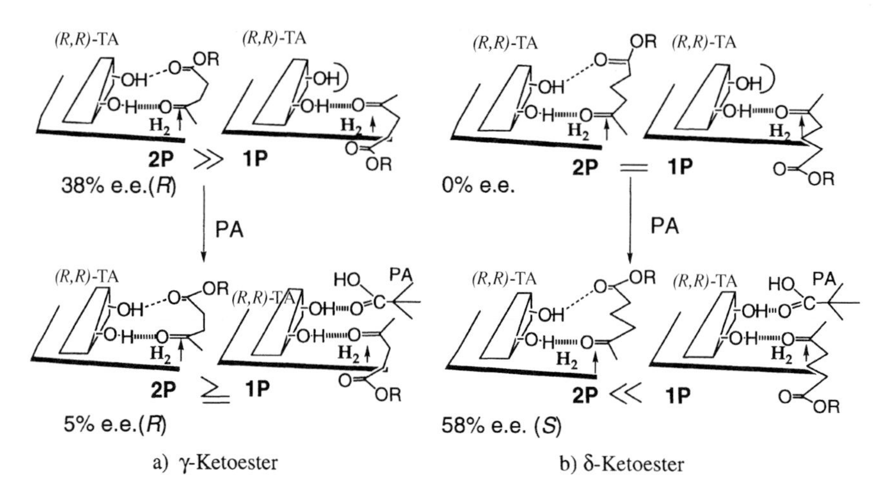

Figure 8.20. Schematic illustration of the relative contribution of 2P and 1P in the hydrogenation of δ- and γ-ketoesters, and the effect of PA in the system. Reprinted by permission from Tai A, *Shokubai (Catalysts & Catalysis)*, **41**, 270 (1999).

In early work, it was observed that the enantiodifferentiating hydrogenation of higher homologues of MAA such as methyl 3-oxopentanoate, methyl 3-oxodecanoate, and methyl 3-oxotetradecanoate afforded better e.e.'s, with a slower reaction rate than for MAA (Table 8.8). The increased e.d.a. is not easily explained based on the 2P model alone, but becomes clear with the extended stereochemical model. Figure 8.21 shows sketches of the 2P and 1P adsorption modes for (a) MAA and (b) its long-chain homologue. For MAA, a good 2P fit is expected, but the 1P adsorption is also possible because of the small size of the methyl group at the acyl side. For the long-chain homologues, the contact between site 2 and the long alkyl group at the acyl side impedes a 1P adsorption mode. This increases the contribution of the 2P mode and eventually gives a higher (*R*)-excess than for MAA. Even the rate differences between MAA and its long chain analogues can be rationalized. The molecular size of MAA is almost the same as that of the active site provided by adsorbed TA, so that almost all sites can simultaneously interact with MAA. At contrast, adsorption of a higher β-ketoester with its long alkyl chain may block access to adjacent sites. As a result, the overall rate is higher for MAA than for long-chain β-ketoesters.

The extended stereochemical model also indicates that MAA – which has long been considered to be the standard β-ketoester – may not be the best substrate for TA-MNi. For MAA/TA-MNi, the *i* factor never reaches 100% because of a small 1P contribution. β-Ketoesters carrying a relatively large alkyl group at the acyl side should be more favored substrates due to steric repulsion, as previously noted for e.g. methyl 3-oxodecanoate. Of course, if steric repulsion plays some role, the e.d.a. is expected to be temperature-dependent. Accordingly, a series of β-ketoesters was hydrogenated over TA-NaBr-MRNiU at 60 and 100 °C (Table 8.14). TA-NaBr-MRNiU is an excellent catalyst for low-temperature hydrogenation, because it is highly active.

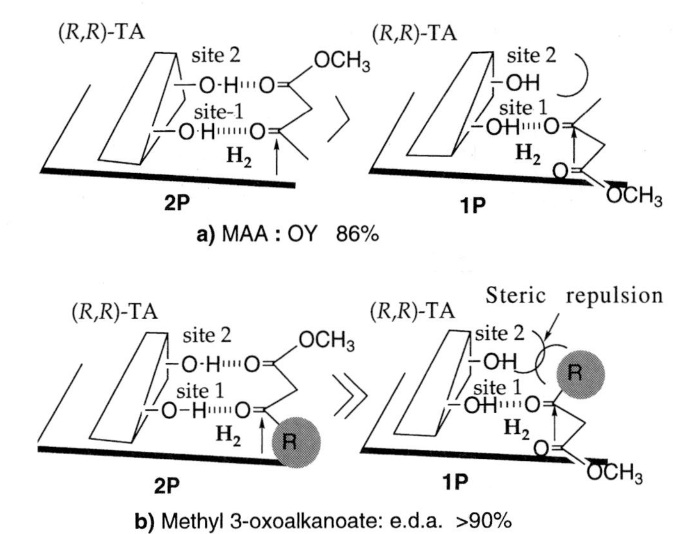

Figure 8.21. Schematic illustration of the relative contribution of 2P and 1P in the hydrogenation of MAA and its homologs, and effect of PA in the system. Reprinted by permission from Tai A, *Shokubai (Catalysts & Catalysis)*, **41**, 270 (1999).

Except for the acetoacetates, all substrates gave a systematically higher e.d.a. at 60 °C than at 100 °C (Table 8.14, entries 4–12). For the acetoacetates (Table 8.14, entries 1–3), neither the type of alkoxy group nor the temperature has much effect on the e.d.a., indicating that the alkoxy group is not located close to the Ni surface. In going from a methyl group at the acyl side (MAA) to an ethyl group (methyl 3-ketopentanoate), the e.d.a. increases (Table 8.14, entries 1 *vs.* 4). Further increase of the chain length at the acyl side has little effect (Table 8.14, entries 5, 6). Branching of the alkyl group (isopropyl, cyclopentyl, or neopentyl) further increases the e.d.a. (Table 8.14, entries 7–9), but a too bulky alkyl group slows down the hydrogenation itself (Table 8.14, entries 10, 11). Thus, as predicted by the extended stereochemical model, an alkyl group of medium-size bulkiness gives the best e.d.a.'s. An outstanding result is the 96% e.e. in the hydrogenation of methyl 4-methyl-3-oxo-pentanoate (Table 8.14, entry 7).

The isopropyl group of methyl 4-methyl-3-oxo-pentanoate may even be a little too large. Hence we decided to test a cyclopropyl group (*c*-Pr) which has a size between that of an ethyl and an isopropyl group. With methyl 3-cyclopropyl-3-oxopropanoate,

MeO (substrate) → (*R*,*R*)-TA-MRNiU / H_2, 60 °C, 24–48 h → MeO (product, OH)

100% yield
> 98% e.e.

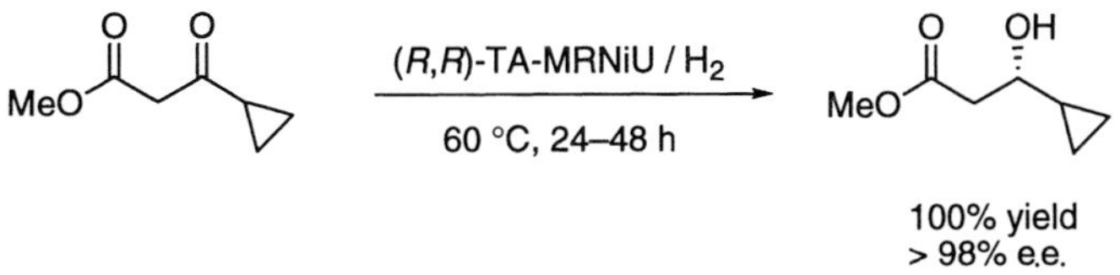

Figure 8.22. Almost perfect enantiodifferentiation in the hydrogenation of methyl 3-cyclopropyl-3-oxopropanoate over TA-MRNiU.

Table 8.14. Enantioface-differentiating hydrogenation of various β-ketoesters over TA-NaBr-MRNiU.[a]

	Substrate	T (°C)	E.d.a. (%)
1	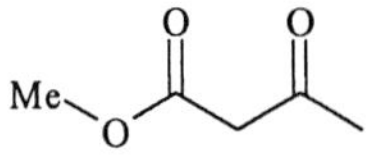 Me	100	86
		60	86
2	iPr	100	85
		60	87
3	tBu	100	84
		60	88
4	Me	100	91
		60	94
5	Me	100	90
		60	93
6	Me	100	87
		60	90
7	Me	100	88
		60	96
8	Me	100	90
		60	95
9	Me	100	84
		60	96
10	Me	100	80
			Slow reaction
11	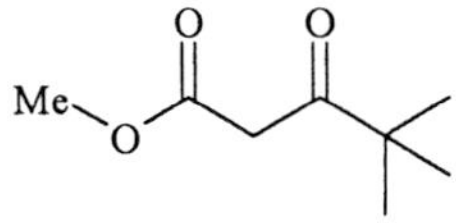 Me	100	No reaction
12	Me	100	95.9
		60	98.6
		40	98.2

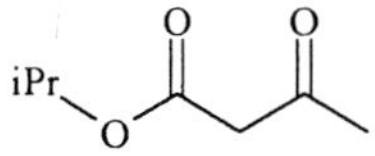

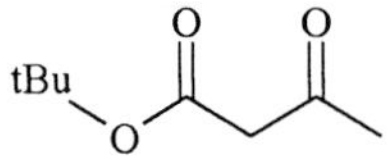

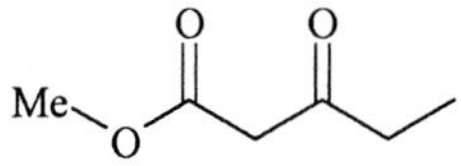

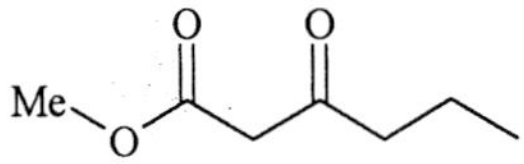

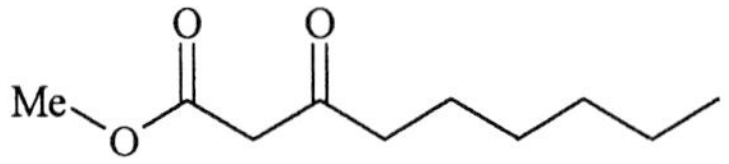

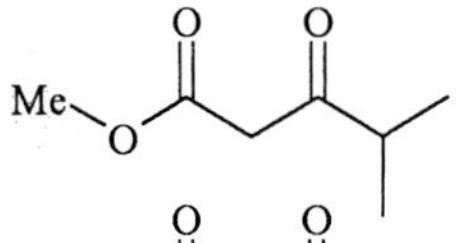

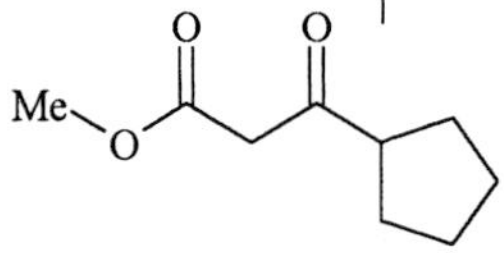

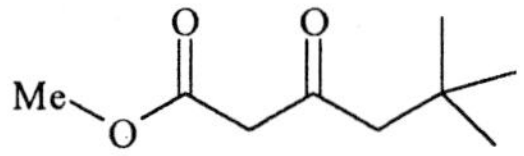

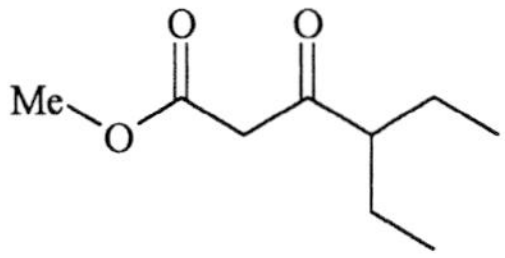

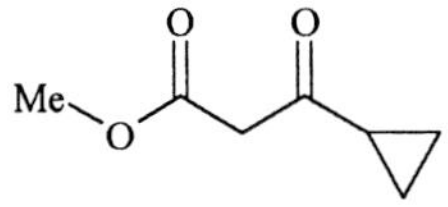

[a] Reaction conditions: 0.9 g MN, 4–10 g substrate, 10 ml THF, 0.2 ml AcOH, initial H_2 pressure: 100 atm.

the hydrogenation proceeded smoothly, giving more than 98% e.d.a. (Table 8.14, entry 12) over TA-NaBr-MRNiU (Fig. 8.22) [76, 77]. Fortunately, the cyclopropyl group remains intact in the hydrogenation conditions. The optically active hydrogenation product is a useful synthetic block, since it can be converted to various functional groups by ring opening based on well-documented cyclopropane chemistry. An optical purity higher than 98% is an extremely favorable starting point for pharmaceutical and agrochemical synthesis, as further optical enrichment is superfluous.

8.3.4 Conclusions of the Model Studies

The above-mentioned e.e. of more than 98% strikingly illustrates that even a simple compound, such as TA adsorbed on Ni, can discriminate between the enantiofaces of a substrate with more than 99% accuracy. Therefore, the present study counters the *a priori* idea that a heterogeneous asymmetric catalyst cannot yield an excellent e.d.a., because the surface is not uniform, or because steric constriction of a stereo-controlling active site is hard to achieve on a solid. In terms of the "catalyst region model", the contribution of the N (non-enantiodifferentiating) regions in TA-NaBr-MRNiU is less than 2%. While we had this catalyst in hand for some time, it was the prediction based on the "extended stereochemical model" that led us to using the very best substrate. This, again, underlines the importance of sufficient tuning of the reaction system.

Reaction and stereochemical models have been proposed by several research groups [2c, 21, 24, 78, 79]. However, in comparison with our work, these models are less instructive, because they insufficiently incorporate information from organic stereochemistry. Moreover, our work allows isolation of optically pure compounds in high yields, which is the eventual goal of synthetic organic chemists.

In conclusion, our models allow simulating the functions of MNi, and predicting ways to improve the e.d.a. of the catalyst, without claiming to imply a real mechanism. Full understanding of MNi will require further significant advances in the physicochemical approach.

Acknowledgements

The authors thank Dr. Yoshiharu Izumi, Professor emeritus of Osaka University, for his encouragement throughout this study, and Professor Tadao Harada of Ryukoku University and Professor Tsutomu Osawa of Toyama University for their cooperation.

References

[1] Izumi Y, Imaida M, Fukawa H, and Akabori S, *Bull. Chem. Soc. Jpn*. **36**, 21 (1963).

[2] a) Izumi Y, *Angew. Chem., Int. Ed. Engl.* **10**, 871 (1971). b) Fish M J and Ollis D F, *Cat. Rev. Sci. Eng.*, **18**, 259 (1978). c) Izumi Y, *Adv. Cat.*, **32**, 215 (1983). d) Sachtler W M H, in Augusttine R L (ed.), Catalysis for Organic Reactions, Marcel Dekker, New York, 1985, p. 189. e) Tai A and Harada T, in Iwasawa Y (ed.), Tailored Metal Catalysts, D. Reidel, Dordrecht, 1986, p. 265. f) Webb G and Wells P B, *Catal. Today,* **12**, 371 (1992). g) Osawa T, Harada T, and Tai A, *Catal. Today*, **37**, 465 (1997).

[3] Smith G V and Musoiu M, *J. Catal.*, **60**, 184 (1979).

[4] Wittmann G G, Bartók B, Bartók M, and Smith G V, *J. Mol. Catal.* **60**, 1 (1990).

[5] Tanabe T, Okuda K, and Izumi Y, *Bull. Chem. Soc. Jpn.*, **46**, 514 (1973).

[6] Harada T, Onaka S, Tai A, and Izumi Y, *Chem. Lett.*, 1131 (1977).

[7] Hoek A and Sachtler W M H, *J. Catal.* **58**, 276 (1979).

[8] Keane M A and Webb G, *J. Catal.*, **136**, 1 (1992).

[9] Klabunovskii E I, Russ. *J. Phys. Chem.*, **47**, 765 (1973).

[10] Murakami S, Harada T, and Tai A, *Bull. Chem. Soc. Jpn.*, **53**, 1356 (1980).

[11] Harada T, Yamamoto M, Onaka S, Imaida M, Tai A, and Izumi Y, *Bull. Chem. Soc. Jpn.,* **54**, 2323 (1981).

[12] Nitta Y, Sekine F, Imanaka T, and Teranishi S, *Bull. Chem. Soc. Jpn.*, **54**, 980 (1981).

[13] Harada T, Tai A, Yamamoto M, Ozaki H, and Izumi Y, Proc. 7th Int. Congr. Catal., Tokyo, 364 (1980).

[14] Gross L H and Rys P, *J. Org. Chem.*, **39**, 2429 (1974).

[15] Harada T, Hiraki Y, Izumi Y, Muraoka J, Ozaki H, and Tai A, Proc. 6th Int. Congr. Catal., London, 1204 (1976).

[16] Nitta Y, Sekine F, Imanaka T, and Teranishi S, *J. Catal.*, **74**, 382 (1982).

[17] Nitta Y, Yamanishi O, Sekine F, Imanaka T, and Teranishi S, *J. Catal.*, **79**, 475 (1983).

[18] Hoek A and Sachtler W M H, *J. Catal.*, **58**, 276 (1976).

[19] Fu L, Kung H H, and Sachtler W M H, *J. Mol. Catal.*, **42**, 29 (1987).

[20] Osawa T, Mita S, Iwai A, Takayasu T, Harada T, and Matsuura I, in Delmon B, et al. (ed.) Preparation of Catalysts VII, Elsevier, Amsterdam, 1998, p. 313.

[21] Hoek A, Woerde H M, and Sachtler W M H, Proc. 7th Int. Congr. Catal., Tokyo, 376 (1980).

[22] Nitta Y, Kawabe M, and Imanaka T, *Applied Catal.,* **30**, 141 (1997).

[23] Nitta Y, Utsumi T, Imanaka T, and Teranishi S, *J. Catal.*, **101**, 376 (1986).

[24] Klabunovskii E I, Vedenypin A A, Karpeiskaya E I, and Pavlov V A, Proc. 7th Int. Congr. Catal., Tokyo, 390 (1980).

[25] Zubareva and N D and Klabunovskii E I, *Izv. Akad. Nauk SSSR., ser. Khim* 1172 (1988).

[26] Orito Y, Niwa S, and Imai S J. *Yukigouseikagaku (Syn. Org. Chem. Jpn).*, **34** 236 (1976).

[27] Izumi,Y, Imaida M, Harada T, Tanabe T, Yajima S, and Ninomiya T, *Bull. Chem. Soc. Jpn.,* **42**, 241 (1969).

[28] Tanabe T and Izumi Y, *Bull. Chem. Soc. Jpn.* **46**, 1550 (1973).

[29] Lipyart E N, Pavlov V A, and Klabunovskii E I, *Kinet. Katal.,* **12,** 1491 (1971).

[30] Bennett A, Christie S, Keane M A, Peacock R D, and Webb G, *Catal. Today*, **10**, 363 (1991).

[31] Yasumori I, Inoue Y, and Okabe K, in Delmon B (ed.), Catalysis, Heterogeneous, Homogeneous, Elsevier, Amsterdam, 1975.

[32] Ozaki H, *Bull. Chem. Soc. Jpn.*, **51**, 257 (1978).

[33] Ozaki H, Tai A, Kobatake S, Watanabe H, and Izumi Y, *Bull. Chem. Soc. Jpn.*, **51**, 3559 (1978).

[34] Nitta Y, Sekine F, Sasaki J, Imanaka T, and Teranishi S, *Chem. Lett.*, 541 (1981).

[35] Keane M A, *Langmuir*, **13**, 41 (1997).

[36] Keane M A, *J. Chem. Soc., Faraday Trans.*, **93**, 2001 (1997).

[37] Tanabe T, *Bull. Chem. Soc. Jpn.*, **46**, 1482 (1973).

[38] Richards D R, Kung H H, and Sachtler W M H, *J. Mol. Catal.*, **36,** 329 (1986).

[39] Hatta A and Suetake W, *Bull. Chem. Soc. Jpn.*, **48**, 2428 (1975).

[40] Hatta A and Suetake W, *Bull. Chem. Soc. Jpn.,* **48**, 3441 (1975).

[41] Inoue Y, Okabe K, and Yasumori I, *Bull. Chem. Soc. Jpn.*, **54**, 613 (1981).
[42] Yasumori I, *Pure Appl. Chem.* **50**, 971 (1978).
[43] Izumi Y, *Proc. Jpn. Acad.*, **53**, 38 (1977).
[44] Fish M J and Ollis D F, *J. Catal.*, **50**, 353 (1977).
[45] Keane M and Webb G, *J. Catal.*, **136**, 1 (1992).
[46] Nitta Y, Imanaka T, and Teranishi S, J. *Catal.*, **80**, 31 (1983).
[47] Harada T, Imachi Y, Tai A, and Izumi Y, in Imelik B (ed) Metal-Support and Metal-Additive Effects in Catalysis, Elsevier, Amsterdam 1982, p 377.
[48] Harada T and Izumi Y, *Chem. Lett.*, 1195 (1978).
[49] Botelaar L J and Sachtler W M H, *J. Mol. Catal*, **27**, 387 (1984).
[50] Brunner H, Muschiol M, Wischert T, Wiel J, and Heraeus W C, *Tetrahedron: Asymmetry*, **1**, 159 (1990).
[51] Tai A, Kikukawa T, Sugimura T, Inoue Y, Abe S, Osawa T, and Harada T, *Bull. Chem. Soc. Jpn.*, **67**, 2473 (1994).
[52] Tai A, Kikukawa T, Sugimura T, Inoue Y, Osawa T, and Fujii S, *Chem. Commun.*, 795 (1991).
[53] Osawa T, Mita S, Iweai A, Miyazaki T, Takayasu O, Harada T, and Matsuura I, *Chem. Lett.*, 1131 (1997).
[54] Tai A, Imachi Y, Harada T, and Izumi Y, *Chem. Lett.*, 1651 (1981).
[55] Tai A, Tskioka K, Imachi Y, Oaki H, Harada T, and Izumi Y, Proc. 8th Int. Congr. Catal., Berlin, V-513 (1984).
[56] Tai A, Harada T, Tsukioka K, Osawa T, and Sugimura T, Proc. 9th Int. Congr. Catal., Calgary, 1082 (1988).
[57] Tai A, Watanabe H, and Harada T, *Bull. Chem. Soc. Jpn.*, **52**, 1468 (1979).
[58] Tatsumi S, *Bull. Chem. Soc. Jpn.*, **41**, 408 (1968).
[59] Gronrgen J A,and Sachtler W M H, *J. Catal.* **38**, 276 (1975).
[60] Wittmann G, György G, and Bartók M, *Helv. Chim. Acta.*, **73**, 635 (1990).
[61] Tai A, Harada T, Hiraki Y, and Murakami S, *Bull. Chem. Soc. Jpn.*, **56**, 1414 (1983).
[62] Hiraki Y, Ito K, Harada T, and Tai A, *Chem. Lett.*, 131 (1981).
[63] Nakahata M, Imaida M, Ozaki H, Harada T, and Tai A, *Bull. Chem. Soc. Jpn.*, **55**, 2186 (1982).
[64] Kikukawa T, Iizuka Y, Sugimura T, Harada T, and Tai A, *Chem. Lett.*, 1269 (1987).
[65] Tai A, Ito K, and Harada T, *Bull. Chem. Soc. Jpn.*, **54**, 223 (1981).
[66] Ito K, Harada T, Tai A, and Izumi Y, *Chem. Lett.*, 1049 (1980).
[67] Ito K, Harada T, and Tai A, *Bull. Chem. Soc. Jpn.*, **53**, 3367 (1980).
[68] Osawa T and Harada T, *Bull. Chem. Soc. Jpn.*, **57**, 1618 (1984).
[69] Osawa T, Harada T, and Tai A, *J. Catal.*, **121**, 7 (1990).
[70] Osawa T, *Chem. Lett.*, 1609 (1985).
[71] Osawa T, Harada T, and Tai A, *J. Mol. Catal.*, **121**, 7, (1990).
[72] Harada T and Osawa T, in Jannes G and Dubois V (eds) Chiral Reaction in Heterogeneous Catalysis, Plenum Press, New York, 1995, p. 83.
[73] Osawa T, Tai A, Imachi Y, and Takasaki S, in Jannes G and Dubois V (eds) Chiral Reaction in Heterogeneous Catalysis, Plenum Press, New York, 1995, p. 75.
[74] Osawa T, Harada T, Tai A, Takayasu O, and Matsuura I, in Blaser H U, Baiker A, Prins R (eds), Heterogeneous Catalysis and Fine Chemicals IV, Elsevier, Amsterdam, 1997, p. 199.
[75] Sugimura T, Osawa T, Nakagawa S, Harada T, and Tai A, Proc. 11th Int. Congr. Catal., Baltimore, 1996, p 231.
[76] Nakagawa S, Sugimura T, Tai A, *Chem., Letts.*, 859 (1997).
[77] Nakagawa S, Sugimura T, Tai A, *Chem., Letts.*, 1257 (1998).
[78] Gronrgen J A,and Sachtler W M H, Proc. 6th Int. Congr. Catal., London (1976), p. 1014
[79] Yasumori I, Yokozeki M, Inoue Y, *Faraday Dicuss. Chem. Soc.* **72**, 385 (1981).

9 Catalytic Hydrogenation, Hydroformylation and Hydrosilylation with Immobilized P- and N-ligands

Daniel J. Bayston and Mario E. C. Polywka

9.1 Introduction

Homogeneous catalysts are known to exhibit high selectivity and activity in a variety of asymmetric transformations under relatively mild reaction conditions [1]. Amongst the numerous types of asymmetric catalysis, perhaps the most impressive systems are those for the hydrogenation of carbon-carbon and carbon-oxygen double bonds with transition metal diphosphine complexes [2]. Many of these catalysts exhibit extremely high enantioselectivities at low catalyst to substrate ratios and, thus, lend themselves to efficient industrial processes. However, it has long been realized that one of the major drawbacks of homogeneous catalysis is the need for separation of the catalytic species from the reaction mixture at the end of the process. Clearly, such a transition metal catalyst is often very expensive, must not be found as a contaminant in pharmaceutical products, and is often destroyed in the attempted separation from such a product. The required separation of catalyst from product can itself be a costly process. Therefore, there is a continuing need for the immobilization of catalytically active species in order to facilitate this procedure [3].

Various methods are available for the immobilization of catalysts. It is not the purpose of this review to describe each method in detail but to give an overview helping to clarify the nature of some of the catalysts referred to later. Immobilization may be divided up into three distinct areas:

1. By formation of a covalent bond with the ligand,
2. By ion-pair formation,
3. By entrapment.

Binding of a ligand to a solid support *via* a covalent bond has become the most often employed method of heterogenization of an enantioselective catalyst. The solid supports themselves may be either organic polymers such as polystyrenes, or inorganic polymers, often silica based. Clearly the support must be functionalized in such a way as to be able to form a covalent bond, and in the case of silica, this is done by reaction with a free hydroxyl group in the polymeric silica network. The simplest method available is a simple grafting of a suitably functionalized monomeric ligand onto a support. This method benefits from being able to use a well-characterized monomer in the grafting with a well-characterized, often commercially available polymer, leading to an immobilized ligand of known origin. Furthermore, the same functionalized monomer may be used to graft onto a variety of supports, and therefore, this method is widely applicable. An alternative method of immobilization is to perform a copolymerization reaction with a monomer functionalized with, for example, a styrene unit. This method has the drawback that a particular monomer may only be suited to a certain type of polymer. A third, less desirable method, is to prepare the ligand on the polymer support. While this method may be performed in a combinatorial manner to rapidly produce a whole array of ligands, it is hardly possible to purify any intermediates in the synthesis, and it is therefore difficult to determine the exact composition of the polymer.

Ion-pair formation clearly can only be used for an ionic catalyst, which is a severely limiting factor in the application of this method. In fact, this technique has found use to date only in rhodium-mediated catalysis. This method uses a homogeneous phase cationic catalyst complexed with an immobilized anion such as a commercially available polymeric resin with a sulfonic acid residue as the counterion. The simplicity of this approach renders it particularly attractive, especially when one considers that the ligand does not need to be modified in any way for it to be immobilized. This means, of course, that a variety of different ligands can rapidly be investigated as an immobilized ligand.

The method of entrapment of a catalyst within a polymer has recently been applied to several catalysts. Two types of entrapment have been used. Either a catalyst can be entrapped in a polymer by simple steric restrictions blocking the exit of a catalyst from the polymer, or, electronic interactions between the catalyst and the polymer matrix hold the catalyst in place. Although entrapment is still in its infancy compared with conventional grafting of ligands to a support, some extremely interesting results have already been obtained which will be described later.

9.2 Asymmetric Hydrogenation with Immobilized Catalysts

9.2.1 Immobilized DIOP Derivatives

Historically, one of the first supported enantioselective catalysts was designed for hydrogenation. In the early seventies, Kagan reported about the synthesis and use of DIOP for homogeneous hydrogenation of α-acylaminoacrylic acids [4]. His group subsequently synthesized an insoluble analog for heterogeneous rhodium-catalyzed asymmetric hydrogenation and hydrosilylation reactions. The synthesis of this supported ligand involved attachment of diol **2** by acetalization, to an insoluble aldehyde **1**, derived from Merrifield resin (Scheme 9.1). The polystyrene bound chiral ditosylate **3** was treated with lithium diphenylphosphide to provide the immobilized DIOP ligand **4** and further reacted with $[RhCl(C_2H_4)_2]_2$ to give an active hydrogenation catalyst.

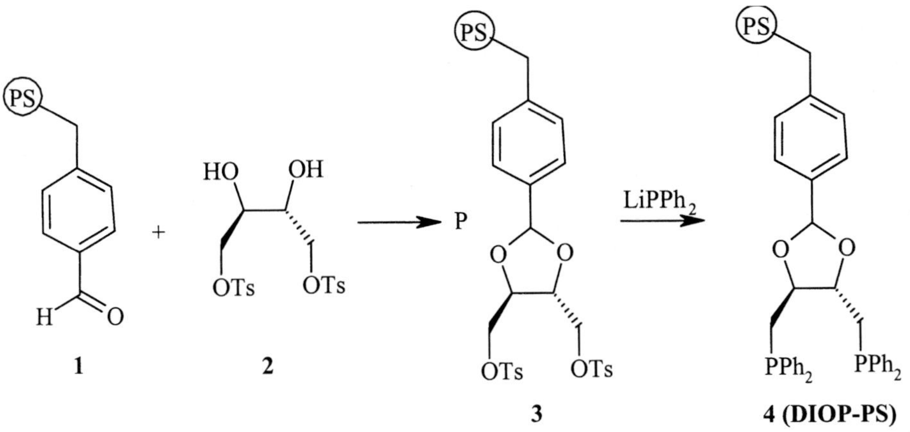

PS HO OH + OTs OTs H O 1 2 → P PS O O OTs OTs 3 $LiPPh_2$ → PS O O PPh_2 PPh_2 4 (DIOP-PS)

Scheme 9.1

The catalytic activity was tested in the hydrogenation of simple olefins: α-ethylstyrene was hydrogenated in an e.e. of only 1.5% (c.f. 15% with soluble Rh-DIOP catalyst) and methyl atropate in an e.e. of 2.5% (c.f. 7% with soluble Rh-DIOP catalyst) with benzene as solvent. α-Acetamidocinnamic acid could not be hydrogenated with this insoluble Rh-DIOP catalyst. The likely cause of the low level of activity was that the solvent, benzene, was unsuitable for the support used. Since the dehydroamino-acid was insoluble in this medium, ethanol was used as a co-solvent which caused the polymer to contract, thus hindering access of the substrate to the active catalytic site. This solvent dependence is a characteristic feature of immobilized catalysis, and matching the solvent to the polymer used is often necessary to obtain good reaction rates and selectivities.

The immobilized DIOP ligand has also been synthesized by radical copolymerization reactions (Scheme 9.2). Thus, styrene monomer **5** was co-polymerized with hydroxyethyl methacrylate (HEMA) and ethylene glycol dimethacrylate to obtain a cross-linked polar polymer **6** more suitable as a catalyst support for use with polar solvents such as ethanol [5]. The hydrogenation of α-acetamidocinnamic acid **7** (Scheme 9.3) with the Rh(I)-DIOP-HEMA catalyst in a benzene/ethanol solvent mixture gave the product **8** in an e.e. of 86% (Table 9.1, entry 3) compared with 81% for the homogeneous Rh(I)-DIOP catalyst (Table 9.1, entry 1). Although the optical purities of the hydrogenation product were comparable to those obtained with the soluble rhodium catalyst, the heterogeneous catalyst system showed a slower reaction rate, presumably due to a slower diffusion rate of substrate into the polymer.

The polymethylvinyl ketone (MVK) catalyst system **12** was designed to incorporate a chiral alcohol unit in the polymer close to the catalytic site so as to effect an interaction (Scheme 9.4) [6]. Thus, hydrogenation of α-acetamidocinnamic acid **7** in benzene/ethanol gave **8** in an e.e. of 83% (Table 9.1, entry 4), comparable to the soluble Rh(I)-DIOP catalyst (Table 9.1, entry 1). Similarly, hydrogenation of α-acetamidoacrylic acid **9** with the immobilized Rh(I)-DIOP-MVK catalyst gave **10** in an e.e. of 75% in a benzene/ethanol mixture (Table 9.1, entry 7), again comparable to the solu-

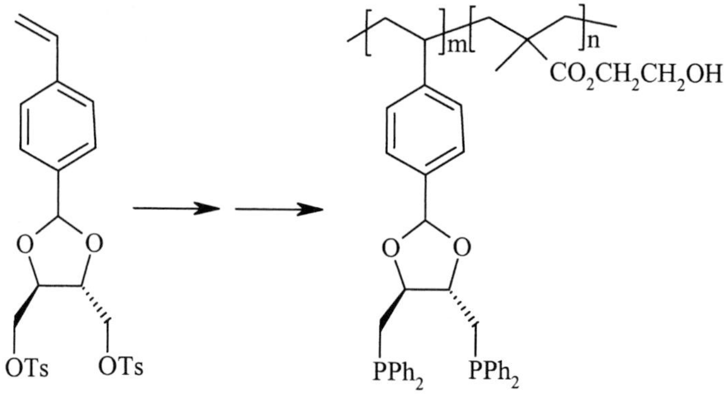

5 → → **6 DIOP-HEMA**

Scheme 9.2

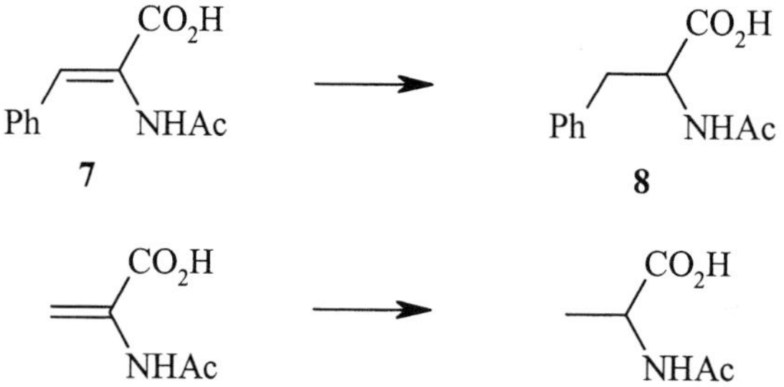

7 → **8**

9 → **10**

Scheme 9.3

Table 9.1. Asymmetric hydrogenation of α-acylaminoacrylic acids with DIOP catalysts.

Entry	Catalyst	Substrate	Solvent	E.e. %
1	Rh(I)-DIOP	**7**	benzene/EtOH	81
2	Rh(I)-DIOP-PS	**7**	benzene/EtOH	no reaction
3	Rh(I)-DIOP-HEMA	**7**	benzene/EtOH	86
4	Rh(I)-DIOP-(*R*)-MVK	**7**	benzene/EtOH	83
5	Rh(I)-DIOP	**9**	benzene/EtOH	70
6	Rh(I)-DIOP	**9**	THF	7
7	Rh(I)-DIOP-(*R*)-MVK	**9**	benzene/EtOH	75
8	Rh(I)-DIOP-(*R*)-MVK	**9**	THF	40
9	Rh(I)-DIOP-(*S*)-MVK	**9**	THF	24

ble Rh(I)-DIOP system (70% e.e., Table 9.1, entry 5). However, upon switching to tetrahydrofuran as solvent, a decrease in e.e. to 7% was observed for the soluble catalyst (Table 9.1, entry 6), while the Rh(I)-DIOP-(*R*)-MVK catalyst yielded **10** in an e.e. of 40% (Table 9.1, entry 8), indicating that an interaction with the free alcohol of the polymer was occurring. Furthermore, use of Rh(I)-DIOP-(*S*)-MVK, containing the opposite chiral alcohol, gave **10** in an e.e. of 24% (Table 9.1, entry 9), a clear indication of a matched/mismatched system. The use of a protic solvent clearly swamped the effect of the polymer hydroxyl groups.

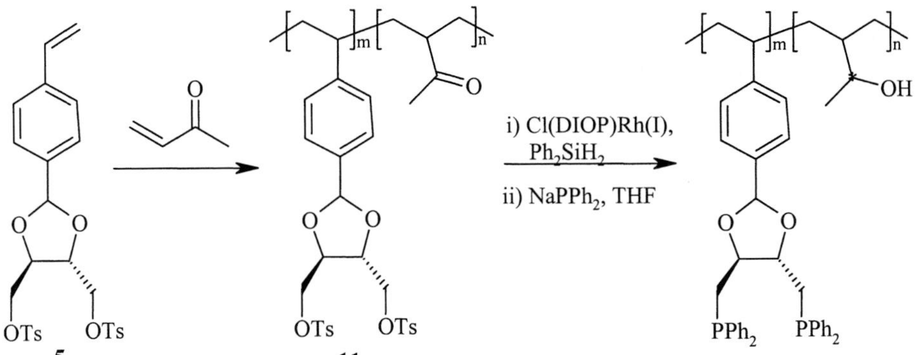

i) Cl(DIOP)Rh(I), Ph_2SiH_2

ii) $NaPPh_2$, THF

OTs OTs **5**

OTs OTs **11**

PPh_2 PPh_2 **12 DIOP-MVK**

Scheme 9.4

9.2.2 Immobilized BPPM Derivatives

More effective immobilized hydrogenation catalysts have been prepared based on the BPPM ligand **13**.

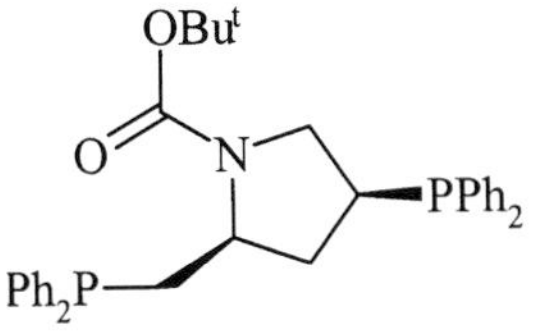

OBu^t PPh_2 Ph_2P **13** BPPM

Again, radical copolymerization reactions were utilized to synthesize immobilized diphosphine ligands. Thus, styrene monomer **14** was co-polymerized with either methyl methacrylate, *tert.*-butyl methacrylate or hydroxyethyl methacrylate (Scheme 9.5) [7]. A route to the opposite enantiomer of the styrene monomer **14** has also been reported, thus giving access to either enantiomer of the immobilized BPPM-type catalyst [8]. The use of these catalysts has been demonstrated in the rhodium-catalyzed asymmetric hydrogenation of both carbon-carbon and carbon-oxygen double bonds. For example, a catalyst prepared from mixing PPM-HEMA **17** and $[Rh(COD)Cl]_2$, efficiently hydrogenated α-acetamidocinnamic acid **7** in alcoholic solvents in the presence of triethylamine in an e.e. of 91%, according to Stille, but in only 23% according to Achiwa [7, 8]. The only apparent difference in ligand preparation of these two catalysts is simply that in the former procedure, a cross-linking agent was utilized in the polymerization step, whereas this was absent in the latter. Although not postulated by either author, such a difference in enantioselectivity may be accounted for by site isolation. The non-cross-linked polymer may be free to form more of an inactive dimeric species than the cross-linked analog.

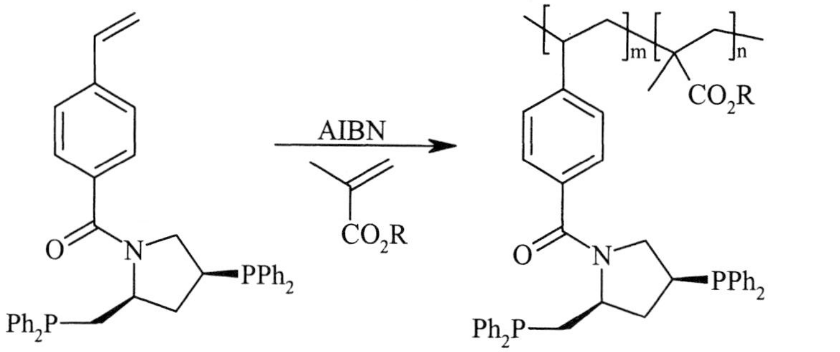

14

R=Me, **PPM-MMA (15)**
R=Bu^t, **PPM-tBu (16)**
R=CH_2CH_2OH, **PPM-HEMA (17)**

Scheme 9.5

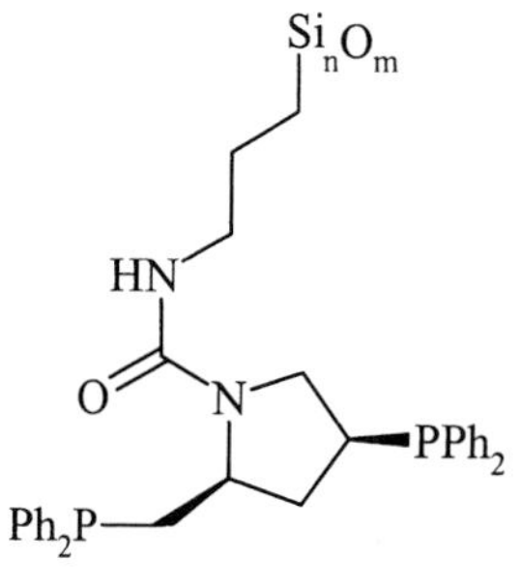

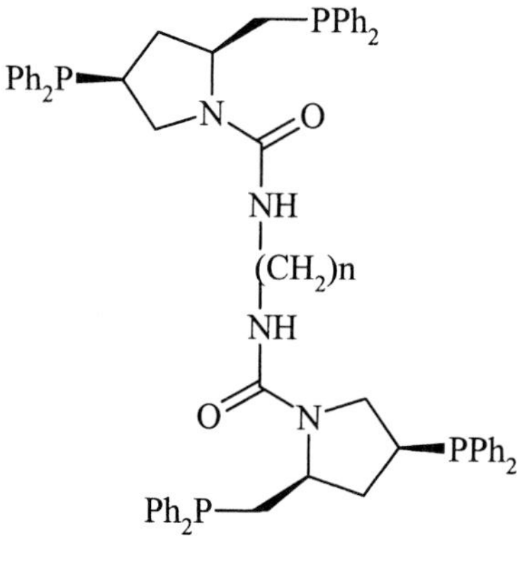

18 Si-PPM

n=6, **PPM-C6-PPM (19)**
n=12, **PPM-C12-PPM (20)**

21 **22**

Figure 9.1. BPPM type ligands.

The role of site isolation has recently been investigated for these types of immobilized ligand [9]. A series of BPPM-type ligands (Figure 9.1) were tested in the rhodium-catalyzed hydrogenation of methyl-α-acetamidocinnamate **21** in order to assess whether the possibility of dimer formation was detrimental to catalytic behavior (Table 9.2). The insoluble Si-PPM ligands **18** were prepared by grafting of a monomer onto silica gel. The *bis*-PPM ligands **19** and **20** were designed as soluble ligands with either a C6 or C12 spacer unit. Table 9.2 gives the results obtained with either a neutral rhodium species, $[Rh(COD)Cl]_2$ or a cationic one, $[Rh(COD)_2]BF_4$. It was observed that the neutral catalysts were several times less active than the cationic ones and also lost activity when their loading was increased (Table 9.2, entries 4–6). The increase in loading would lead to adjacent catalytic sites that could adversely interact with each other, causing lower activity. A lower catalyst loading on the support would decrease the proportion of active sites close enough to interact, thus making the overall activity higher. The experiments with the bis-PPM ligands (Table 9.2, entries 2 and 3) showed that by increasing the local concentration of catalytic sites, the activity decreased accordingly, thus reinforcing the theory of site isolation. The cationic rhodium complexes did not show such a pronounced site isolation phenomenon as the neutral ones, and it was concluded that these complexes have little tendency to interact with each other. The Si-PPM ligands were concluded to be amongst the easiest to prepare and the most efficient in terms of catalyst activity, being comparable to their homogeneous counterparts.

Table 9.2. Asymmetric hydrogenation of methyl *α*-acetamidocinnamate, **21** with PPM catalysts.

Entry	Loading[a]	Ligand	Rh scource	Rate[b]	E.e. %
1		BPPM	$[Rh(COD)Cl]_2$	9	93
2		PPM-C12-PPM	$[Rh(COD)Cl]_2$	3.9	92.6
3		PPM-C6-PPM	$[Rh(COD)Cl]_2$	1.8	92.4
4	0.016	Si-PPM	$[Rh(COD)Cl]_2$	12.5	84.8
5	0.11	Si-PPM	$[Rh(COD)Cl]_2$	3.9	85
6	0.2	Si-PPM	$[Rh(COD)Cl]_2$	1.2	82.2
7		BPPM	$[Rh(COD)_2]BF_4$	18	94.8
8		PPM-C12-PPM	$[Rh(COD)_2]BF_4$	18	95.5
9		PPM-C6-PPM	$[Rh(COD)_2]BF_4$	18	95.5
10	0.11	Si-PPM	$[Rh(COD)_2]BF_4$	13	93.5
11	0.2	Si-PPM	$[Rh(COD)_2]BF_4$	11	91.2

[a] mmol lig/g support, [b] the rate is given as max. turnover frequency [min^{-1}].

A study was also performed to investigate the effect of site isolation on the asymmetric iridium-catalyzed hydrogenation of imine **23** to **24** (Scheme 9.6). While the homogeneous Ir-BPPM catalyst was deactivated after 26% conversion, the heterogeneous counterparts were much more active. It is evident from Table 9.3 that the rate of hydrogenation decreased with increasing catalyst loading. The deactivation was suspected to be attributable to the formation of inactive hydride-bridged iridium dimers. Hence, site isolation of the catalyst on a low-loading silica polymer increased the rate of hydrogenation.

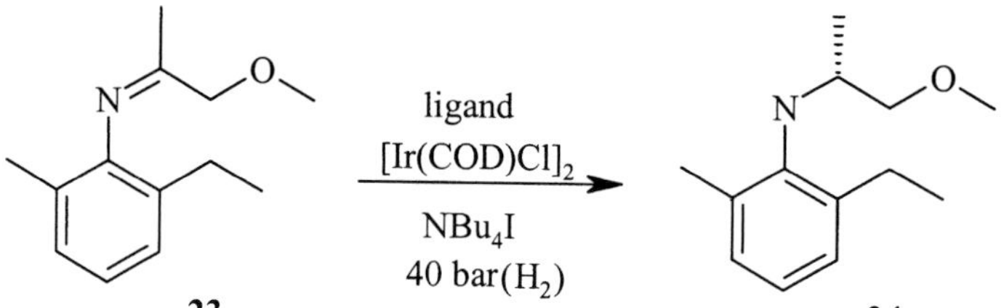

ligand
$[Ir(COD)Cl]_2$
NBu_4I
40 bar(H_2)

23 **24** **Scheme 9.6**

Table 9.3. Iridium-catalysed hydrogenation of imine **23**.

Ligand	Loading [a]	Rate [b]	E.e.
BPPM		0 [c]	45.2
Si-PPM	0.016	5.1	55.2
Si-PPM	0.058	2.4	55.5
Si-PPM	0.19	0.45	4.8

[a] mmol lig/g support, [b] the rate is given as max. turnover frequency [min^{-1}], [c] catalyst deactivated after 26% conversion.

9.2.3 Immobilized BINAP Derivatives

Of the many catalysts developed for transition metal-catalyzed asymmetric hydrogenations, the Ru-BINAP complexes have become the most extensively applied, both in academia and industry [10]. Their ability to promote highly enantioselective transformations over a wide range of substrates with high substrate to catalyst ratios ensures a continuing interest in this area and has promoted various attempts to immobilize this catalyst. A novel method of immobilization of the BINAP framework (**25**) has been reported by Davis who employed a supported aqueous phase (SAP) catalyst [11]. This SAP consists of a thin water film that lies on a high-surface-area hydrophilic support, a controlled pore glass, and contains the water soluble BINAP-$(SO_3Na)_4$ ruthenium complex (**26**). The hydrophilicity of the ligand and support creates interaction energies sufficient to maintain the immobilization. This SAP was designed to conduct catalytic reactions at an interface.

The usefulness of this novel catalyst was tested in the asymmetric hydrogenation of alkene **27** to Naproxen **28** (Scheme 9.7). Using a homogeneous [Ru(BINAP)-C_6H_6Cl]Cl catalyst in methanol gave the product in e.e.'s of up to 96%. Under identical reaction conditions, the homogeneous [Ru(BINAP-$(SO_3Na)_4C_6H_6Cl$]Cl catalyst system also produced Naproxen with an e.e. of 96% when using methanol as solvent, but only 70% when using water/ethyl acetate as solvent. For the SAP catalyst to be effective, clearly methanol could not be used as solvent, since leaching of the catalyst from the support would occur. Therefore, it was necessary to use a hydrophobic solvent such as ethyl acetate. In this solvent, it was found that the SAP-[Ru(BINAP-$(SO_3Na)_4C_6H_6Cl$]Cl catalyst was active for the hydrogenation of **27** only when the SAP was pretreated with water, yielding Naproxen with an e.e. of 70%. Although the

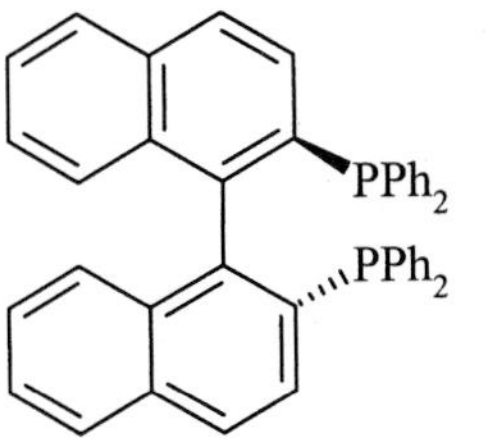

NaO_3S SO_3Na PPh$_2$ PPh$_2$ P P SO_3Na SO_3Na

25 BINAP

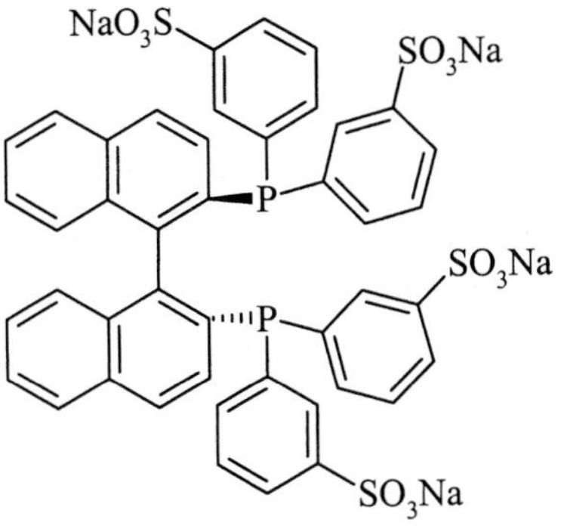

26 BINAP-4SO_3Na

reaction rate was much greater than with the biphasic homogeneous [Ru(BINAP-$(SO_3Na)_4C_6H_6Cl$]Cl catalyst, it was still seven times lower than with the homogeneous [Ru(BINAP)C_6H_6Cl]Cl in methanol. The water dependence of the SAP catalyst was ascribed to its need for mobility, which would be severely restricted under anhydrous conditions. However, it was postulated that in the presence of water, a loss of chloro-ligand by aquation was responsible for the decline in enantioselectivity. Therefore, a change of the hydrophilic liquid phase from water to ethylene glycol was assessed, using cyclohexane/chloroform as the hydrophobic solvent. In this case, the Ru-Cl bond was expected to be left intact and, indeed, Naproxen with an e.e. of 96% was obtained in the hydrogenation of **27,** though still with a lower overall activity than with the homogeneous catalyst. After the hydrogenation reaction the catalyst was removed by filtration and reused with little loss in activity and no detectable loss of catalyst to the filtrate.

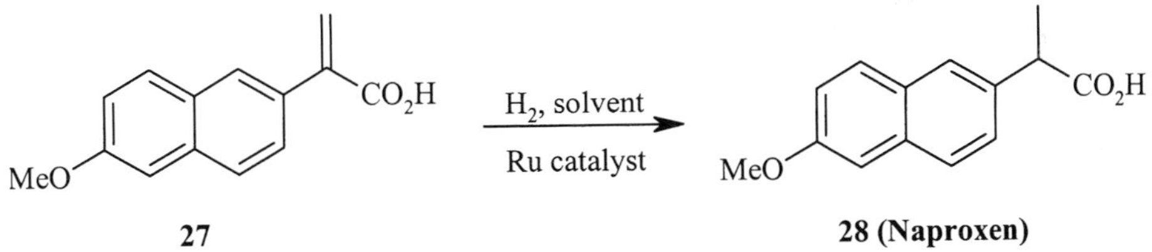

MeO CO_2H H_2, solvent Ru catalyst MeO CO_2H

27 **28 (Naproxen)**

Scheme 9.7

A further novel method of immobilization of catalysts has been described by Vankelecom et al. who have encapsulated both epoxidation and hydrogenation catalysts in a membrane [12]. Thus, [Ru(BINAP)(*p*-cymene)Cl]Cl was occluded in an elastomeric-type polydimethylsiloxane membrane and held in place merely by steric restrictions. This catalyst was shown to be active for the hydrogenation of methyl acetoacetate **29** (Scheme 9.8) in a variety of highly polar solvents. The best e.e.'s obtained were approx. 70% by using polyethyleneglycol as solvent. Although this result is inferior to the homogeneous Ru-BINAP system (>98% in MeOH), this may be due to the use of highly polar solvents, a necessity for compatibility with the catalytic membrane.

29 → 30

Scheme 9.8

A third approach to immobilization of the BINAP framework followed a more traditional technique of grafting a monomer onto a functionalized cross-linked polystyrene (Scheme 9.9) [13]. In this case enantiomerically pure BINOL **31** was transformed to the monoester derivative by Friedel-Crafts chemistry to give **32**. A monosubstituted derivative was chosen rather than a *bis*-substituted BINOL so as to impart maximum flexibility and mobility to the ligand and reduce unnecessary cross-linking. Functional group transformations provided diphosphine **33** with a pendant acid linker. Standard peptide coupling to cross-linked aminomethylated polystyrene furnished the required immobilized BINAP ligand **34**. The active hydrogenation catalyst was formed according to the method of Genet [14] by stirring with (COD)Ru(*bis*-methallyl) and HBr to generate what is believed to be a [Ru(PS-BINAP)Br_2] catalyst. This immobilized catalyst was assessed in the asymmetric hydrogenation of various carbon-carbon and carbon-oxygen double bonds. Best results were obtained in the hydrogenation of β-ketoesters such as **29** or **35** (Scheme 9.10), giving the corresponding β-hydroxyesters in an e.e. of 97% and 88%, respectively (the olefin of β-ketoester **35** was also hydrogenated). It is evident that the loss of C_2 symmetry was not detrimental to the enantioselectivity of the catalyst, although immobilization was shown to somewhat reduce its activity. However, the catalyst was shown to be active over several reuses with only a slight drop in activity and enantioselectivity.

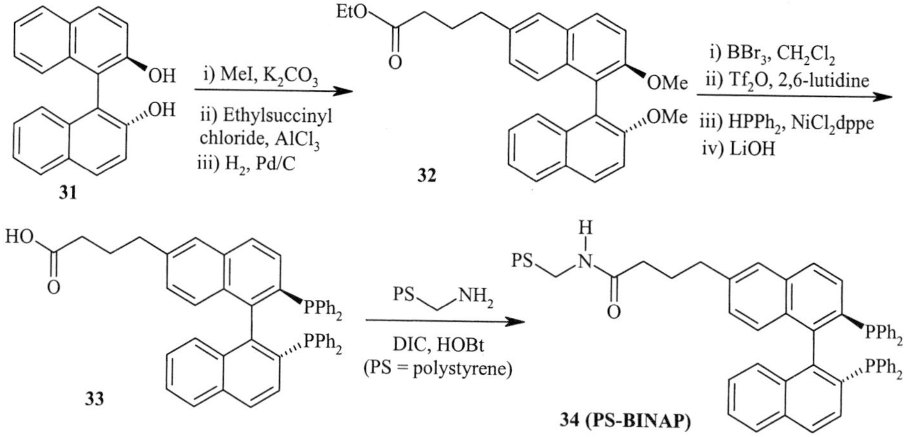

31 → 32: i) MeI, K_2CO_3; ii) Ethylsuccinyl chloride, $AlCl_3$; iii) H_2, Pd/C

32 → 33: i) BBr_3, CH_2Cl_2; ii) Tf_2O, 2,6-lutidine; iii) $HPPh_2$, $NiCl_2$dppe; iv) LiOH

33 → 34 (PS-BINAP): PS–CH_2–NH_2, DIC, HOBt (PS = polystyrene)

Scheme 9.9

35 → 36

Scheme 9.10

9.2.4 Immobilization of Catalysts on Cation-Exchange Resins

A conceptually simple method for the immobilization of a variety of ligands has been developed which takes advantage of the ionic nature of certain rhodium catalysts. Thus, a cationic catalyst may be fixed on a support containing an ionic functional group such as a sulfonated resin [15]. This technique was first applied by Mazzei, who reported the intercalation of the cationic rhodium catalyst **37** onto a series of mineral clays, particularly from the smectite group [16]. Such clays have a layer structure composed of alternating layers of cations and anionic silicate sheets. The interlayer distance of these sheets may be increased sufficiently, by swelling in either water or an alcohol, to allow the ion exchange of the large rhodium complex cation on the smectite and effecting immobilization. This supported rhodium catalyst was tested in the asymmetric hydrogenation of α-acetamidoacrylic acid **7**, yielding **8** in an e.e. of 72% when using hectorite as the support (Scheme 9.3). The activity of the catalyst reportedly decreased upon reuse, although no leaching of rhodium was detected. These types of support were claimed to possess better thermal and mechanical stabilities than conventional organic polymeric supports, important properties of a reusable catalyst.

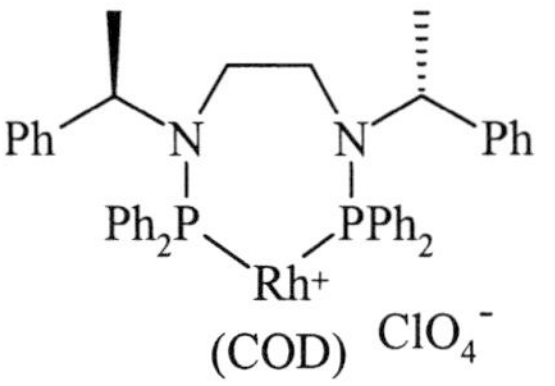

37

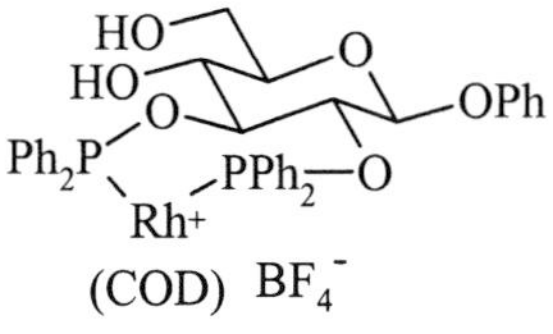

38

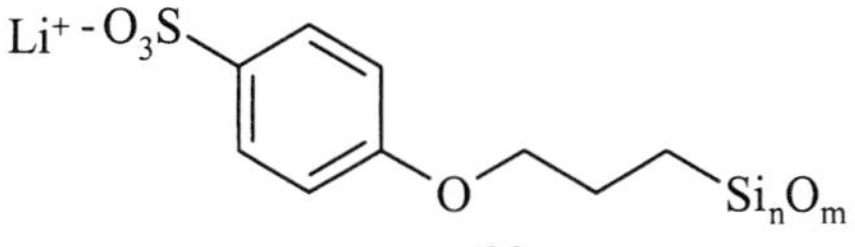

39

The ion-exchange principle has been extensively investigated by Selke and co-workers for the carbohydrate-based rhodium catalyst **38** [17]. This group used either a cross-linked polystyrene or a silica support, **39**, both with pendant sulfonic acid groups to exchange with the rhodium cation. Each support was found to have its own benefits; both gave product with e.e.'s of ~95% in the hydrogenation of prochiral α-acetamido acrylates with similar activities, even though the inorganic silica support was unable to swell. The volume consistency of the silica support enabled its use in either polar or non-polar solvents and therefore enabled the reduction of substrates with varying solubility. This support would thus be more suitable to be used in a column or continuous flow apparatus than the flexible polystyrene support. However, the silica support was less effective at immobilizing the ionic complex catalyst and caused rather high degrees of rhodium leaching. Preliminary results indicated that the leaching of rhodium might be overcome by simple pretreatment of the support with aniline. In contrast, the cross-linked polystyrene was shown to leach rhodium to a lesser extent with reuse but was ineffective in solvents that did not swell the polymer. Both of the supported catalysts were reused for 7 runs with decreasing activity.

40 DIPAMP

41 Me-DUPHOS

Perhaps the most promising example of this type of immobilization technique is a more complex system comprising a support, an anchoring agent and a chiral ligated metal species [18]. In this case the support, a metal oxide such as a silica, alumina or zeolite, was treated with an anchoring agent. A variety of anchoring agents were used and were chosen from the heteropolyacids: phosphotungstic acid, phosphomolybdic acid or silicotungstic acid. The supported anchoring agent was then treated with a chiral metal complex such as Rh(COD)(DIPAMP)BF_4, **40,** or Rh(COD)(Me-DUPHOS)Cl, **41,** to form an immobilized hydrogenation catalyst. The support-anchor and anchor-metal interactions that held the catalyst together were proposed to be simply attractive Van der Waals forces or ion-exchange forces, rather than covalent bonds. The catalysts were suitable for use in alcoholic solvents and were activated either at first use in a reactor or by a reduction step such as a prehydrogenation. An example of the employment of this type of catalyst was the hydrogenation of methyl-α-acetamidoacrylate **21**. Table 9.4 shows the results obtained with two of these supported catalysts. Rather surprizingly, it was observed that the supported Rh-DIPAMP catalyst showed higher activity and selectivity than the homogeneous analog. Furthermore, both the rates of hydrogenation and enantioselectivity increased upon reuse of the catalyst. This phenomenon was also observed for the Rh-Me-DUPHOS catalysts.

Table 9.4. Hydrogenation of methyl-α-acetamidoacrylate, **21**.

Catalyst	Use	Supported[a]		Soluble	
		Rate[b]	E.e. %	Rate[b]	E.e. %
Rh(DIPAMP)	1	0.32	90	0.25	76
Rh(DIPAMP)	3	1.67	92	–	–
Rh(Me-DUPHOS)	1	1.8	83	3.3	96
Rh(Me-DUPHOS)	3	4.4	95	–	–

[a] Alumina-phosphotungstic acid, [b] moles H_2/mole Rh/min.

It is believed that the increased activity and selectivity of the immobilized catalysts arise from a change in the steric environment of the active metal in the supported moiety. The catalysts were found to be extremely stable, even surviving prolonged exposure to air. Thus, these catalysts appear to be amongst the very best immobilized catalysts for this type of transformation. It will be interesting to follow further developments in this field to see how general these catalysts will be.

9.2.5 Transfer Hydrogenation with Immobilized Catalysts

The immobilized transition-metal diphosphine catalysts described so far all used molecular hydrogen for the reduction of either carbon-carbon, carbon-nitrogen or carbon-oxygen double bonds. In the majority of cases, high pressure is essential for any reaction to occur. However, the technique of transfer hydrogenation obviates the need for specialized high-pressure equipment and uses either isopropanol or formic acid:triethylamine as the source of hydrogen. An immobilized transfer hydrogenation catalyst was reported in 1995 and utilized a tetraamine-rhodium complex to study molecular imprinting effects [19]. This rhodium complex **45** (Scheme 9.11), formed by simple copolymerization with a diisocyanide (in the absence of Na-phenylethanolate), was found to catalyze the transfer hydrogenation of acetophenone in isopropanol with an e.e. of 33%. However, diamine **42** could be co-polymerized in the presence of sodium (*S*)-phenylethanolate **43**, as shown in Scheme 9.11 to give polymer **44**. After addition of isopropanol to remove the chiral (*S*)-phenylethanol template the resulting polymer **45** reduced acetophenone **46** in an e.e. of 43% (Scheme 9.12), an increase of 10% from the non-templated complex. This increase in e.e. was ascribed to a favorable molecular imprinting effect of the (*S*)-phenylethanolate, thereby creating chiral pockets within the polymer network.

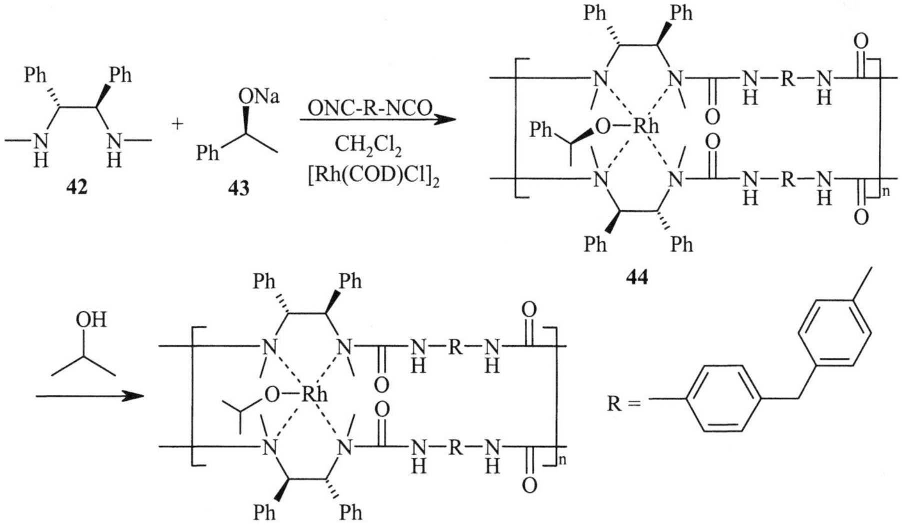

Scheme 9.11

46 → 47

Scheme 9.12

Other immobilized chiral diamines have found varying success in the transfer hydrogenation of acetophenone derivatives. For example, the chiral rhodium diamine complexes **48** and **49**, hybrid silsesquioxane materials prepared by sol-gel hydrolysis of the parent monomers with $Si(OEt)_4$, gave **47** in an e.e. of only 80% at 30% conversion in the isopropanol-promoted transfer hydrogenation of acetophenone [20]. More effective catalysts based on Noyori's Ts-DPEN ligand **50** have recently been reported by two separate groups [21]. The polystyrene bound ligand **51**, prepared by a radical copolymerization of the corresponding monomer, was used in both an iridium and ruthenium-catalyzed transfer hydrogenation of acetophenone with isopropanol as hydrogen donor [22]. Interestingly, it was found that the heterogeneous iridium catalyst prepared from $[Ir(COD)Cl]_2$ gave slightly higher e.e.'s than the homogeneous monomeric counterpart, i.e. 92% compared to 89%, respectively. This was attributed to a postulated chiral microenvironment in the polymer matrix, which enhanced the enantioselectivity for the heterogeneous catalyst. The equivalent ruthenium complexes, prepared from $[Ru(p\text{-cymene})Cl_2]_2$, were found to be less active, less selective but more stable upon reuse.

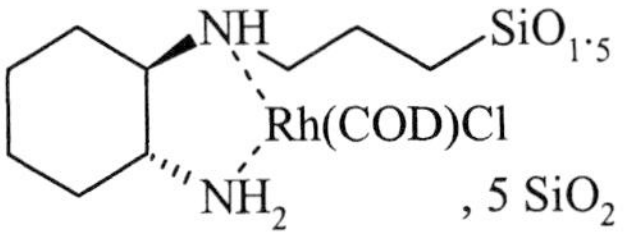

48 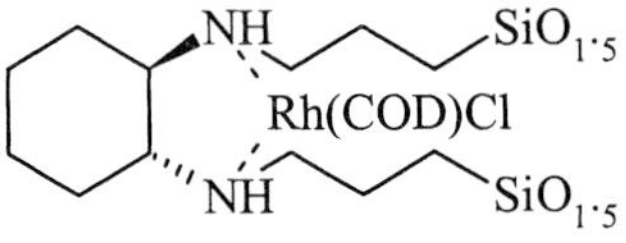 **49**

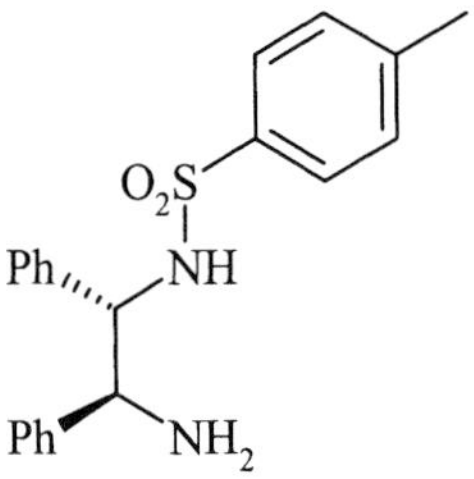

50 Ts-DPEN 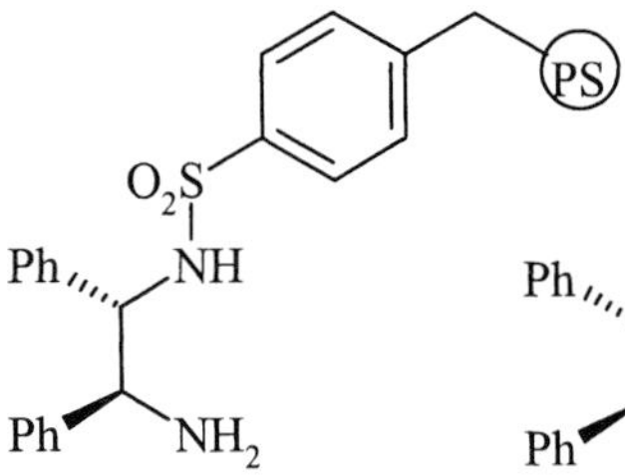 **51** 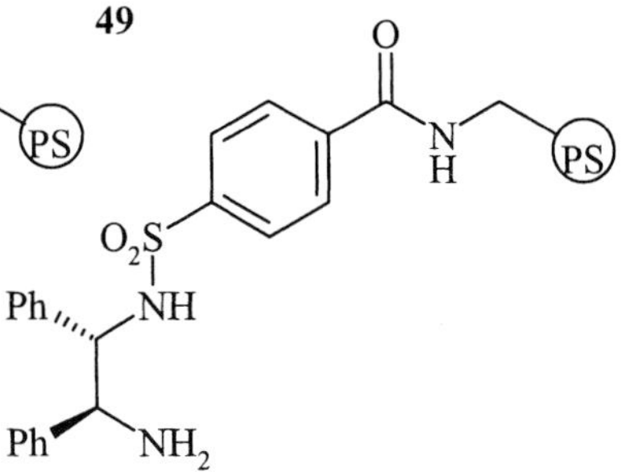 **52**

The polystyrene bound ligand **52**, prepared by grafting the parent acid onto cross-linked aminomethylated polystyrene, was also used in the ruthenium-catalyzed transfer hydrogenation of acetophenone using formic acid:triethylamine as the hydrogen donor [23]. The complex prepared from **52** and $[Ru(p\text{-cymene})Cl_2]_2$ afforded **47** with e.e.'s of up to 99% when using a suitable solvent to swell this type of polymer, such as dichloromethane. However, when neat formic acid:triethylamine was used, it was found that a more suitable support was a polystyrene containing a polar polyethyleneglycol linker. In this case, e.e.'s of up to 97% were obtained in the transfer hydrogenation of acetophenone. Upon reuse throughout 3 runs, these catalysts showed decreased activity, accompanied by a slight decrease in enantioselectivity.

9.2.6 Other Immobilized Ligands for Asymmetric Hydrogenation

A variety of other ligands (**53–56**) for the transition-metal-catalyzed asymmetric hydrogenation of prochiral substrates have been developed (Figure 9.2). The polystyrene-bound dimenthylphosphine ligand **53** was found to be active in the rhodium-catalyzed hydrogenation of α-acetamidocinnamic acid (**7**) [24]. The results obtained were highly solvent-dependent, as can be expected for this type of support, with the best obtained e.e. being 58% in a benzene:ethanol solvent mixture. In contrast, the use of a silica support for the dimenthylphosphine ligand allowed the sole use of alcoholic solvents and led to e.e.'s in the vicinity of 80% for the same substrate [25]. Regardless of the support, it was necessary to use a phosphine:rhodium ratio of at least 2:1 to obtain optimum e.e. values, a clear indication of the *bis*-phosphine rhodium catalyst being more effective than the monophosphine complex. In both of these cases, a high rhodium leaching was observed which restricted the effective reusability of the catalysts.

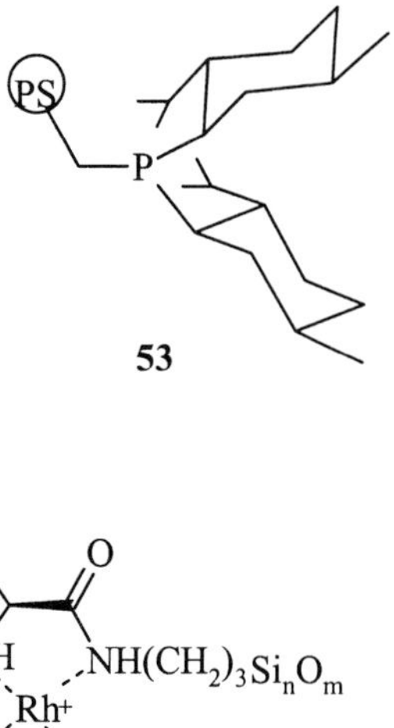

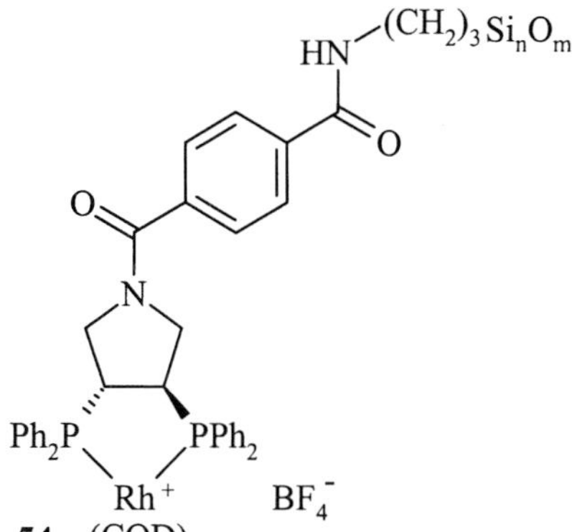

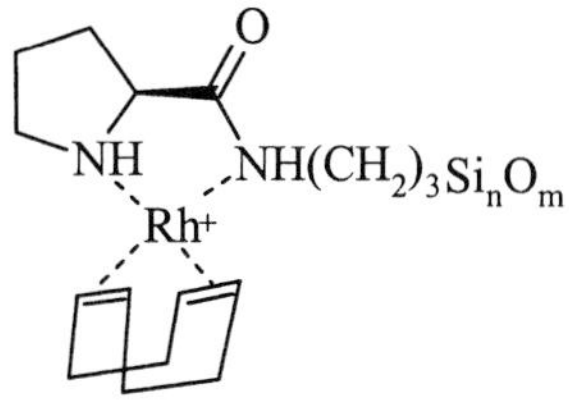

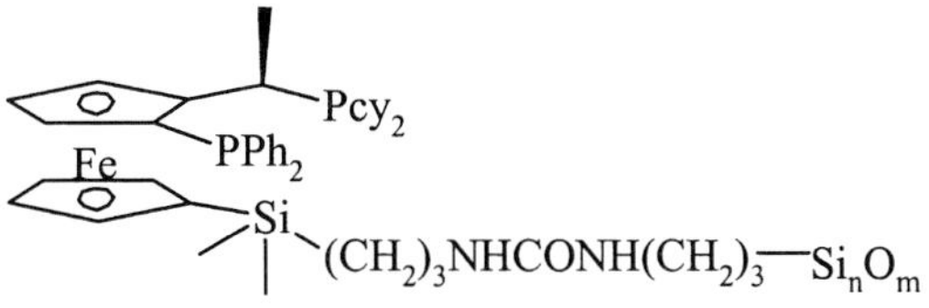

Figure 9.2. Ligands for the transition-metal-catalyzed asymmetric hydrogenation of prochiral substrates.

A particularly selective hydrogenation catalyst is the C_2 symmetric diphosphine rhodium complex **54** developed by Nagel and Kinzel [26]. This complex was found to give e.e.'s of up to 100% for the hydrogenation of methyl *α*-acetamidocinnamate **57** (Scheme 9.13), clearly indicating that this is one of the best immobilized ligands yet developed, at least for this particular substrate. An interesting feature of this catalyst was the dependence of e.e. on the pore size of the silica support. A pore size of 4 nm resulted in the aforementioned e.e. of 100%, whereas a pore size of 10 nm, under identical reaction conditions, gives a hydrogenation product having only 89%. No explanation was given for this phenomenon, although no such effect was observed for *α*-acetamidocinnamic acid as substrate, which afforded **58** with an almost constant e.e. of 93%, regardless of the pore size of the silica-bound catalyst.

CO_2Me NHAc Ph **57** → CO_2Me NHAc Ph **58**

Scheme 9.13

Another extremely selective catalyst for the hydrogenation of *α*-acetamidocinnamic acid **7** has been reported by Corma et al. [27]. In this case a proline derivative was anchored on a zeolite with a well-defined supermicropore (pore diameter of 1.2–3 nm). Both the heterogeneous and homogeneous counterpart of these rhodium complexes **55** were reported to be stable in air for long periods of time, unlike phosphine-metal species which are sensitive to oxygen, especially the homogeneous systems. Enantiomeric excesses of up to 98% were obtained with this catalyst, an enhancement of 15% compared with the homogeneous analog. This constitutes a particularly rare case of a heterogeneous catalyst being more selective than its homogeneous counterpart. Increased selectivities were postulated to arise from steric effects of the support at the site of reaction in the confined spaces of the micropores where the metal complex is anchored. Furthermore, it was reported that the catalysts were reused several times without loss of activity or rhodium content.

The immobilized ferrocene-based ligand **56** deployed in the Ir-catalyzed hydrogenation of imines has recently appeared in the patent literature [28]. Although the synthetic sequence for obtaining this novel ligand is rather long, it is of great interest due to the paucity of catalysts able to effectively reduce carbon-nitrogen double bonds, whether they be heterogeneous or homogeneous [29]. As its iridium complex, this ligand has been shown to be particularly effective for the asymmetric hydrogenation of imine **23** (Scheme 9.6), giving amine **24** in an e.e. of 80%. Moreover, analogous homogeneous catalysts tend to lose their activity during the hydrogenation reaction due to formation of inactive chloro-bridged iridium dimers. Such inactivation is avoided with this heterogeneous catalyst, presumably attributed to a site isolation effect, and reuse of the catalyst appears to be possible with little reported loss of activity. Amine **24** is an important intermediate in the industrial synthesis of chloroacetanilide-type herbicides, and therefore, this type of catalysis raises immediate industrial interest.

A whole library of diphosphine ligands has been synthesized by combinatorial methods based on the monomeric phosphines **59** and **60** [30]. The authors claim that a combinatorial approach may be a valuable tool for the discovery of new asymmetric catalysts partially because of the limitation of rational design which plagues this field, and partially due to the lengthy process of preparing and testing individual catalysts. A 63-member library was synthesized and tested on a solid support. This library was made up of a peptide sequence **61** containing up to 11 amino acids, including the amino acids **59** and **60** (Aib = α-aminoisobutyric acid). It was assumed that the peptides would adopt a helical structure and thus allow coordination of both phosphines to a metal center. The library of supported ligands was tested in the rhodium-catalyzed asymmetric hydrogenation of methyl α-acetamidoacrylate, **21**. Although the best e.e. obtained was only 18%, the exercise demonstrated a novel method of ligand design. Moreover, it will be interesting to observe any future developments in this area.

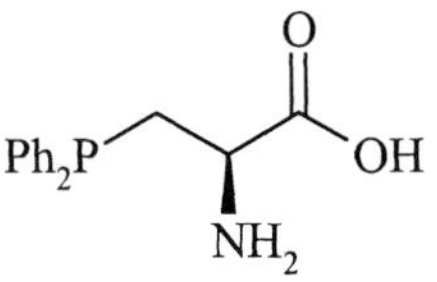

Ph_2P O OH NH_2 **59**

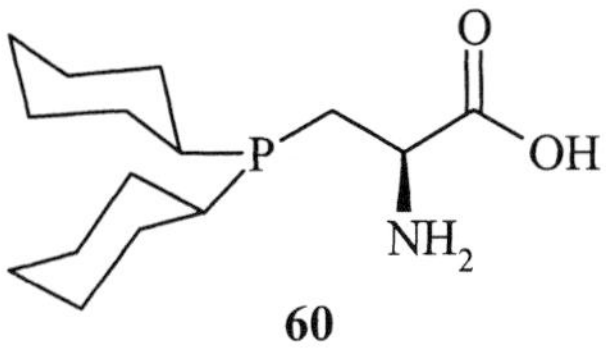

P O OH NH_2 **60**

Ac-Ala-Aib-Ala-[ ]-Ala-Aib-Ala-NH_2

61

9.3 Enantioselective Hydroformylation with Immobilized Catalysts

9.3.1 Immobilized DIOP Derivatives

The immobilized catalysts that have been tested for the asymmetric hydroformylation reaction generally contain the same ligands as for the aforementioned hydrogenation. Again, one of the first catalysts tested was a metal catalyst containing a DIOP-type ligand. The ligands PS-DIOP, **64**, and PS-DIPHOL, **62**, were prepared by nucleophilic displacement with a phosphine nucleophile onto the ditosylate **63**, which was prepared by copolymerization of the styrene-containing monomer (Scheme 9.14).

The active hydroformylation catalyst was prepared by adding $[HRh(CO)(PPh_3)_3]$ to the polymeric diphosphine ligands. Styrene (**65**) was chosen as a model substrate for the test reactions (Scheme 9.15). For both of the catalysts, the selectivity to aldehydes was almost complete, with no alcohols or ethylbenzene observed in any case. Furthermore, the branched/normal ratio of aldehydes (**66**/**67**) for the Rh-PS-DIOP catalyst was about 6, compared with a ratio of about 2 for the homogeneous Rh-DIOP catalyst (Scheme 9.15). Use of the Rh-PS-DIPHOL catalyst led to an increase in ratio of up to 20. Unfortunately, the e.e.'s obtained were poor. In the homogeneous case, the best e.e. obtained was only 25%, whereas for Rh-PS-DIOP, the best e.e. was 11%,

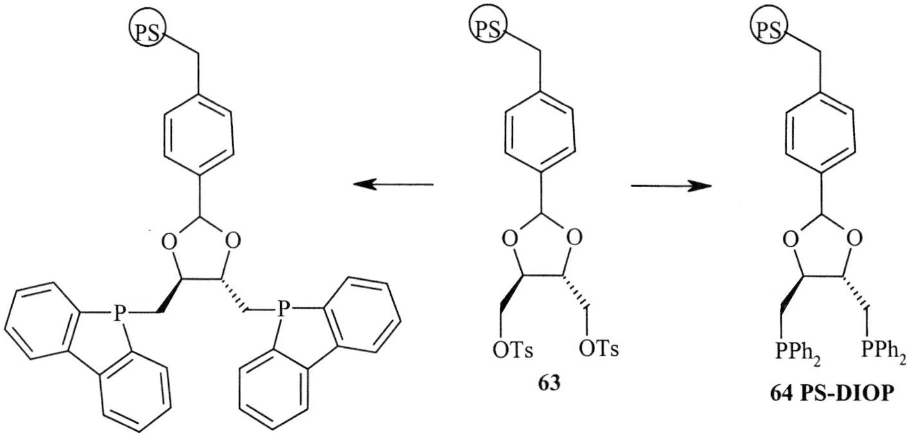

PS PS PS
O O O O O O
P P
OTs OTs
63
PPh_2 PPh_2
64 PS-DIOP

62 PS-DIPHOL

Scheme 9.14

and for Rh-PS-DIPHOL, only 6%. The inability of the polymer-bound catalysts to reproduce the e.e.'s found with their homogeneous counterparts was ascribed to the product hydrotropaldehyde racemizing under the reaction conditions. This postulate was supported by the asymmetric hydroformylation of cis-2-butene **68**, where the product was less likely to racemize. In this case, the branched aldehyde product, **69**, was obtained in an e.e. of 27% for both the homogeneous and heterogeneous Rh-DIOP catalysts (Scheme 9.16). Over a range of substrates, the polymer-supported catalysts again yielded a two- to three-fold increase in the branched/normal ratio of aldehydes. This phenomenon has also been observed for achiral, immobilized chelating diphosphine hydroformylation catalysts.

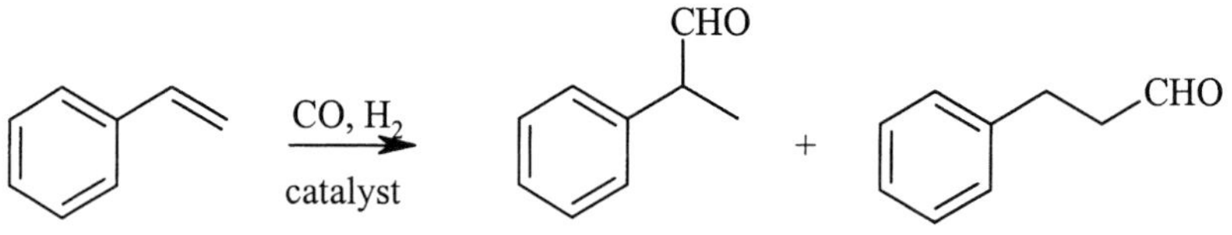

CO, H_2
catalyst
CHO
CHO
65 **66** **67** **Scheme 9.15**

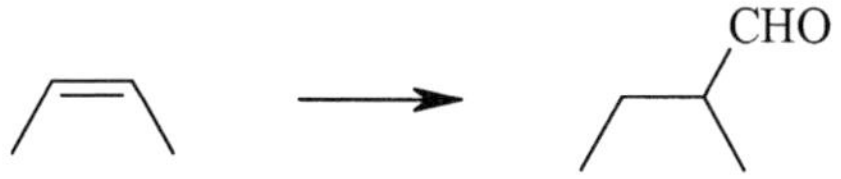

CHO
68 **69** **Scheme 9.16**

Further studies on the hydroformylation of styrene with a PS-DIOP-$PtCl_2$-$SnCl_2$ catalyst have shown that similar optical yields in the range of 28% e.e. can be obtained. The branched/normal ratio of aldehydes appeared to be less affected by immobilization in platinum-tin complexes; the ratio is approximately 0.5 but varies according to exact reaction conditions.

9.3.2 Immobilized BPPM Derivatives

Somewhat higher optical yields have been obtained in the hydroformylation of styrene with homogeneous and heterogeneous BPPM-$PtCl_2$-$SnCl_2$ catalysts as was reported in a thorough piece of work by Stille. The immobilized ligands were prepared by radical copolymerization of the monomeric α,β-unsaturated amide **71** (Scheme 9.17). Two different types of polymer were synthesized: HEMA-PPM **70** prepared by copolymerization with hydroxyethylmethacrylate, and PS-PPM **72** prepared by copolymerization with styrene and divinylbenzene. The enantioselectivities obtained in the asymmetric hydroformylation of styrene were consistently around 70% for the homogeneous and both heterogeneous catalysts (Table 9.5). The HEMA-PPM-$PtCl_2$-$SnCl_2$ catalyst (Table 9.5, entry 2) appeared to be slightly less active than the homogeneous catalyst but significantly more active than the catalyst immobilized on polystyrene. However, the PS-PPM-$PtCl_2$-$SnCl_2$ catalyst gave a higher branched/normal ratio of aldehydes (Table 9.5, entry 3). No explanation was given for this phenomenon.

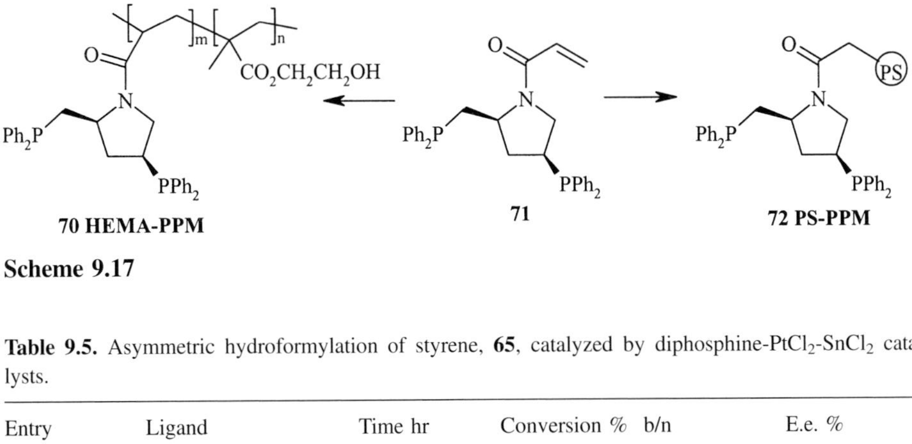

O m n $CO_2CH_2CH_2OH$ N Ph_2P PPh_2 **70 HEMA-PPM** O N Ph_2P PPh_2 **71** O PS N Ph_2P PPh_2 **72 PS-PPM**

Scheme 9.17

Table 9.5. Asymmetric hydroformylation of styrene, **65**, catalyzed by diphosphine-$PtCl_2$-$SnCl_2$ catalysts.

Entry	Ligand	Time hr	Conversion %	b/n	E.e. %
1	BPPM	6	52	0.46	68
2	HEMA-PPM	24	45	0.45	70
3	PS-PPM	90	40	0.60	73
4	PS-PPM[a]	90	38	0.53	73

Reactions run at 60 °C at 2200 psi, $H_2/CO=1$. [a] Reused catalyst.

In the asymmetric hydroformylation of olefins the question still arose as to whether racemization of the product aldehydes occurs under the reaction conditions. To answer this, it was deemed necessary to remove the aldehyde from the reaction mixture as soon as it was formed. The most practical method of doing this was to convert the aldehyde to a substrate that could no longer be racemized. Running the hydroformylation reaction in triethyl orthoformate immediately converted any aldehyde into the non-racemizable acetals **73** and **74** (Scheme 9.18). Although the rate of reaction was considerably slower in the presence of triethyl orthoformate (only 22% conversion after 10 days with PS-PPM-$PtCl_2$-$SnCl_2$), the product acetal was found to possess an e.e. of >98%. This was clear proof that racemization of the aldehydes under hydroformylation conditions was occurring in the absence of triethyl orthoformate. Even with the homogeneous PPM-$PtCl_2$-$SnCl_2$ catalyst, the reaction rates in the presence of triethyl orthoformate were at least an order of magnitude lower, thus giving impractically long reaction times.

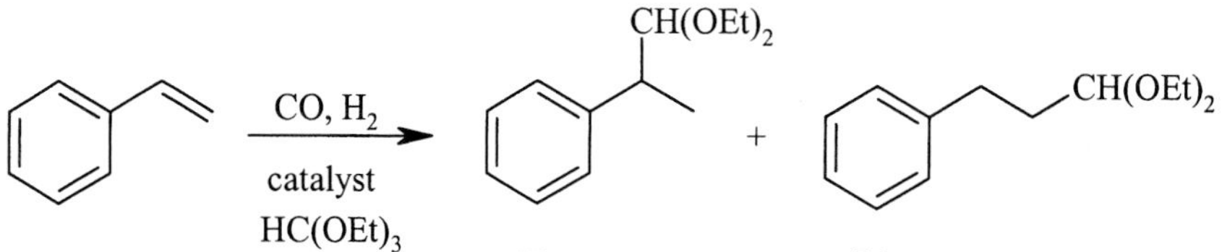

Scheme 9.18

9.3.3 Immobilized Phosphine-Phosphite Derivatives

A more recent report has given a detailed account on a particularly active and enantioselective catalyst for asymmetric hydroformylation reactions based on a phosphine-phosphite-rhodium system. The ligand, itself, is based on a *bis*-binaphthyl skeleton with each phosphorous species possessing complementary donating properties (Figure 9.3).

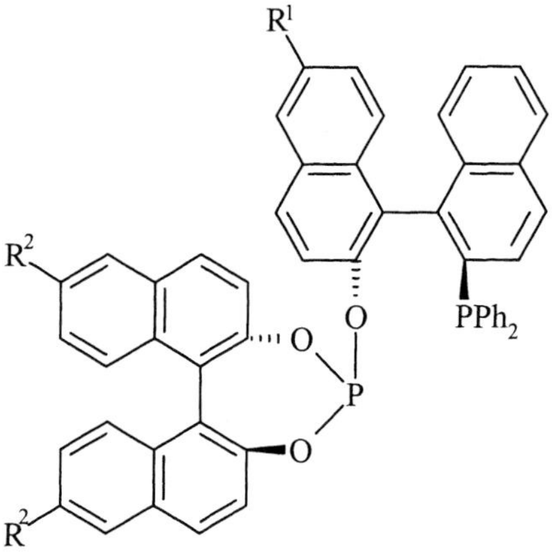

	R^1	R^2
75a	H	H
75b	-CH=CH_2	H
75c	H	-CH=CH_2
75d	-CH=CH_2	-CH=CH_2

Figure 9.3. BINAPHOS-related ligands.

The monomers **75 b**, **75 c** and **75 d** were co-polymerized with 1,2-, 1,3- and 1,4-divinylbenzene containing 3- and 4-ethylstyrene. The amount of divinylbenzenes present in the mixture was 55% thus causing uniquely high amounts of cross-linking in the immobilized ligands. Most conventional polystyrene-bound ligands described so far have possessed less than 10% cross-linking. Higher amounts of cross-linking were traditionally thought to suppress reactivity of the polymer-supported catalyst due to the inability of such a polymer to swell sufficiently to allow reasonable diffusion of reagents to the active sites. This was shown not to be significant for the polymer-supported BINAPHOS ligand as can be seen in the $Rh(acac)(CO)_2$ promoted hydroformylation of styrene (Table 9.6).

Table 9.6. Asymmetric hydroformylation of styrene, **65**, catalyzed by immobilized BINAPHOS catalysts.

Entry	Catalyst	b/n	E.e.
1	Rh(acac)-**75a**	89:11	92
2	PS-**75b**-Rh(acac)	84:16	89
3	PS-**75c**-Rh(acac)	89:11	89
4	PS-**75d**-Rh(acac)	88:12	68
5	PS-[Rh(acac)(**75d**)]	87:13	85

The rhodium complexes of both **75 b** and **75 c** gave e.e.'s of 89% of hydrotropaldehyde with a b/n-ratio of 84:16 and 89:11, respectively (Table 9.6, entries 2 and 3). This compares favorably with the homogeneous catalyst (Table 9.6, entry 1). However, catalyst PS-**75 d**-Rh(acac) resulted in an e.e. of only 68% but in a similar b/n-ratio (Table 9.6, entry 4). A postulated cause for the lower selectivity of this catalyst was the polymerization of the monomer. It is conceivable that the copolymerization reaction results in ligand PS-**75 d** being frozen as a mixture of conformers, some of which are not suitable to coordinate to rhodium. This may result in formation of a less stable rhodium species, for example, with monodentate coordination. The existence of different active species may explain the lowered enantioselectivity for this catalyst. Further evidence to support this theory was obtained by co-polymerizing preformed rhodium-ligand monomer, Rh-**75 d**. The resulting catalyst, PS-[Rh(acac)(**75 d**)] showed increased activity, with hydrotropaldehyde being obtained with an e.e. of 85% and with a b/n-ratio of 87/13 (Table 9.6, entry 5). The conformation of this catalyst should be fixed to a suitable form for efficient bidentate coordination to rhodium. Ligands PS-**75 b** and PS-**75 c** can keep their flexibility in the polymer matrix, because only one of the two binaphthyl groups is bound to the polymer backbone.

The polymer-bound Rh-BINAPHOS catalysts were also active for the hydroformylation of vinyl acetate **76** (Scheme 9.19). For example, use of PS-**75 b**-Rh(acac) as catalyst resulted in a branched to normal b/n-ratio of 85/15 with an e.e. of 91% for the branched isomer **77**. Again, this compares well with the homogeneous catalyst, affording a 92% e.e. and a b/n-ratio of 84/16.

This polymer-supported rhodium-phosphine-phosphite complex serves as the first example of a truly effective and practical asymmetric hydroformylation catalyst.

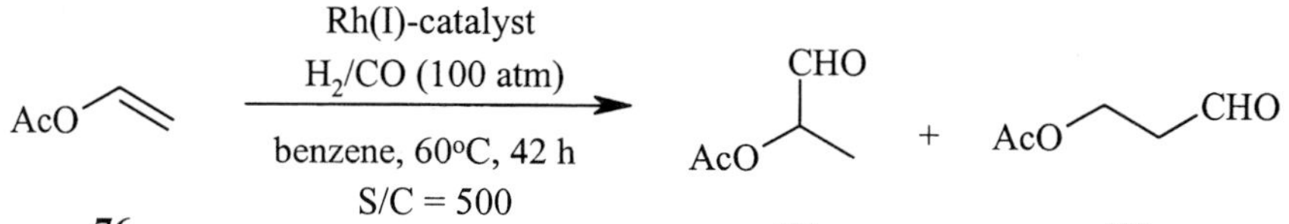

Rh(I)-catalyst
H_2/CO (100 atm)
benzene, 60°C, 42 h
S/C = 500

AcO 76 → AcO 77 (CHO) + AcO 78 (CHO)

Scheme 9.19

9.4 Catalytic Asymmetric Hydrosilylation with Immobilized Ligands

Comparatively little work has been done in the area of asymmetric hydrosilylation with supported catalysts, and all efforts have been directed at hydrosilylation of ketones rather than olefins. Most of the research has involved the use of the Rh-DIOP catalyst **79**, prepared as described for the hydrogenation [4]. Acetophenone has been used as the typical substrate and was reacted with the polymer-bound rhodium catalyst in the presence of a diaryl silane to obtain the intermediate reduced species **80** (Scheme 9.20). Hydrolysis then provided the enantiomerically enriched alcohol **47**. It was found that when using freshly prepared immobilized catalyst, the e.e. values of **47** (29%) were comparable to those of the homogeneous Rh-DIOP catalyst. Interestingly, the enantiomeric excess of **47** slightly depended upon the ratio of immobilized ligand to rhodium. A 1/1 diphosphine/Rh ratio led to an e.e. of 20.5% of **47,** whereas a 4/1 ratio gave 23%. The best result of 29% was obtained from a 2/1 diphosphine/ rhodium ratio. No satisfactory explanation for these effects could be found. It is especially interesting to note that the use of a >2/1 diphosphine/rhodium ratio in the homogeneous catalyst results in the formation of an inactive dimer, whereas the immobilized catalyst remains active at this level. Clearly, this is another example of the benefits of site isolation.

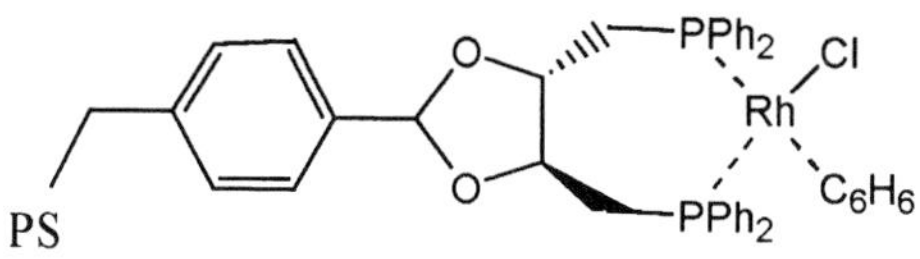

PS — PPh_2, Cl, Rh, C_6H_6, PPh_2

79

46 (O) —catalyst, H_2SiAr_2→ 80 ($OSiHAr_2$) —H^+→ 47 (OH)

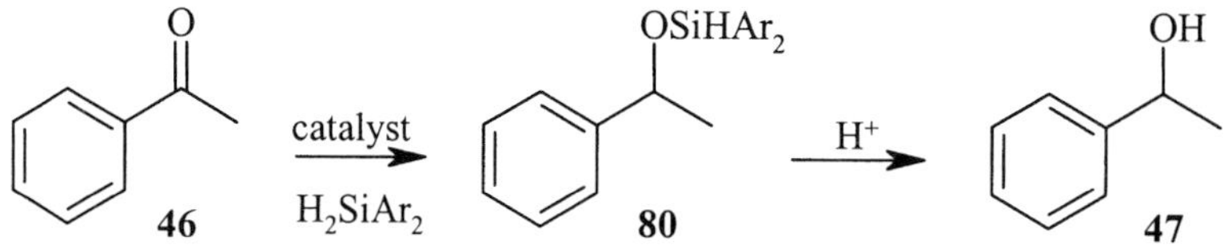

Scheme 9.20

The same ligand has also been attached to a silica support [36]. This Rh-Si-DIOP catalyst also hydrosilylated acetophenone but gave **47** in an e.e. of only 20% and in only 22% yield, thus being far less active than the polystyrene supported analog.

Another silica supported ligand is ligand **81** [37]. In the rhodium-catalyzed hydrosilylation of acetophenone, **47** was obtained in an e.e. of 19% with **81**.

Ph
P
Si_mO_n
Ment
81

9.5 Conclusions

In conclusion, immobilized asymmetric hydrosilylation of both ketones and olefins has been largely overlooked since the first results in this area were published. With the success of both hydrogenation and hydroformylation catalysts in recent years, it is apparently only a matter of time before the area of hydrosilylation is studied in greater detail.

References

[1] Noyori, R. *Asymmetric Catalysis in Organic Synthesis*; John Wiley & Sons: New York, **1994**.

[2] Ojima, I. (ed.) *Catalytic Asymmetric Synthesis*; VCH: New York, **1993**.

[3] a) Jannes, G. and Dubois, V. (ed.) *Chiral Reactions in Heterogeneous Catalysis*; Plenum Press: New York and London, **1995**; b) Hetflejs, J. *Studies in Surface Science and Catalysis*, **1985**, *27*, 497; c) Blaser, H. U. and Pugin, B. *Supported Reagents and Catalysts in Chemistry (Spec. Publ. R. Soc. Chem.)*, **1998**, *216*, 101; d) Baiker, A. *Curr. Opin. Solid State Mater. Sci.* **1998**, *3(1)*, 86.

[4] a) Dumont, W., Poulin J-C., Dang, T-P. and Kagan, H. B. *J. Am. Chem. Soc.* **1973**, *95(25)*, 8295; b) Ohkubo, K., Haga, M., Yoshinaga, K. and Motozato, Y. *Inorg. Nucl. Chem. Lett.* **1980**, *16(3)*, 155; c) Ohkubo, K., Fujimori, K. and Yoshinaga, K. *Inorg. Nucl. Chem. Lett.* **1980**, *15(5)*, 231.

[5] a) Takaishi, N., Imai, H., Bertelo, C. A. and Stille, J. K. *J. Am. Chem. Soc.* **1976**, *98*, 5400; b) Takaishi, N., Imai, H., Bertelo, C. A. and Stille, J. K. *J. Am. Chem. Soc.* **1978**, *100*, 264.

[6] Masuda, T. and Stille, J. K. *J. Am. Chem. Soc.* **1978**, *100*, 268.

[7] a) Achiwa, K. *Chem. Lett.* **1978**, *8*, 905; b) Achiwa, K. *Heterocycles* **1978**, *9(11)*, 1539.

[8] Baker, G. L., Fritschel, S. J., Stille, J. R. and Stille, J. K. *J. Org. Chem.* **1981**, *46*, 2954.

[9] a) Pugin, B. and Muller, M. *Stud. Surf. Sci. Catal. (Heterogeneous, Catalysis & Fine, Chemical III)* **1993**, *78*, 107; b) Pugin, B. *J. Mol. Catal.* **1996**, *107*, 273.

[10] a) Ohta, T., Miyake, T., Seido, N., Kumobayashi, H. and Yakaya, H., *J. Org. Chem.* **1995**, *60*, 357; b) Mashima, K., Kusano, K., Ohta, T., Noyori, R. and Takaya, H. *J. Chem. Soc., Chem. Commun.* **1989**, 1208; c) Kitamura, M., Tokunaga, M. and Noyori, R. *J. Org. Chem.* **1992**, *57*, 4053; d) Miyashita, A., Yasuda, A., Takaya, H., Toriumi, K., Ito, T., Souchi, T. and Noyori, R. *J. Am. Chem. Soc.* **1980**, *102*, 7932; e) Noyori, R., Ohkuma, T., Kitamura, M., Takaya, H., Sayo, N., Kumobayashi, H. and Akutagawa, S. *J. Am. Chem. Soc.* **1987**, *109*, 5856.

[11] a) Wan, K. T. and Davis, M. E. *J. Catal.* **1994**, *148*, 1; b) Wan, K. T. and Davis, M. E. *Nature* **1994**, *370*, 449.
[12] Vankelecom, I. F. J., Tas, D., Parton, R. J., Van der Vyver, V. and Jacobs, P. A. *Angew. Chem. Int. Ed. Engl.* **1996**, *35(12)*, 1346.
[13] Bayston, D. J., Fraser, J. L., Ashton, M. R., Baxter, A. D., Polywka, M. E. C. and Moses, E. *J. Org, Chem.* **1998**, *63(9)*, 3137.
[14] a) Genet, J. P., Pinel, C., Ratovelomanana-Vidal, V., Mallart, S., Pfister, X., Cano De Andrade, M. C. and Laffitte, J. A. *Tetrahedron: Asymmetry* **1994**, *5*, 665; b) Genet, J. P., Mallart, S., Pinel, C., Juge, S. and Laffitte, J. A. *Tetrahedron: Asymmetry* **1991**, *2*, 43; c) Genet, J. P., Pinel, C., Ratovelomanana-Vidal, V., Mallart, S., Pfister, X., Bischoff, L., Cano De Andrade, M. C., Darses, S. and Laffitte, J. A. *Tetrahedron: Asymmetry* **1994**, *5*, 675; d) Genet, J. P., Ratovelomanana-Vidal, V. and Cano De Andrade, M. C. *Tetrahedron Letters* **1995**, *36*, 2063. e) Genet, J. P., Ratovelomanana-Vidal, V., Cano De Andrade, M. C., Pfister, X., Guerreiro, P. and Lenoir J. Y. *Tetrahedron Letters* **1995**, *36*, 4801.
[15] Brunner, H., Bielmeier, E. and Wiehl, J. *J. Organomet. Chem.* **1990**, *384(1–2)*, 223.
[16] Mazzei, M., Marconi, W. and Riocci, M. *J. Mol. Catal.* **1980**, *9*, 381.
[17] a) Selke, R., Haupke, K. and Krause, H. W. *J. Mol. Catal.* **1989**, *56(1–3)*, 315; b) Selke, R. and Capka, M. *J. Mol. Catal.* **1990**, *63(3)*, 319; c) Selke, R. *J. Mol. Catal.* **1986**, *37(2–3)*, 227.
[18] Augustine, R. L. *WO 9828074*.
[19] Gamez, P., Dunjic, B., Pinel, C. and Lemaire, M. *Tetrahedron Lett.* **1995**, *36(48)*, 8779.
[20] Adima, A., Moreau, J. J. E. and Wong Chi Man, M. *J. Mater. Chem.* **1997**, *7(12)*, 2331.
[21] a) Ohkuma, T., Ooka, H., Hashiguchi, S., Ikariya, T. and Noyori R. *J. Am. Chem. Soc.* **1995**, *117*, 2675; b) Hashiguchi, S., Fujii, A., Takehara, J., Ikariya, T. and Noyori, R. *J. Am. Chem. Soc.* **1995**, *117*, 7562; c) Fujii, A., Hashiguchi, S., Uematsu, N., Ikariya, T. and Noyori, R. *J. Am. Chem. Soc.* **1996**, *118*, 2521; d) Matsumura, K., Hashiguchi, S., Ikariya, T. and Noyori, R. *J. Am. Chem. Soc.* **1997**, *119*, 8738; e) Uematsu, N., Fujii, A., Hashiguchi, S., Ikariya, T. and Noyori, R. *J. Am. Chem. Soc.* **1996**, *118*, 4916.
[22] Ter Halle, R., Schulz, E. and Lemaire, M. *Synlett.* **1997**, 1257.
[23] Bayston, D. J., Travers, C. T. and Polywka, M. E. C. *Tetrahedron: Asymmetry* **1998**, *9(12)*, 2015.
[24] Krause, H. W. *React. Kinet. Catal. Lett.* **1979**, *10(3)*, 243.
[25] Kinting, A., Krause, H. and Capka, M. *J. Mol. Catal.* **1985**, *33*, 215.
[26] Nagel, U. and Kinzel, E. *J. Chem Soc., Chem. Commun.* **1986**, 1098.
[27] Corma, A., Iglesias, M., del Pino, C. and Sanchez, F. *J. Chem. Soc., Chem. Commun.* **1991**, 1253.
[28] Pugin, B. WO 97/02232; Pugin, B. WO 96/32400.
[29] Burk, M. J., Casy, G. and Johnson, N. B. *J. Org. Chem.* **1998**, *63*, 6084.
[30] Gilbertson, S. R. and Wang, X. *Tetrahedron Lett.* **1996**, *37(36)*, 6475.
[31] a) Stille, J. K. *Chem. Ind.* **1985**, *22*, 23; b) Bayer, E. and Schurig, V. *Chemtech* **1976**, 212.
[32] Fritschel, S. J., Ackerman, J. J. H., Keyser, T. and Stille, J. K. *J. Org. Chem.* **1979**, *44(18)*, 3152.
[33] a) Parrinello, G., Deschenaux, R. and Stille, J. K. *J. Org. Chem.* **1986**, *51*, 4189; b) Pittman, C. U., Kawabata, Y. and Flowers, L. I. *J. Chem. Soc., Chem. Commun.* **1982**, 473.
[34] a) Stille, J. K. and Parrinello, G. *J. Mol. Catal.* **1983**, *21*, 203; b) Stille, J. K. *J. Macromol. Sci., Chem.* **1984**, *A21(13–14)*, 1689; c) Parrinello, G. and Stille, J. K. *Polym. Prepr. (Am. Chem. Soc., Div. Polym. Chem.)* **1986**, *27(2)*, 9; d) Parrinello, G. and Stille, J. K. *J. Am. Chem. Soc.* **1987**, *109(23)*, 7122.
[35] Nozaki, K., Itoi, Y., Shibahara, F., Shirakawa, E., Ohta, T., Takaya, H. and Hiyama, T. *J. Am. Chem. Soc.* **1998**, *120*, 4051.
[36] Kolb, I., Cerny, M. and Hetflejs, J. *React. Kinet. Catal. Lett.* **1977**, *7*, 199.
[37] Capka, M. *Collect. Czech. Chem. Commun.* **1977**, *42(12)*, 3410.

10 Catalytic Heterogeneous Enantioselective Dihydroxylation and Epoxidation

P. Salvadori, D. Pini, A. Petri, and A. Mandoli

10.1 Introduction

The oxidation of C=C groups affords interesting oxygenated compounds such 1,2-diols and epoxides and is therefore one of the most powerful reactions in organic synthesis. By performing the reaction enantioselectively on prochiral substrates, optically active products can be obtained. This opens synthetic routes to many natural and biologically active chiral substances. In this respect, catalytic enantioselective procedures provide the most relevant approach [1].

Homogeneous catalysis often allows the use of a single active species, which can be adjusted to solve specific problems by adopting suitable reaction conditions. However, the separation of products, the recovery and the reuse of the catalyst can adversely affect the cost of processes. Heterogeneous catalysis is an obvious way to overcome such difficulties. With respect to enantioselective reactions, a severe limitation of traditional heterogeneous catalysts is that a catalytic surface mostly contains different types of catalytically active centers, with different and sometimes opposite stereoselectivity. Heterogenization of a homogeneous catalyst represents a modern way to combine the advantages of heterogeneous and homogeneous catalysis in enantioselective processes.

Within this context of heterogenization, the present chapter reviews the main contributions in enantioselective dihydroxylation and epoxidation.

10.2 Asymmetric Dihydroxylation

Within the last decade, the asymmetric dihydroxylation (AD) of alkenes catalyzed by osmium tetroxide (OsO_4) and chiral ligands has become one of the most important enantioselective reactions in organic chemistry [2]. As yet, the *Cinchona* alkaloid derivatives dihydroquinidine (DHQD) **1** and dihydroquinine (DHQ) **2**, first used as chiral ligands in AD by the Sharpless group, are the only auxiliaries able to furnish diols with both high chemical yield and enantioselectivity [3].

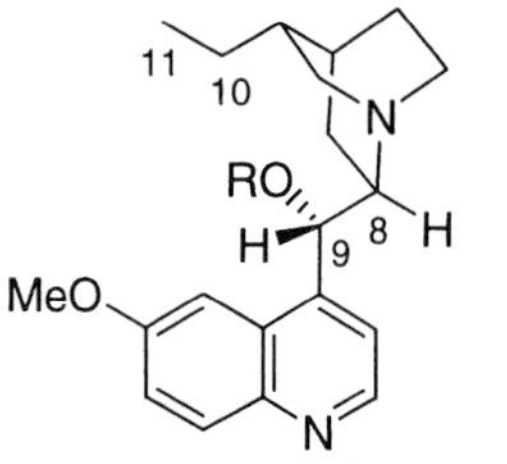

Dihydroquinidine (R=H)
DHQD
1

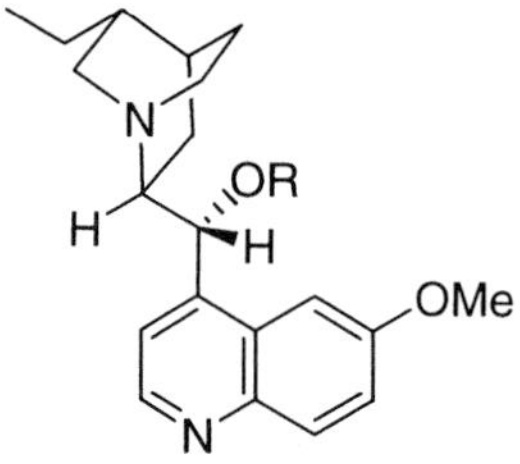

Dihydroquinine (R=H)
DHQ
2

These diastereomeric ligands behave in AD like enantiomers (Scheme 10.1): thus, the use of quinidine derivatives **1** leads to an excess of a diol that is the enantiomer of the diol produced in excess when using the quinine derivative **2**.

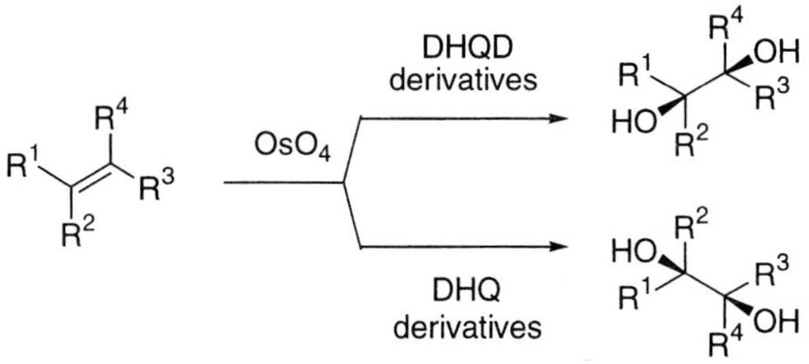

Scheme 10.1

Originally, stoichiometric amounts of the OsO_4 and the alkaloid derivatives (**1** and **2**, with R = $COCH_3$) were used, but this approach was uneconomical. Later it was discovered that the asymmetric process becomes catalytic if a stoichiometric oxidant like N-methylmorpholine-N-oxide (NMO) is used together with OsO_4 [4]. However, the enantioselectivity of the catalytic reaction was lower than that of the stoichiometric reaction. The presence of a second catalytic cycle, which furnishes the diol with little or no e.e., was found to be the cause of these different e.e. values [5]. The enantioselectivity was remarkably increased by eliminating this cycle, either by very slow addition of the olefin to the reaction mixture, or by performing the reaction in a biphasic system (tBuOH:H_2O) with $K_3Fe(CN)_6$ as stoichiometric co-oxidant in the presence of an excess of K_2CO_3 [6, 7].

The nature of the C9 substituent of the *Cinchona* alkaloid backbone was found to be of great importance [8]. Over 500 derivatives have been tested in catalytic AD reactions, and apart from the 4-chlorobenzoate ester (CLB) **3** used in the original catalytic procedure, the most remarkable results have been obtained with the phenanthryl ether (PHN) **4**, with ethers of phthalazine (PHAL) **5** and its analogs (DP-PHAL, DPP) **6–7**, with ethers of diphenylpyrimidine (PYR) **8** and of anthraquinone derivatives

(AQN) **9**. With these ligands, it is possible to obtain diols with high ee's (in many cases >90%) from almost any class of alkenes [9–13].

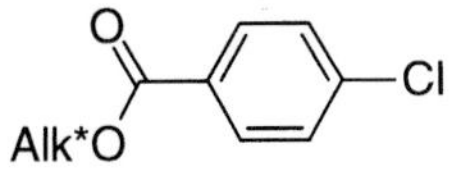

chlorobenzoate
(CLB)
3

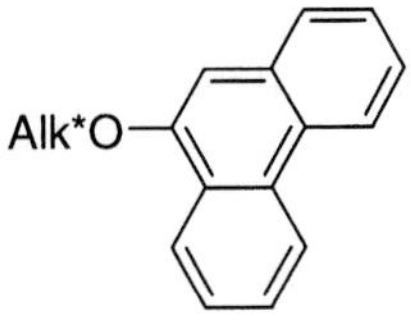

phenanthryl ether
(PHN)
4

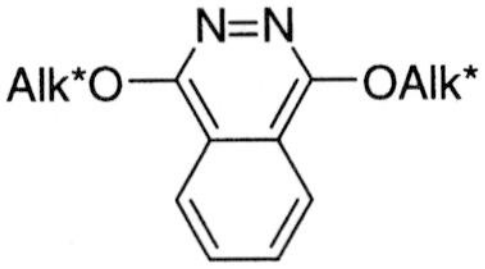

phthalazine
(PHAL)
5

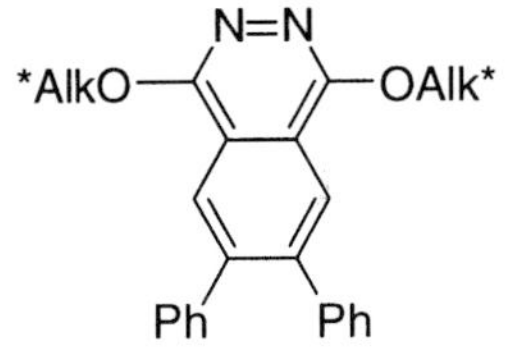

diphenylphthalazine
(DP-PHAL)
6

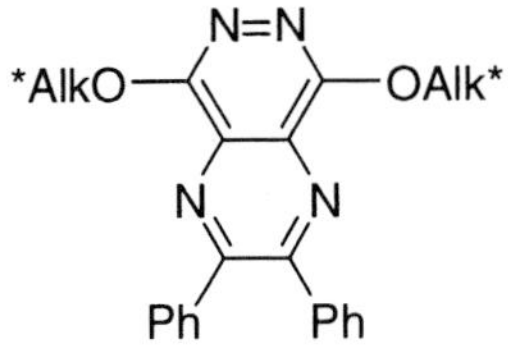

diphenylpyrazinopyridazine
(DPP)
7

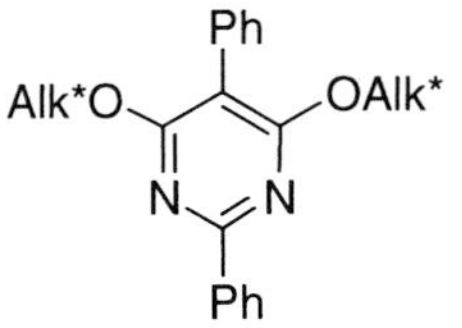

diphenylpyrimidine
(PYR)
8

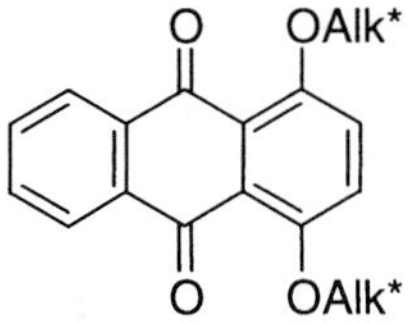

anthraquinone
(AQN)
9

Alk* = (DHQD) or (DHQ)

In parallel to these studies, *Cinchona* alkaloid derivatives have been immobilized on insoluble supports [14, 15]. Typical supports are linear or cross-linked organic polymers and inorganic matrices such as silica gel.

10.2.1 Use of Functionalized Polymers: Insoluble Polymer-Bound Catalysts for AD (IPB-AD)

A well-designed insoluble polymeric catalyst should meet the following requirements: no product contamination, easy recovery by filtration, stability in the reaction conditions, compatibility with the reaction medium, swelling effect, recyclability, and high enantioselectivity. One of the most challenging problems in the design of polymer-bound catalysts lies in the selection of the proper support which can offer the desirable swelling characteristics in the reaction solvent. In addition, the complexation sites must be spaced and suitably "diluted" in the polymeric backbone. There are two general strategies in synthesizing chiral polymeric ligands: the chemical modification of preformed polymers by a chiral reagent, or the copolymerization of a monomer containing the desired chiral group with an achiral monomer used as diluent and cross-linking agent. In particular, the latter approach offers several advantages such as the control of the chiral monomer concentration and the possibility of obtaining materials with different properties by varying polymerization parameters. This can be illustrated by the stepwise development of IPB-AD catalysts.

Initial attempts: copolymerization with acrylonitrile

Polymer-bound *Cinchona* alkaloids were already known before the advent of the highly effective catalytic AD process 16]. Initially, the hydroxyl group and the quinuclidine tertiary amino group were used as sites for anchoring to the polymer backbone [16]. These polymeric ligands however, showed low enantioselectivity since these functional groups appeared to play an important role in the chiral discrimination. Accordingly, the alkaloids were anchored to the polymer through the C10-C11 double bond, leaving the hydroxyl group available for chemical modification and the quinuclidine ring nitrogen available for coordination on the Os.

Copolymerization of 9-O-acylquinine derivatives with acrylonitrile in the presence of AIBN as radical initiator, using different molar ratios of the comonomers, furnished insoluble copolymers with varying extent of alkaloid incorporation (y) [17].

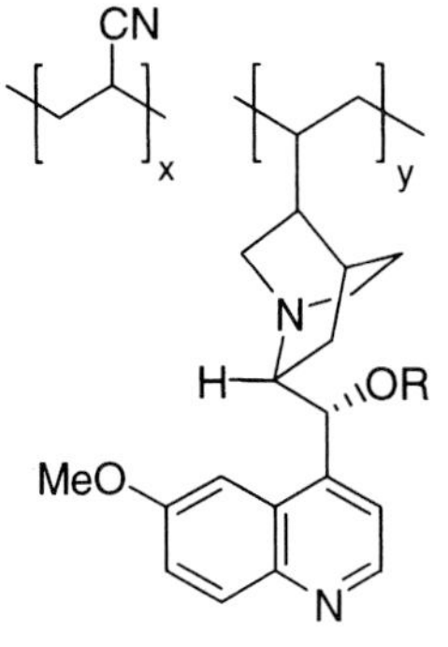

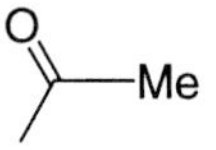

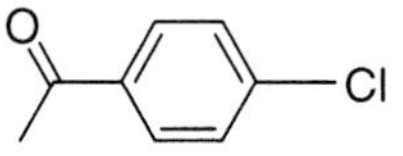

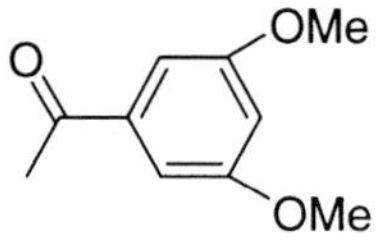

These crude copolymers were continuously extracted with different solvents (acetone, methanol) in order to obtain fractions insoluble in the AD reaction medium. Polymer **10a** containing 10 mol% of the acetate derivative, was used in the AD of *trans*-stilbene and other olefins with NMO as co-oxidant. Diols were obtained in good chemical yields but with low enantioselectivity (6–30%). An interesting trend emerged from these studies: only by using copolymers having a low alkaloid content (10–15% by mol), high chemical yields could be obtained. In contrast, with high alkaloid content copolymers, no diol was formed. This result was explained by the probable proximity of two quinuclidine moieties in polymers with high alkaloid concentrations; these units are able to bind OsO_4 to form very stable bi-chelate complexes, which are known to inhibit the hydrolysis of the intermediate osmate ester and, hence, the formation of the diol [18]. Enantioselectivities with various C9 modifiers (up to 46%) were lower than those provided by the homogeneous process with the same ligand and olefin. The OsO_4/polymeric alkaloid complexes could be simply and quantitatively recovered at the end of the reaction by filtration or centrifugation and reused several times without significant loss of activity or enantioselectivity.

Effects of a spacer between polymer and alkaloid

At about the same time, copolymers were prepared starting from acrylonitrile and quinidine derivatives that were spaced from the main polymeric chain [19].

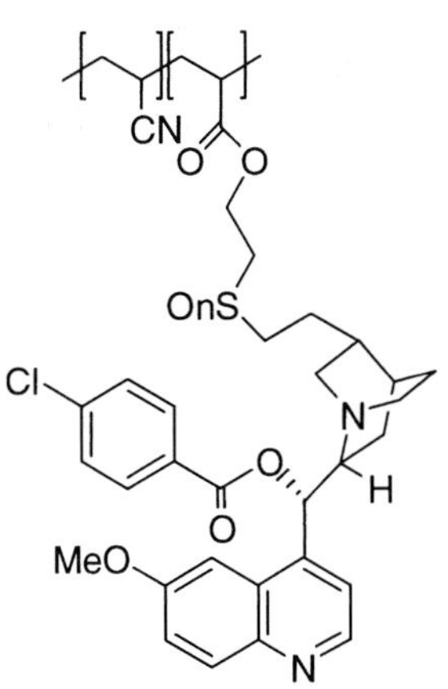

11a n = 1

11b n = 2

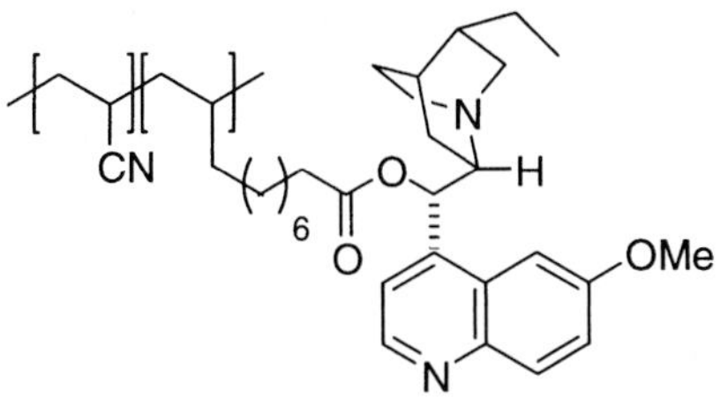

12

In the AD of *trans*-stilbene with the co-oxidants NMO and $K_3Fe(CN)_6$, high e.e.'s (from 78 to 87%) were obtained using polymers **11 a, b–12**, in which the alkaloid moiety is spaced from the polyacrylonitrile chain by different groups. Thus, a spacer seems to strongly increase the enantioselectivity in the heterogeneous AD. In a similar approach, the linear polymer **14** was synthesized from acrylonitrile and the styrene-containing monomer **13**; in **13**, a spacer is placed between the *para*-chlorobenzoate derivative of quinine (Q)-CLB and the easily polymerizable styrene moiety (Scheme 10.2) [20].

With polymer **14**, the dihydroxylation of *trans*-stilbene leads to higher e.e.'s than with **11** and **12**, even if the values are still lower than those of the homogeneous process. Control experiments confirmed that the enantioselectivity of the heterogeneous reaction was lowered by the formation of racemic diol via non-enantioselective pathways, due to achiral complexation of OsO_4 to N-methylmorpholine (formed when using NMO as co-oxidant) or to the nitrile groups of the polyacrylonitrile copolymers [15].

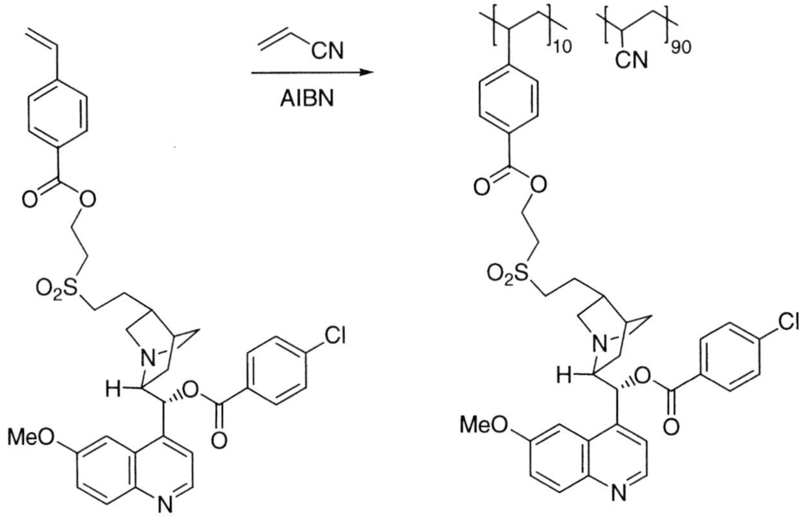

13 14

Scheme 10.2

Linear and cross-linked polystyrene; swellable vs. non-swellable polymers

Linear polystyrenic copolymers **15–17** were obtained from quinidine derivatives [21,22]. While initial catalytic results were rather disappointing [21], it was later reported that in the AD of *trans*-stilbene, homopolymer **15** exhibits excellent enantioselectivity with $K_3Fe(CN)_6$ as the oxidant (ee = 89%); lower e.e.'s were observed with NMO [22].

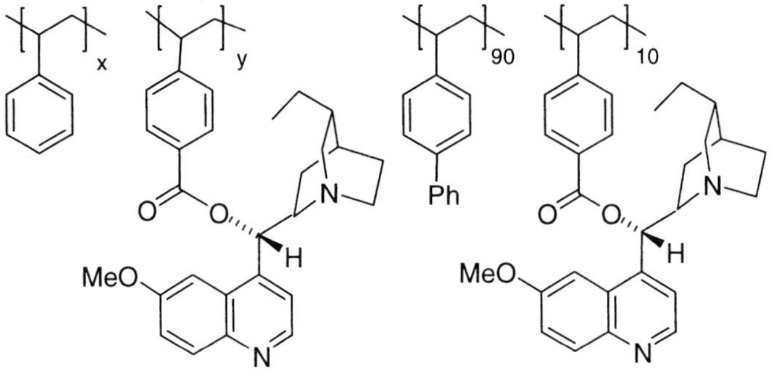

15 x = 0 ; y = 100
16 x = 90 ; y = 10

17

A cross-linked polystyrenic polymer **18** was synthesized by copolymerization of the chiral monomer **13** with styrene and divinylbenzene as cross-linking agent [23].

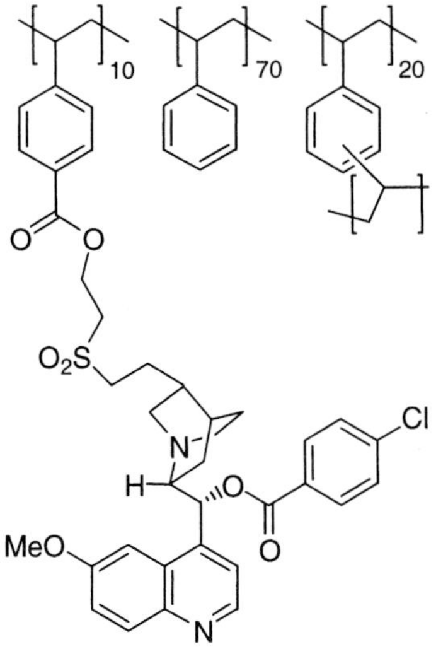

18

AD of *trans*-stilbene with NMO as co-oxidant was fast and gave a high level of enantioselectivity. On the other hand, formation of the diol was not at all observed with $K_3Fe(CN)_6$ in *t*BuOH:H_2O. This result was related to the different behavior of the copolymer in the reaction solvents used. In *t*BuOH:H_2O, the material is in a collapsed state, which prevents the contact between substrate and catalyst. In contrast, the reaction was fast with NMO in acetone:H_2O (10:1). In this solvent, the polymer was quite swollen, and the enantioselectivities of a series of diols compared well to those reported for the soluble ligand.

In an effort to circumvent such swellability problems, a polymeric support was sought that could swell in polar protic solvents. Polymer **19** was then obtained by copolymerization with hydroxyethylmethacrylate (HEMA) and with ethylene glycol dimethacrylate (EGDMA) as a cross-linking agent [24].

Thanks to the alcohol groups, polymer **19** swelled very well both in acetone:H_2O and *t*BuOH:H_2O, in contrast to the polystyrenic analogue **18**. In AD with both NMO and $K_3Fe(CN)_6$, the reaction proceeded with good enantioselectivity (up to 95%) and high chemical yields with aliphatic and aromatic olefins, both terminal and internal. With this polymer **19**, the heterogeneous catalytic AD was successfully expanded to trisubstituted and aliphatic monosubstituted alkenes, using both co-oxidants. An important conclusion from these results is that the polymeric support must be compatible with the reaction conditions to obtain e.e.'s comparable to those achieved in the homogeneous phase.

Similar results in terms of chemical yield and enantioselectivity were obtained in AD of two *trans*-disubstituted aromatic alkenes with copolymers **20–21**, synthesized from a benzoate monomer and methyl methacrylate (MMA) or HEMA [22]. As previously noted, an increase of the polymer polarity was found to favor the compatibility of the polymeric support with the reaction medium.

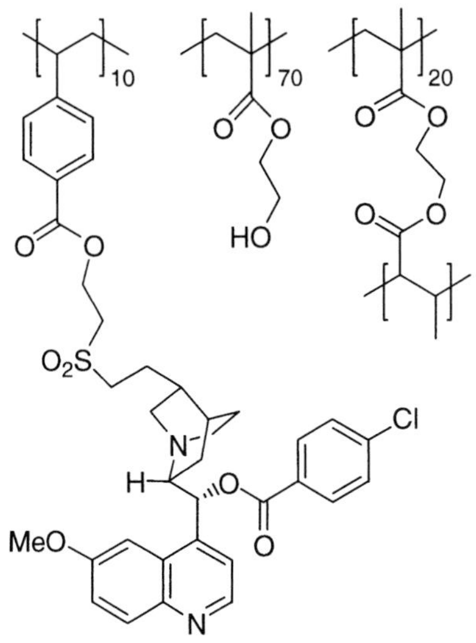

19

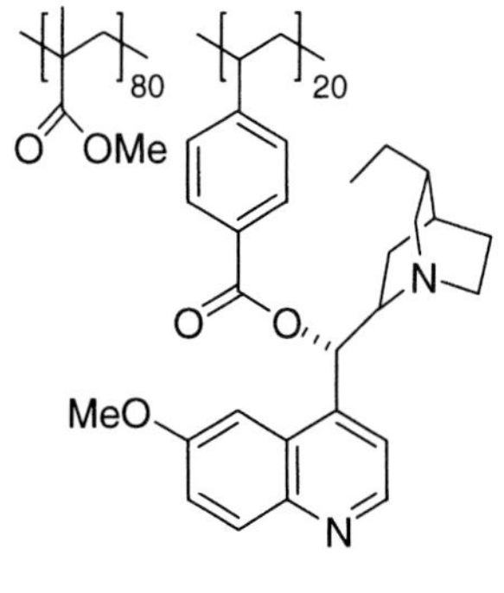

20

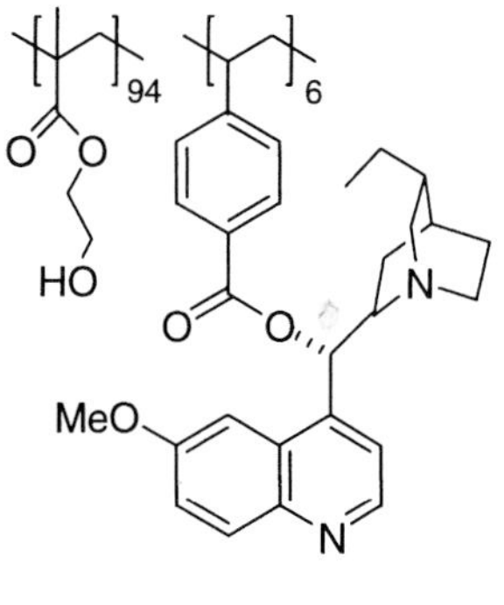

21

Variation of the O9 substituent on the alkaloid

From the observations in homogeneous catalytic AD, it was expected that replacing the 4-chlorobenzoate with a different substituent would further improve the enantioselectivity in the heterogeneous process [2]. Polyhydroxymethacrylic copolymers **22** and **23** were obtained from the corresponding chiral monomers containing the PHN and PHAL derivative of quinidine (QD) [25, 26].

These copolymers were used in AD with $K_3Fe(CN)_6$ as co-oxidant, reaching high e.e.'s (in many cases >80%) comparable to those reported for the soluble ligands. A significant improvement was obtained with mono- and disubstituted aliphatic alkenes. It should be pointed out that with polymer **23**, four of the six possible substitution

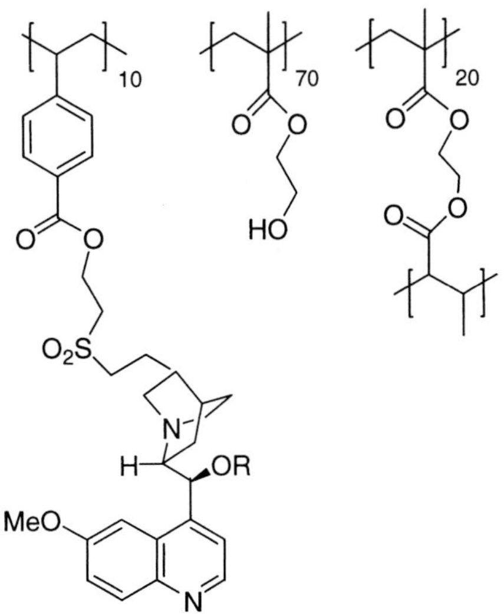

22 R =

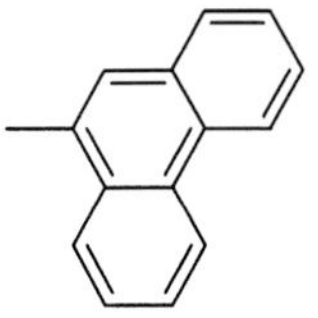

23 R =

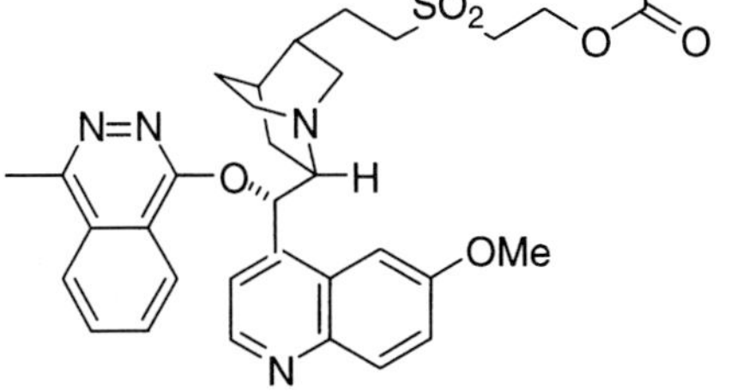

classes of olefins could be dihydroxylated in the heterogeneous phase to give diols with synthetically useful enantioselectivities.

Kinetic studies were performed with these heterogeneous AD catalysts [15, 26]. The dihydroxylation rate and enantioselectivity with both co-oxidants and insoluble ligands were compared to those obtained in the homogeneous phase, using a soluble analogue of the chiral monomer. These experiments show that the behavior of an alkaloid anchored on a polyhydroxymethacrylic support is highly similar to that of a soluble alkaloid ligand. Hence, it seems that similar reaction mechanisms are operative in the homogeneous and heterogeneous catalysts.

Direct copolymerisation of the C=C groups in *bis-Cinchona* alkaloid derivatives **24–26** with monomers such as EGDMA, MMA and HEMA has been reported by different groups [27–29]. In all cases, high enantioselectivities were claimed for AD of *trans*-stilbene and other olefins.

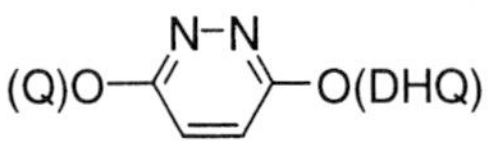

(Q)O–(pyridazine)–O(DHQ)

24

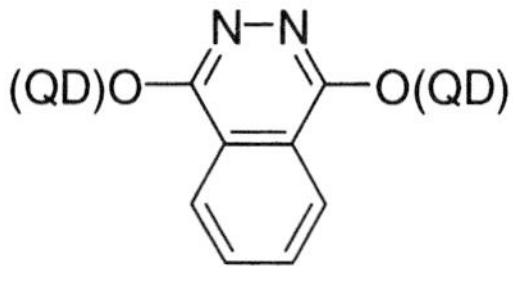

(QD)O–(phthalazine)–O(QD)

25

(QD)O–(2,5-diphenylpyrimidine)–O(QD)

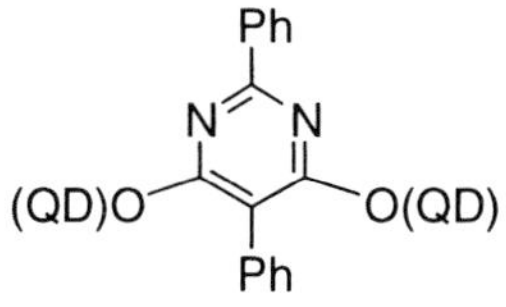

26

However, it was later demonstrated that by using the published procedures, one does not obtain copolymers, but rather poly(methyl methacrylate), poly(hydroxy-ethyl methacrylate) and poly(ethylene glycol dimethacrylate) with some physically trapped unreacted chiral monomers [15, 26, 30]. Indeed, it is well-known that allylic carbon-carbon double bonds (like in **24–26**) do not easily participate in a free radical vinyl copolymerization, and might even retard or inhibit the formation of polymers. Therefore, it is vital to carefully check materials obtained by copolymerization, since the observed catalytic reaction may be due to leached species that were either weakly bound to the support or simply physically trapped. A simple procedure is the continuous extraction of the crude copolymers with methanol or any other solvent in which the chiral monomers are soluble; leaching of the chiral monomer or of part of it should be assessed by direct analysis of the washings.

Soluble polymers

In addition to the insoluble polymeric ligands discussed above, various approaches have been developed, using soluble polymer-bound ligands or polymer-supported reagents for AD [31 a–d]. Regarding the soluble polymers, different groups prepared *Cinchona* alkaloid derivatives linked to polyethylene glycol monomethyl ether [31 a–c]. These polymeric ligands are completely soluble in the AD solvent mixture. Hence, the e.e.'s are very high and the reaction is fast. However, an additional step is required to isolate the products and to recover the ligands, since the polymers must be precipitated from the reaction mixture by addition of a solvent in which they are insoluble.

10.2.2 Use of Inorganic Supports

As a more attractive approach to heterogeneous catalytic AD on a large scale, the immobilization of pyridazine derivative **25** on silica gel was investigated [32]. The insoluble ligand **27** was used with different alkenes and $K_3Fe(CN)_6$. While the yields were comparable to those obtained in the homogeneous reaction, the e.e.'s of the diols were lower, especially for aliphatic mono- and disubstituted olefins. In addition, reuse of silica gel-bound ligand indicated some leaching of OsO_4.

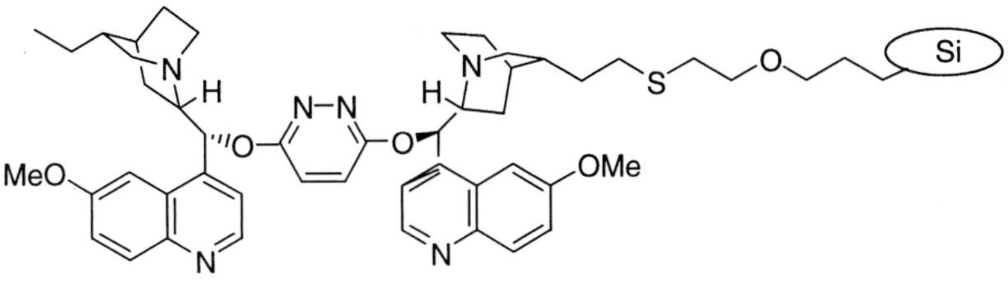

27

In related work, the silica gel-supported alkaloid **28** was used with *trans*-1,2-disubstituted alkenes and 1-phenylcyclohexene with $K_3Fe(CN)_6$ as the oxidant [33]. While this procedure leads to diols in high chemical yields and enantioselectivities, there is appreciable leaching of the chiral ligand and OsO_4 from the insoluble material, probably due to the strongly alkaline reaction conditions [34].

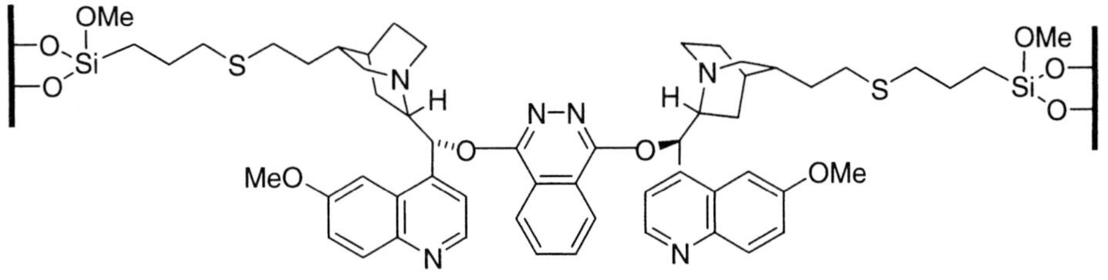

28

Bis(9-O-dihydroquinidinyl)pyrazinopyridazine **29 a–c** and *bis*(9-O-dihydro-quinidinyl)pyrimidine **30 a, b** were immobilized on different commercial functionalized silicas [35].

High enantioselectivities were achieved especially with ligands having ether and ester linkers. Lower asymmetric induction was observed with the amido-linked ligand **29 b**. It was proposed that residual amino groups on the silica surface or the amido linker, itself, may have a negative influence. Repeated use of the immobilized ligands is possible only after the addition of small quantities of the osmium salt after each run; otherwise, a considerable decrease in the chemical yield is observed. The silica-supported alkaloid **29 a** was used seven times in the AD of styrene without significant loss of enantioselectivity, while with **29 c,** lower e.e.'s were obtained after each run, presumably because part of the ligand was lost in the alkaline reaction conditions.

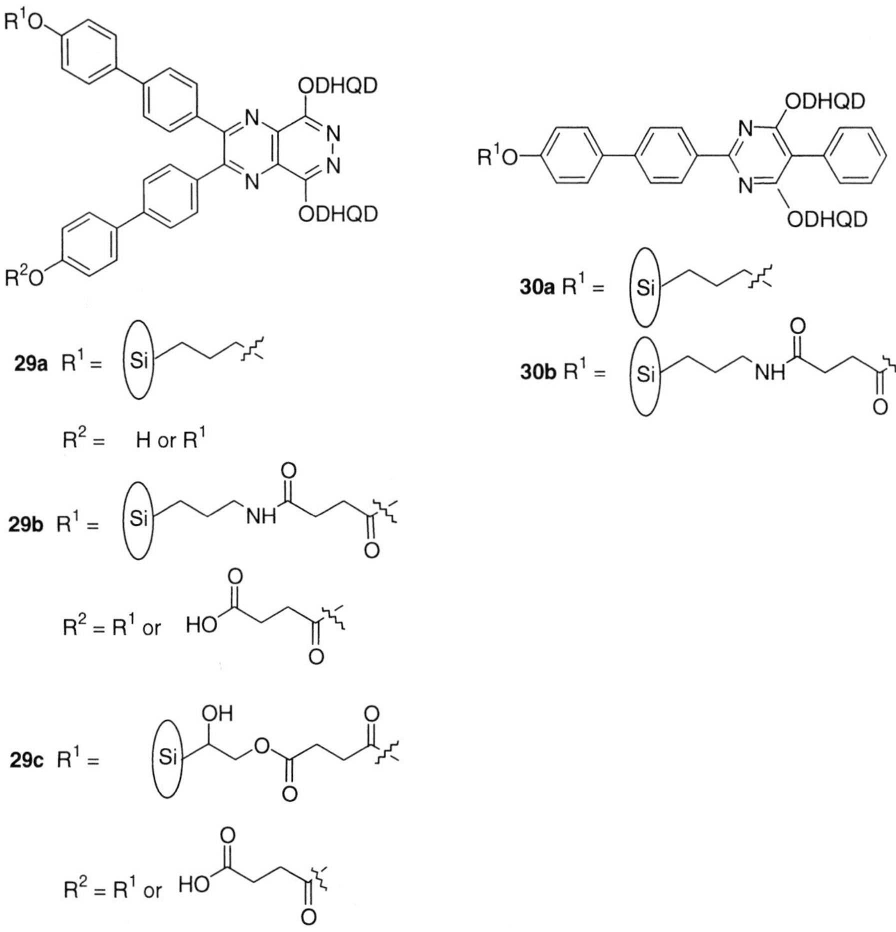

R¹O ODHQD N N N N ODHQD R²O

29a R^1 = Si

R^2 = H or R^1

29b R^1 = Si NH O O

R^2 = R^1 or HO O O

29c R^1 = Si OH O O O

R^2 = R^1 or HO O O

R¹O N N ODHQD ODHQD

30a R^1 = Si

30b R^1 = Si NH O O

10.3 Heterogeneous Catalytic Asymmetric Epoxidation of Carbon-Carbon Double Bonds

Several procedures are available for the highly enantioselective homogeneous catalytic epoxidation of C=C bonds, not only for unfunctionalized olefins, but also for substrates containing hydroxyl and carbonyl functional groups. While heterogeneous variants of all relevant homogeneous systems have been reported, the results lag behind in comparison with the best homogeneous results. Globally, the field of heterogeneous enantioselective epoxidation is less mature than the asymmetric dihydroxylation considered above. Nevertheless, recent reports have opened interesting perspectives, as will become clear in the remaining part of this chapter.

10.3.1 Epoxidation of Unfunctionalized Alkenes with Mn(salen) Catalysts

After extensive work, mainly from Katsuki's and Jacobsen's research groups, manganese(III) complexes derived from chiral tetradentate *bis*(salicylaldiminate) ligands (generally known as 'salen') emerged in the last decade as the most effective catalysts for the asymmetric epoxidation of unfunctionalized olefins, with **31** being one of the most effective complexes (Scheme 10.3) [36–38].

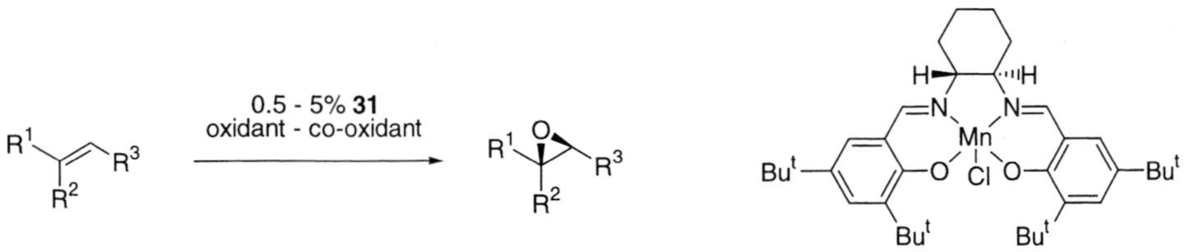

Scheme 10.3

The discrimination between the enantiotopic faces of prochiral alkenes was found to depend strongly on the substitution pattern of the salen, as elucidated by a modular variation study of the amine and salicylidene fragments of the ligand. Furthermore, the nature of the terminal oxidant and the presence of co-oxidant additives proved to be important in order to optimize activity and selectivity. A combination of catalyst **31,** sodium hypochlorite, *meta*-chloroperbenzoic acid (*m*-CPBA) or iodosylbenzene (PhIO) as stoichiometric oxidant, and a tertiary amine-N-oxide as co-oxidant leads to e.e. values in the range of 75 to 98% for a variety of terminal, *cis*-1,2-disubstituted and trisubstituted olefins [36]. Although the debate on the exact mechanism of the oxo transfer step is still open, there is general agreement that the reaction involves the oxidation of **31** by the stoichiometric oxidant to afford a Mn(V) oxo complex. In the catalytic cycle, the metal center shuttles between the oxidation states +III and +V [36, 39]. However, the simultaneous presence of Mn(III) and Mn(V) oxo species is inherently harmful for the catalysis, because of the formation of inactive μ-oxo-bridged dimers, (salen)Mn(IV)-O-Mn(IV)(salen). This is one of the reasons for the generally limited turnover number encountered with this class of catalysts [40]. In this respect, it is evident that site isolation of the Mn complex on a surface or within the pores of an insoluble support may increase catalyst life and turnover number. This perspective has stimulated the preparation of supported chiral Mn(salen) complexes [39].

10.3.1.1 Organic Insoluble Polymer-Bound Jacobsen-Type Catalysts (IPB-AE)

Salen ligands with two polymerizable groups

In a first approach, a chiral Mn(salen) complex, bearing two spaced pendant vinyl groups, was polymerized in a radical reaction with ethylene glycol dimethacrylate, leading to copolymer **32** [41]. Due to the high cross-linking density, the material was insoluble and swelled little in organic solvents. Nevertheless, when toluene was used as a porogenic agent in the course of the copolymerization, a macroporous texture was obtained with a high surface area and an easy access for substrate and terminal oxidant. In the epoxidation of representative alkenes with PhIO in acetonitrile, use of **32** led to activities and chemoselectivities comparable to those of soluble catalysts [41].

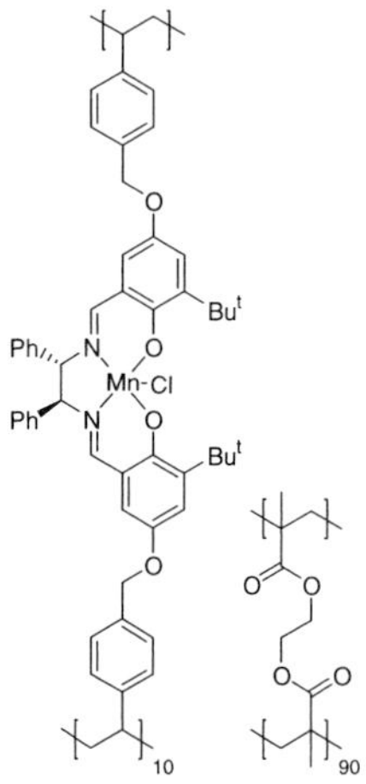

32 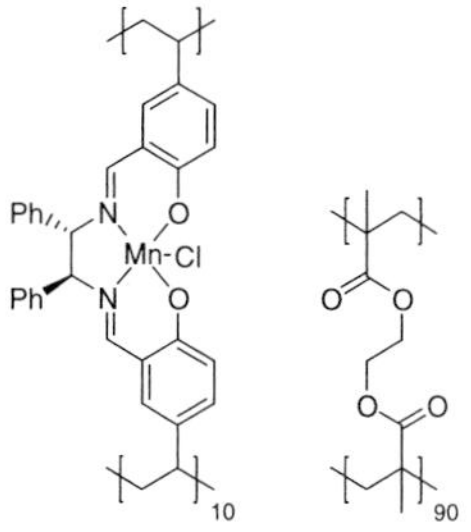 **33** 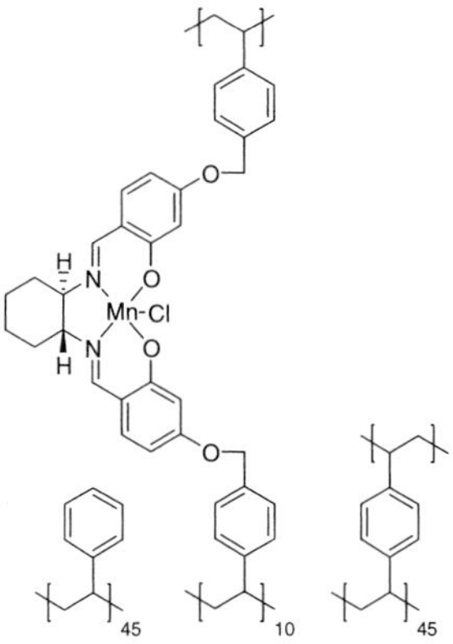 **34**

Unfortunately, enantioselectivity values were much lower, with a maximum e.e. of 30% in the case of dihydronaphthalene, even in the presence of pyridine-N-oxide as co-oxidant. Interestingly, activity and stereoselectivity were unchanged over 5 recycles. Speculating that the steric hindrance of the chirally imprinted macromolecular chain would force the substrate to follow a well-defined approach trajectory to the catalytic center, the same authors and others subsequently evaluated the non-spaced supported complexes **33** and **34**. These complexes lack the bulky 3,3′ substituents which are normally required for high enantioselectivity of the salen complex [42, 43]. However, the low e.e. values (<15%) recorded with such catalysts discouraged further exploitation of this concept.

As the limited e.e. values observed with **32** could be traced to the rigidity of the heavily cross-linked polymeric matrix, and possibly to unfavorable stereoelectronic effects in the modified salen ligand, it was decided to prepare a terpolymer **35** containing styrene, divinylbenzene and a non-spaced chiral Mn(salen) monomer. This Mn monomer more closely resembles the optimal structure of the soluble catalyst **31** [44].

Dichloromethane proved to be unsuitable as a solvent for reactions with **35**, since colloidal suspensions were formed. However the catalyst was perfectly insoluble in acetonitrile. A reaction with *m*CPBA/NMO and **35** in acetonitrile resulted in satisfactory epoxidation rate and recycling, even if no e.e. value higher than 41% could be recorded. Nevertheless, the styrene epoxide e.e. was higher with **35** than with the highly cross-linked **32**. In order to further reduce the steric hindrance around the catalytic center, the spaced copolymers **36a, b** were synthesized, resulting in a remarkable stereoselectivity improvement for all substrates examined [45]. In the case of indene and *cis*–methylstyrene, e.e. values as high as 60% and 62%, respectively, were recorded using **36a** with 97% chemoselectivity at higher than 90% conversion. Curiously, the oxidation of dihydronaphthalene and 1-phenylcyclohexene was much less clean, with yields lower than 50% and reduced stereoselectivity (e.e. values are respectively 37 and 42% with **36b**).

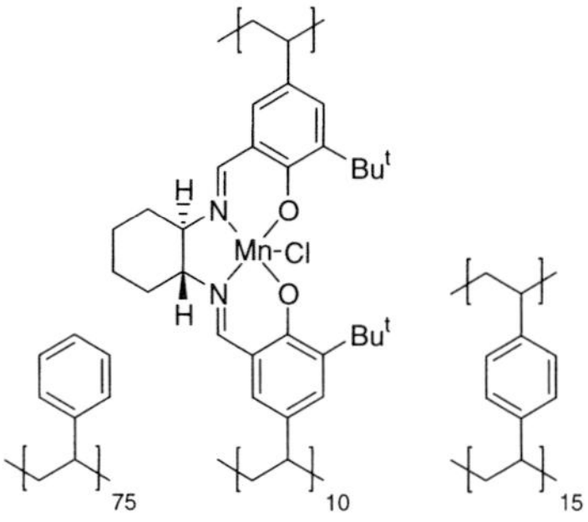

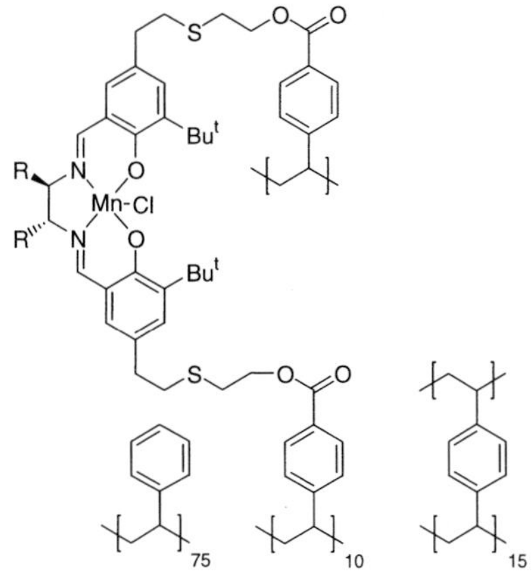

Salen ligands with a single link to the polymer backbone

While the previous examples clearly demonstrate that an increased flexibility around the metal center is beneficial, it is evident that even in **36**, the double attachment of the supported Mn complex to the macromolecular backbone is a potential cause of distortion of the catalyst geometry. Obviously, one of the most appealing solutions to such a problem is the preparation of materials featuring a single linkage between the chiral unit and the macromolecular matrix. However, the synthesis of e.g. monostyryl Mn(salen) comonomers is heavily complicated by the tendency of non-symmetric salen ligands to equilibrate to mixtures also containing C_2-symmetric species.

To overcome these difficulties, a different strategy was developed based on the solid-state sequential construction of the Mn(salen) structure onto functional polymers [46]. Starting with various achiral gel-type or macroporous materials, the supported catalysts **37–41** were prepared.

The catalytic properties of these materials were evaluated mainly in the epoxidation of 1-phenyl-1-cyclohexene with *m*-CPBA/NMO in CH_2Cl_2 at 0 °C. Despite the limited substrate scope, several interesting conclusions emerge from the results. While the supported complexes **39** and **40** proved barely active (<5% almost racemic epoxide within 2 h) and **41** furnished low e.e.'s, fair yields (36–49%) were obtained using **37–38**, with e.e. values in the range of 61–91%. In the case of **40** and **41,** the poor results were ascribed to the heavy perturbation of the optimal salen structure of **31**, leading to modified or excessive steric hindrance around the metal center. A related example in this sense is probably **42**, prepared by starting with a commercial Merrifield resin which gave modest e.e. values with most of the substrates tested [47].

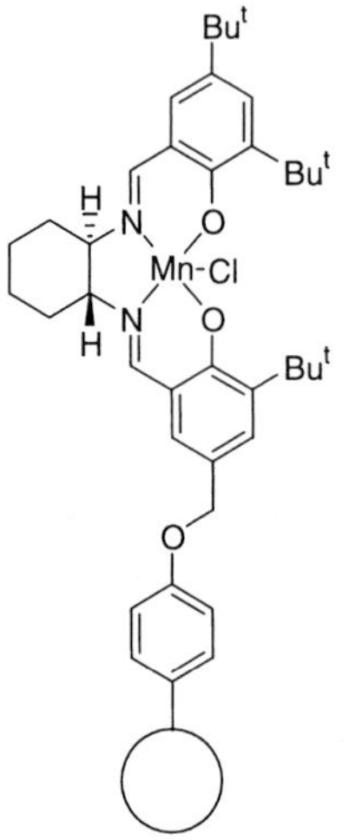

37a styrene/DVB/4-hydroxystyrene (66.4/1.6/32)

37b styrene/DVB/4-hydroxystyrene (44/24/32)

38 MMA/EGDMA/4-hydroxystyrene (19/68/13)

39 styrene/DVB/N-4-hydroxyphenylmaleimide (20/60/20) copolymeric support

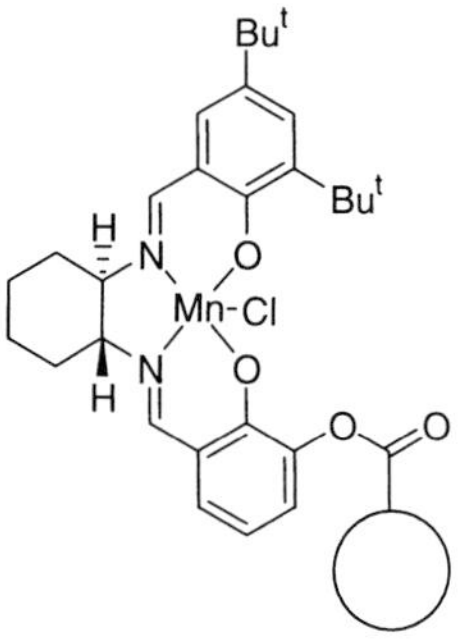

40 MMA/EGDMA/ 3-methacryloyloxysalicylaldehyde (45/30/25) copolymeric support

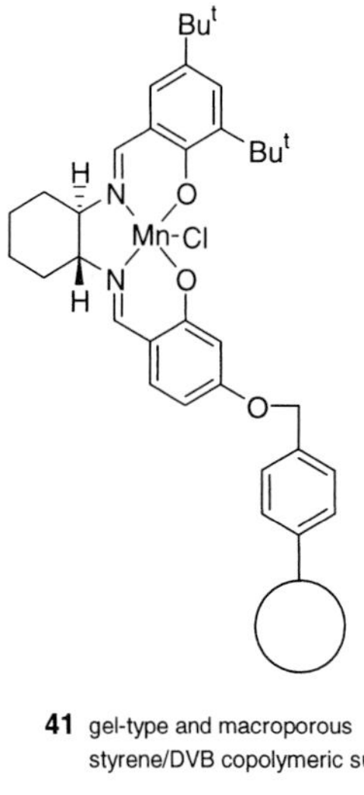

41 gel-type and macroporous styrene/DVB copolymeric supports

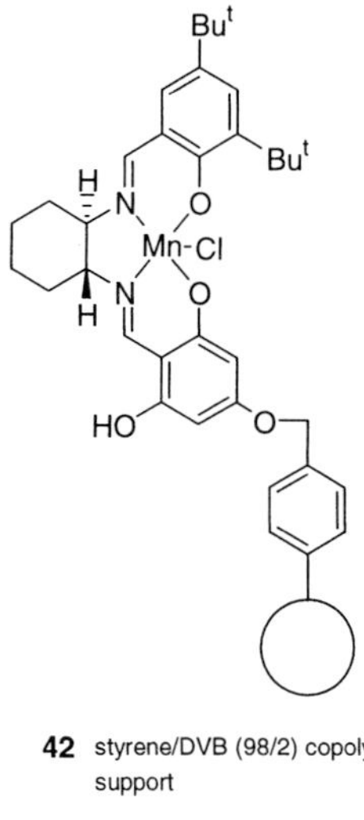

42 styrene/DVB (98/2) copolymeric support

On the other hand, the low activity of material **39** may be attributable to domain formation within the polymer. It is known that styrene and maleimide have a strong tendency to form a 1:1 alternating copolymer, regardless of the polymerization feed composition. Hence, after stepwise derivatization, one obtains a polymer with high local Mn(salen) concentration, eventually leading to formation of inactive μ-dimers. A similar bimolecular deactivation may arise with the gel-type system **37a**, because the macromolecular chains are flexible in this polymer with its low cross-linking density. Even if **37a** still displayed appreciable activity (36% yield), an evident improvement was observed when switching to the more rigid porous materials **37b** and **38** (47 and 49% yield, respectively).

Regarding the stereoselectivity, the e.e. values of the styrene-based materials **37a** and **37b** (61 and 66%, respectively) show that simple engineering of the macromolecular texture does not suffice to attain e.e. values matching those of soluble catalysts. Indeed, the influence of the achiral comonomer and cross-linker seem dominant, with the methacrylate-type copolymer **38** yielding a remarkable e.e. of 91%. While it cannot be excluded that the outstanding performance of **38** results from an apt combination of morphological and microscopic parameters (pore size, surface area, chain mobility, local polarity), this study proves that a highly enantioselective-supported epoxidation catalyst can be developed. A more detailed investigation is therefore desirable, not only to define substrate scope and recycling effectiveness, but also for a comprehensive understanding of the effect of macromolecular properties on catalytic activity and stereoselectivity.

Coordinative immobilization of the Mn-salen complex

The immobilization methods discussed so far relied on the covalent linking of the ligand fragment to the polymeric support. In a different approach, the materials **43a–c** were developed, in which the chiral complexes are anchored through coordination of the metal center to pyridine units [48].

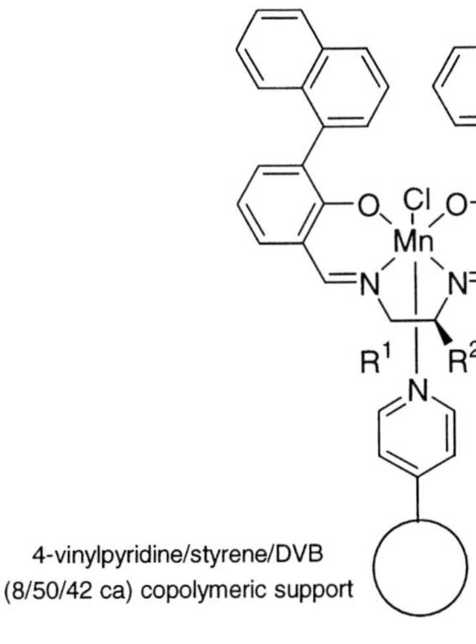

4-vinylpyridine/styrene/DVB
(8/50/42 ca) copolymeric support

43a $R^1, R^2 = (CH_2)_4$

43b $R^1 = H, R^2 = Me$

43c $R^1 = R^2 = Ph$

The supported catalysts were prepared by adsorption of the corresponding Mn(salen) on a 4-vinylpyridine/styrene/divinylbenzene copolymer from a dichloromethane solution. After Soxhlet extraction to remove the unbound complex, the complex loading was only 10 μmol/g, even if the concentration of pyridyl fragments was 380-fold higher. The actual existence of coordinative bonds, with pyridine as axial ligand, was inferred from the blue shift of a UV-Vis absorption band.

43a–c proved active in the epoxidation of styrene and 4-substituted styrenes with iodosylbenzene in CH_2Cl_2 at 4 °C. In 0.5–24 h, conversions in the range of 25–74% were obtained, with complete chemoselectivity. Unfortunately the enantioselectivity did not prove equally good, with e.e. values lower than 50% (note that these substrates are notoriously problematic [36]). At present, it is unclear whether this is a consequence of the anchoring technique or of the poor chiral discrimination capability of the unusual salen ligands in this study. In conclusion, it is interesting to note that the materials **43** could be recycled up to 10 times, demonstrating a quite strong binding of the chiral complex to the polymer.

10.3.1.2 Inorganic Polymer-Supported Jacobsen-Type Catalysts

Recently a 'ship-in-a-bottle' approach was reported for the preparation of Mn(salen) complexes trapped within the cages of an EMT zeolite. The procedure comprises the stepwise treatment of the zeolite with the chiral diamine, a suitable salicylaldehyde (unsubstituted or 3-*tert*-butyl-5-methyl substituted), Mn(II) acetate and finally LiCl under aerobic conditions [49]. The encapsulated catalysts **44a, b** were used for the epoxidation of selected olefins, mostly with the CH_2Cl_2/NaClO biphasic system, at room temperature for 24 h. Because of the restricted access of the relatively large reagents to the cages, some size selectivity effects were observed. Moreover, the enantioselectivity values were comparable to those with the soluble complex **31**. A remarkable e.e. of 88% for *cis*-β-methylstyrene was obtained with **44b**, when the reaction was run in the presence of a catalytic amount of pyridine-N-oxide in dichloroethane. However, both activity (15–47% conversion) and chemoselectivity (58–87%) of the supported catalysts proved significantly lower than with the homogeneous counterpart **31**. Moreover, a substantial worsening of enantioselectivity was found in a recycling attempt.

Shortly after, an independent contribution described the preparation of Y zeolite-encapsulated Mn(salen) **44c** [50]. In comparison with the work discussed above, the choice of a less hindered chiral ligand was dictated by the smaller dimensions of the zeolite Y supercages compared to those of EMT. A range of analytical techniques was employed to characterize the material supporting the structure **44c** with one catalytic center every 5 supercages and a slight excess of uncomplexed ligand. Catalytic tests were conducted with CH_2Cl_2/NaClO at 5 °C for 12–15 h, without any co-oxidant. Under those conditions, conversion values in the 5–40% range were achieved, together with 61–100% epoxide selectivity and an e.e. of 5–58%. Analogously to what was observed with **44a, b**, control experiments demonstrated that the catalytic activity was essentially associated with intrazeolite Mn(salen) chiral units, but unfortunately no attempt to recycle the material was reported.

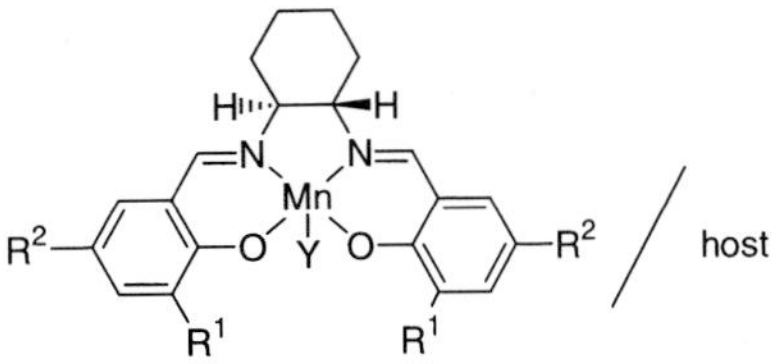

44a	$R^1, R^2 = H$	Y = Cl	host = EMT zeolite
44b	$R^1 = Bu^t, R^2 = Me$	Y = Cl	host = EMT zeolite
44c	$R^1 = R^2 = H$	Y = -	host = Y zeolite
44d	$R^1 = R^2 = Bu^t$	Y = -	host = laponite
44e	$R^1 = R^2 = Bu^t$	Y = Cl	host = Al-MS

In connection with these contributions, it is interesting to discuss the clay-supported catalyst **44d**, prepared by direct ion-exchange of the chiral complex **31** on a synthetic laponite [51]. After moderate enantioselectivities (32–34%) in the first epoxidation run with dihydronaphthalene and PhIO, a marked lowering of catalytic effectiveness was observed in subsequent runs. Careful analysis of the recovered material showed that while the metal leaching was modest (about 10% after 2 runs) and the N/Mn ratio remained constant, the carbon content increased steadily, probably because of the deposition of degradation products from the substrate. Moreover, comparison with IR and thermogravimetric data of analogous non-chiral Mn(salen) complexes suggests that oxidative decomposition of the supported ligand may take place.

The generally overlooked possibility of interactions between a manganese complex and an inorganic host has been addressed in the case of catalyst **31** embedded by impregnation into Al-, Ga- and Fe-substituted mesoporous silicates (MS). Thermoanalytical, UV-Vis and IR spectroscopic data were interpreted in terms of hydrogen bonding between surface silanol groups and salen aromatic rings [52]. The Al-substituted material **44e** was also employed in a limited catalysis study in the presence of NaClO, affording an e.e. of 55% with dihydronaphthalene.

In order to minimize unwanted interactions between the chiral catalyst and the support, an immobilization technique based on 'van der Waals wrapping' of metal complexes within the apolar elastomeric framework of a polysiloxane membrane was proposed, indicating that there might be a potential for the development of membrane reactors [53, 54].

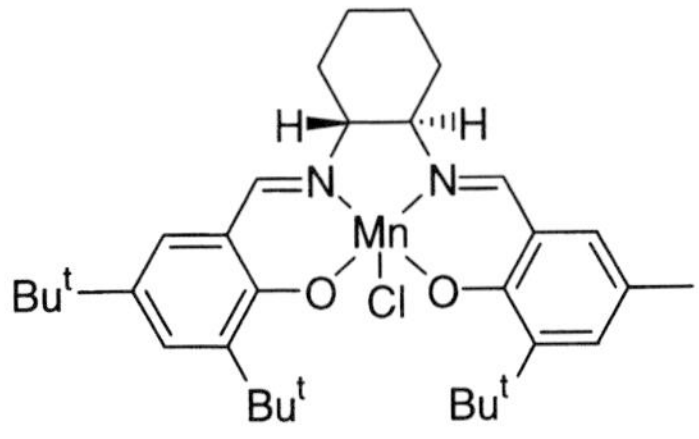

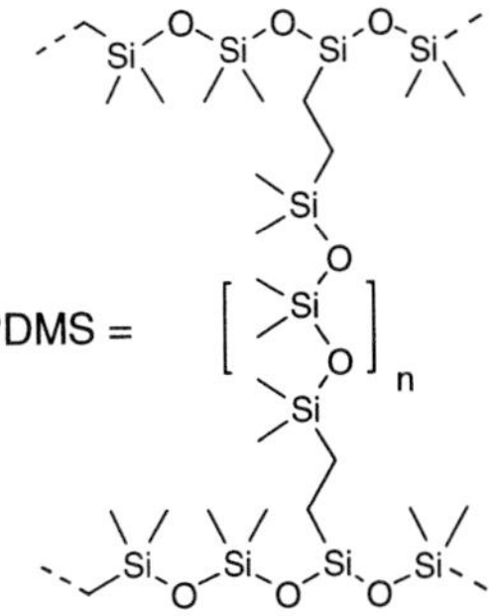

45 : $Y\text{-}Bu^t$ / PDMS

46 : $Y\text{-}CH_2\text{-}Y$ / PDMS

Unfortunately, severe complex leaching (9–52%) was observed during the epoxidation of selected substrates with the supported catalyst **45**, a problem that was only marginally mitigated by the use of a specially designed dimeric ligand (catalyst **46**) [55].

10.3.2 Epoxidation of Allylic Alcohols with Sharpless-Type Ti Catalysts

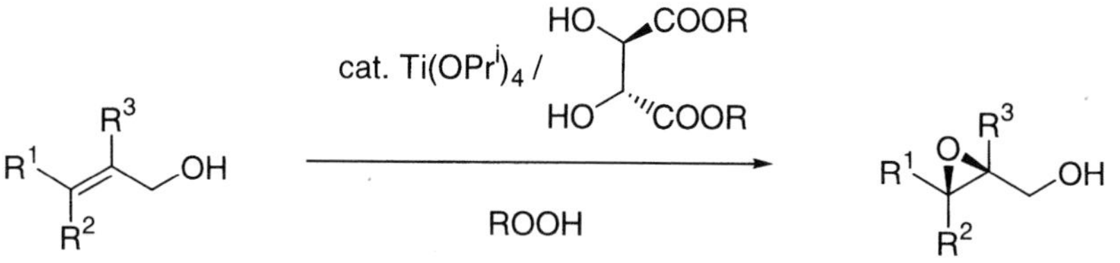
cat. Ti(OPr^i)$_4$ /
HO COOR
HO COOR
ROOH
R1 R2 R3 OH
R1 R2 R3 O OH

Scheme 10.4

Shortly after the introduction by Sharpless's group of the Ti(IV)-catalyzed asymmetric epoxidation procedure for allylic alcohols (Scheme 10.4), papers devoted to the preparation of supported tartrate esters started appearing, in an attempt to develop a heterogeneous variant. An e.e. of 66% was achieved with polystyrene-bound tartrates **47**; e.e. values in the 90–98% range were reported with the combination of a dialkyl tartrate and titanium-pillared montmorillonite (**48**) [56–59].

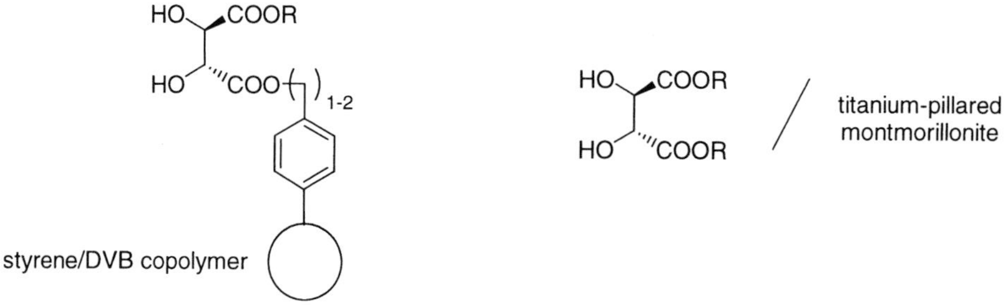
HO COOR
HO COO 1-2
styrene/DVB copolymer
HO COOR
HO COOR
titanium-pillared montmorillonite

47 R = Me, Et **48** R = Et, Pr^i

More recently, good results in terms of enantioselectivity were obtained by employing soluble, linear polyesters prepared by bulk or phase transfer polycondensation of tartaric acid with 1,n-diols [60]. In order to synthesize insoluble materials, forcing reaction conditions were subsequently adopted, resulting in polytartrates **49** with variable degrees (3–15%) of branching/cross-linking [61].

The catalyst resulting from **49** and Ti(OiPr)$_4$ proved insoluble in the reaction solvent (CH_2Cl_2), and was active in the epoxidation of *trans*- and *cis*-allylic alcohols and homoallylic alcohols. However, with the latter two substrate classes, weeks were required to reach acceptable conversions even using stoichiometric amounts or an excess of catalyst [61–63]. On the contrary, with *trans*-allylic alcohols, fair to good isolated yields were attained within hours, with e.e. values mostly in the 60–90% range. In general, it was observed that a branching/cross-linking degree higher than about 10% caused a dramatic degradation of enantioselectivity and, at variance with the homogeneous catalytic system, a 2:1 polymeric ligand to Ti ratio was found to be optimal. Even if the recovered polyester **49** showed unchanged IR characteristics, no recycling attempt was described [61]. Consequently, the possibility of developing a reusable, effective heterogeneous catalyst for the Sharpless epoxidation of allylic alcohols is still awaiting demonstration.

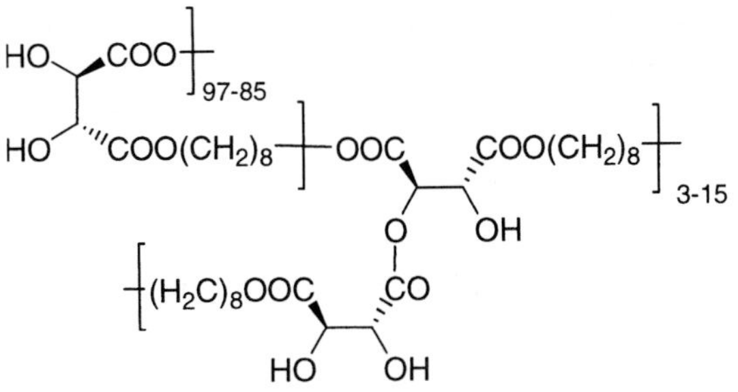

10.3.3 Epoxidation of *α,β*-Unsaturated Ketones under Julià-Colonna Conditions

The groups of Julià and Colonna have shown that polyamino acids can induce good to high e.e. values in the asymmetric epoxidation of *α,β*-unsaturated carbonyl compounds in an organic solvent/alkaline hydrogen peroxide biphasic system. Recently, good results were also obtained in anhydrous conditions (Scheme 10.5) [64].

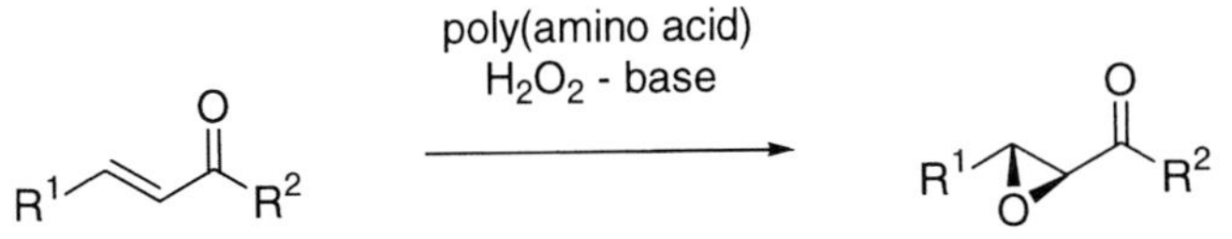

Scheme 10.5

Normally the polyamino acid is insoluble in the reaction medium; it forms a highly swollen gel which complicates the successive workup and discourages recycling of the recovered paste-like material. In addition, a decrease of the catalyst efficiency is sometimes observed, which is explained by partial hydrolysis of the polypeptide chain under the strongly alkaline conditions [65].

To overcome some of these drawbacks, the supported polyamino acids **50 a, b** were introduced. These are obtained by polymerization of the corresponding N-carboxy anhydrides, which is initiated by aminomethylated polystyrene [66]. Materials derived from macroporous, heavily cross-linked polystyrene afforded reduced enantioselectivities, just like materials with a very low or very high functionalization degree. The best results were obtained with polyalanine or polyleucine with an average polymerization degree (4–38), anchored to 2% cross-linked microporous polystyrene. The optimal functionalization degree of the polystyrene is between 20 and 50%.

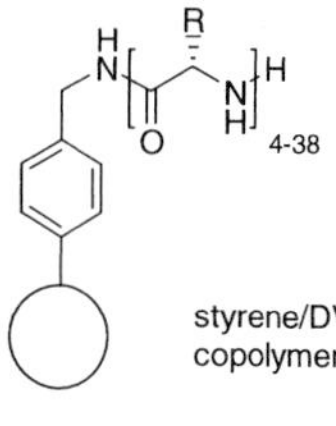

styrene/DVB (98/2) copolymeric support

50a R = Me

50b R = Bu^i

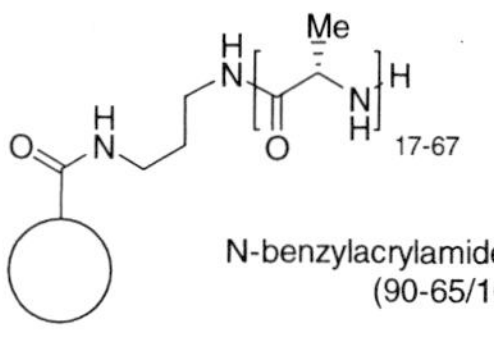

N-benzylacrylamide/N-3-(aminopropyl)methacrylamide (90-65/10-35) copolymeric support

51

By using **50a, b**, a 66–98% yield and an e.e. of 76–99% could be attained in the epoxidation of variously substituted chalcones in toluene/H_2O_2-NaOH. Furthermore, the supported catalyst was easily recovered by filtration and reused up to 12 times, with only a marginal decrease of activity and enantioselectivity. In this regard, it is worth noting that recently non-aqueous systems with urea-H_2O_2 or DABCO-H_2O_2 have been described. This allows the use of **50b** in a fixed-bed reactor [67]. Although only a microscale experiment was described and no recycling attempts were quoted, this opens interesting perspectives for the development of a continuous-flow process.

Shortly after, the polyacrylamide-anchored polyalanine **51** was reported, hereby showing that up to 80% optical purity could be reached in the epoxidation of chalcone in CCl_4/H_2O_2-NaOH [68]. Surprizingly, **51** proved poorly recyclable, with a serious decrease in enantioselectivity already in the first reuse. A concomitant reduction of the optical rotatory power of the supported polyamino acid was observed, suggesting a progressive racemization of the amino acid residues under reaction conditions.

10.4 Conclusions

In this chapter, the main approaches to heterogenization of organic ligands and metal complexes have been discussed in the context of enantioselective dihydroxylation (AD) and epoxidation (AE). In AD, the use of insoluble polymer-bound *Cinchona* alkaloid derivatives has allowed the attainment of high levels of enantioselectivity, in many cases comparable to those obtained in the homogeneous phase with soluble ligands. This approach might be useful for the development of technologies for small- and large-scale production of optically active diols. On the contrary, the use of inorganic matrices as insoluble supports for the chiral ligands has not met with the same success.

In the case of AE catalysts, a varied set of catalyst/support combinations has been described. Because the catalytic systems are quite diverse, each case poses specific problems when the development of a supported variant is attempted. Despite continuing efforts in the field of heterogeneous enantioselective epoxidation catalysis, the current systems mostly do not yet match the homogeneous counterparts, even if recent contributions disclose interesting perspectives.

Note added in proof. After the submission of this manuscript, a few more papers appeared, dealing with the asymmetric epoxidation of alkenes with MCM-41 or silica gel-bound Jacobsen's catalyst and with the epoxidation of enones with silica gel-adsorbed polyleucine [69,70].

References

[1] Ojima I, *Catalytic Asymmetric Synthesis*, VCH-Publishers: New York, 1993.
[2] (a) Lohray B B, *Tetrahedron: Asymmetry* **1992**, *3*, 1317–1349. (b) Kolb H C, Van Nieuwenhze M S, Sharpless K B, *Chem. Rev.* **1994**, *94*, 2483–2547.
[3] Hentges S G, Sharpless K B, *J. Am. Chem. Soc.* **1980**, *102*, 4263–4265.
[4] Jacobsen E N, Markó I, Mungall W S, Scröder G, Sharpless K B, *J. Am. Chem. Soc.* **1988**, *110*, 1968–1970.
[5] Wai J S M, Markó I, Svendsen J S, Finn M G, Jacobsen E N, Sharpless K B, *J. Am. Chem. Soc.* **1989**, *111*, 1123–1125.
[6] Lohray B B, Kalantar T H, Kim B M, Park C Y, Shibata T, Wai J S M, Sharpless K B, *Tetrahedron Lett.* **1989**, *30*, 2041–2044.
[7] (a) Minato M, Yamamoto K, Tsuji J, *J. Org. Chem.* **1990**, *55*, 766. (b) Kwong H, Sorato C, Ogino Y, Chen H, Sharpless K B, *Tetrahedron Lett.* **1990**, *31*, 2999–3002.
[8] Ogino Y, Chen H, Manoury E, Shibata T, Beller M, Lübben D, Sharpless K B, *Tetrahedron Lett.* **1991**, *32*, 5761–5764.
[9] Sharpless K B, Amberg W, Beller M, Chen H, Hartung J, Kawanami Y, Lübben D, Manoury E, Ogino Y, Shibata T, Ukita T, *J. Org. Chem.* **1991**, *56*, 4585–4588.
[10] (a) Sharpless K B, Amberg W, Bennani Y L, Crispino G A, Hartung J, Jeong K, Kwong H, Morikawa K, Wang Z M, Xu D, Zhang X L, *J. Org. Chem.* **1992**, *57*, 2768–2771. (b) Amberg W, Bennani Y L, Chadha R J, Crispino G A, Davis W D, Hartung J, Jeong K-S, Ogino Y, Shibata T, Sharpless K B, *J. Org. Chem.* **1993**, *58*, 844–849.
[11] Wang L, Sharpless K B, *J. Am. Chem. Soc.* **1992**, *114*, 7568–7570.
[12] Becker H, King S B, Taniguchi M, Vanhessche K P M, Sharpless K B, *J. Org. Chem.* **1995**, *60*, 3940–3941.
[13] Becker H, Sharpless K B, *Angew. Chem. Int. Ed. Engl.* **1996**, *35*, 448–451.
[14] Bolm C, Gerlach A, *Eur. J. Org. Chem.* **1998**, 21–27.
[15] Salvadori P, Pini D, Petri A, *SynLett* **1999**, 1181–1190.
[16] (a) Kobayashi N, Iwai K, *J. Am. Chem. Soc.* **1978**, *100*, 7071–7072. (b) Kobayashi N, Iwai K, *J. Polym. Sci., Polym. Chem. Ed.* **1980**, *18*, 223–233.
[17] (a) Pini D, Rosini C, Nardi A, Salvadori P, *Fifth IUPAC Symposium on Organometallic Chemistry Directed Towards Organic Synthesis*, Abstract PS 1–67, Florence (Italy), October 1–6 **1989**. (b) Petri A, *Tesi di Laurea*, University of Pisa, Italy, **1990**. (c) Pini D, Petri A, Nardi A, Rosini C, Salvadori P, *Tetrahedron Lett.* **1991**, *32*, 5175–5178.
[18] Jacobsen E N, Markó I, France M B, Svendsen J S, Sharpless K B, *J. Am. Chem. Soc.* **1989**, *111*, 737–739.
[19] Kim B M, Sharpless K B, *Tetrahedron Lett.* **1990**, *31*, 3003–3006.
[20] Pini D, Petri A, Mastantuono A, Salvadori P, in *Chiral Reactions in Heterogeneous Catalysis*, Jannes G and Dubois V (Eds.), Plenum Press: New York, 1995; p. 155–176.
[21] Lohray B B, Thomas A, Chittari P, Ahuja J R, Dhal P K, *Tetrahedron Lett.* **1992**, *33*, 5453–5456.
[22] Song C E, Roh E J, Lee S, Kim I O, *Tetrahedron: Asymmetry* **1995**, *6*, 2687–2694.
[23] Pini D, Petri A, Salvadori P, *Tetrahedron: Asymmetry* **1993**, *4*, 2351–2354.

[24] Pini D, Petri A, Salvadori P, *Tetrahedron* **1994**, *50*, 11321–11328.
[25] (a) Petri A, Pini D, Rapaccini S, Salvadori P, *Chirality* **1995**, *7*, 580–585. (b) Petri A, Pini D, Salvadori P, *Tetrahedron Lett.* **1995**, *36*, 1549–1552.
[26] Salvadori P, Pini D, Petri A, *J. Am. Chem. Soc.* **1997**, *119*, 6929–6930.
[27] Lohray B B, Nandanan E, Bhushan V, *Tetrahedron Lett.* **1994**, *35*, 6559–6562.
[28] Song C E, Yang J W, Sa H J, Lee S, *Tetrahedron: Asymmetry* **1996**, *7*, 645–648.
[29] Nandanan E, Sudalai A, Ravindranathan T, *Tetrahedron Lett.* **1997**, *38*, 2577–2580.
[30] Canali L, Song C E, Sherrington D C. *Tetrahedron: Asymmetry* **1998**, *9*, 1029–1034.
[31] (a) Bolm C, Gerlach A, *Angew. Chem. Int. Ed. Engl.* **1997**, *36*, 741. (b) Han H, Janda K D, *J. Am. Chem. Soc.* **1996**, *118*, 7632. (c) Han H, Janda K D, *Tetrahedron Lett.* **1997**, *38*, 1527. (d) Han H, Janda K D, *Angew. Chem. Int. Ed. Engl.* **1997**, *36*, 1731–1733.
[32] Lohray B B, Nandanan E, Bhushan V, *Tetrahedron: Asymmetry* **1996**, *7*, 2805–2808.
[33] Song C E, Yang J W, Ha H-J, *Tetrahedron: Asymmetry* **1997**, *8*, 841–844.
[34] Mandoli A, *PhD Thesis*, University of Pisa, Italy, **1998**.
[35] Bolm C, Maischak A, Gerlach A, *Chem. Comm*, **1997**, 2353–2354.
[36] Jacobsen E N, in *Comprehensive Organometallic Chemistry II*, Abel E W, Stone F G A, and Wilkinson E (Eds.), Pergamon: New York, 1995, vol. 12, p. 1097–1135.
[37] Katsuki T, *J. Mol. Catal. A: Chemical* **1996**, *113*, 87–107.
[38] For an example of asymmetric epoxidation with polymer anchored amino alcohol/molybdenum complexes, see: Cazaux I, Caze C, *React. Polym.* **1993**, *20*, 87–97. For the development of a polymer supported combinatorial library of chiral imine ligands, see: Francis M B, Jacobsen E N, *Angew. Chem., Int. Ed. Engl.* **1999**, *38*, 937–941.
[39] Canali L, Sherrington D C, *Chem. Soc. Rev.* **1999**, *28*, 85–93.
[40] Collman J P, Lee V J, Kellen-Yuen C J, Chang X, Ibers J A, Brauman J I, *J. Am. Chem. Soc.* **1995**, *117*, 692–703.
[41] De B B, Lohray B B, Sivaram S, Dhal P K, *Tetrahedron : Asymmetry* **1995**, *6*, 2105–2108.
[42] De B B, Lohray B B, Sivaram S, Dhal P K, *J. Polym. Sci., Part A : Polym. Chem.* **1997**, *35*, 1809–1818.
[43] Breysse E, Pinel C, Lemaire M, *Tetrahedron: Asymmetry* **1998**, *9*, 897–900.
[44] Minutolo F, Pini D, Salvadori P, *Tetrahedron Lett.* **1996**, *37*, 3375–3378.
[45] Minutolo F, Pini D, Petri A, Salvadori P, *Tetrahedron: Asymmetry* **1996**, *7*, 2293–2302.
[46] a) Canali L, Cowan E, Deleuze H, Gibson C L, Sherrington D C, *Chem. Comm.* **1998**, 2561–2562. b) Canali L, Sherrington, D C, Deleuze, H, *React. Funct. Polym.* **1999**, *40*, 155–168.
[47] Angelino M D, Laibinis P E, *Macromolecules* **1998**, *31*, 7581–7587.
[48] Kureshy R I, Khan N H, Abdi S H R, Iyer P, *React. Funct. Polym.* **1997**, *34*, 153–160.
[49] Ogunwumi S B, Bein T, *Chem. Comm.* **1997**, 901–902.
[50] Sabater M J, Corma A, Domenech A, Fornés V, Garcia H, *Chem. Comm.* **1997**, 1285–1286. For a related paper, see: Piaggio P, McMorn P, Langham C, Bethel D, Bulman-Page P C, Hancock F E, Hutchings G J, *New J. Chem.* **1998**, *22*, 1167–1169.
[51] Fraile J H, Garcia J I, Massam J, Mayoral J A, *J. Mol. Catal. A: Chemical* **1998**, *136*, 47–57.
[52] Frunza L, Kosslick H, Landmesser H, Hoft E, Fricke R, *J. Mol. Catal. A: Chemical* **1997**, *123*, 179–187.
[53] Vankelecom I F J, Tas D, Parton R F, Van de Vyver V, Jacobs P A, *Angew Chem., Int. Ed. Engl.* **1996**, *35*, 1346–1348.
[54] Parton R F, Vankelecom I J F, Tas D, Janssen K B M, Knops-Gerrits P-P, Jacobs P A, *J. Mol. Catal. A: Chemical* **1996**, *113*, 283–292.
[55] Janssen K B M, Laquiere I, Dehaen W, Parton R F, Vankelecom I F J, Jacobs P A, *Tetrahedron: Asymmetry* **1997**, *8*, 3481–3487.
[56] Rossiter B E, in *Asymmetric Synthesis*, Morrison J D (Eds.), Academic Press: Orlando, 1985; vol. 5, pp. 193–246.
[57] Farral M J, Alexis M, Tracarten M, *Nouv. J. de Chim.* **1983**, *7*, 449–451.
[58] See also: Suresh P S, Pillai V N R, Srinivasan M, in *Macromol-New Front., Proc. IUPAC Int. Symp Adv. Polym. Sci. Technol.*, Srinivasan K S V (Eds.), Allied Publishers Ltd., New Delhi, vol. 1, 437–440.
[59] Choudary B M, Valli V L K, Durga Prasad A D, *J. Chem. Soc., Chem. Comm.* **1990**, 1186–1187.

[60] Canali L, Karialainen J K, Sherrington D C, Hormi O E O, *J. Chem. Soc., Chem. Comm.* **1997**, 123–124.
[61] Karjalainen J K, Hormi O E O, Sherrington D C, *Tetrahedron: Asymmetry* **1998**, *9*, 1563–1575.
[62] Karjalainen J K, Hormi O E O, Sherrington D C, *Tetrahedron: Asymmetry* **1998**, *9*, 2019–2022.
[63] Karjalainen J K, Hormi O E O, Sherrington D C, *Tetrahedron: Asymmetry* **1998**, *9*, 3895–3901.
[64] Ebrahim S, Wills M, *Tetrahedron: Asymmetry* **1997**, *8*, 3163-3173 and references therein.
[65] Colonna S, Molinari H, Banfi S, Julia S, Masana J, Alvarez A, *Tetrahedron* **1983**, *39*, 1635–1641.
[66] Itsuno S, Sakakura M, Ito K, *J. Org. Chem.* **1990**, *55*, 6047–6049.
[67] Cappi M W, Chen W-P, Flood R W, Liao Y-W, Roberts S M, Skidmore J, Smith J A, Williamson N M, *Chem. Comm.* **1998**, 1159–1160.
[68] Boulahia J, Carrière F, Sekiguchi H, *Makromol. Chem.* **1991**, *192*, 2969–2974.
[69] a) Kim G-J, Shin J-H, *Tetrahedron Lett.* **1999**, *40*, 6827–6830. b) Pini D, Mandoli A, Orlandi S, Salvadori P, *Tetrahedron: Asymmetry* **1999**, *10*, 3883–3886.
[70] a) Geller T, Roberts S M, *J. Chem. Soc., Perkin Trans. 1* **1999**, 1397–1398. b) Carde L, Davies H, Geller T, Roberts S M, *Tetrahedron Lett.* **1999**, *40*, 5421–5424.

11 Enantioselective C-C Bond Formation with Heterogenized Catalysts

S. Abramson, N. Bellocq, D. Brunel, M. Laspéras, and P. Moreau*

11.1. Introduction

C-C bond formation plays a central role in organic synthesis [1]. These reactions can be catalyzed by acids, bases and transition metal complexes *via* various routes such as electrophilic and nucleophilic substitution, addition reactions including alkylation, the Diels-Alder reaction, the Heck reaction, formylation, alkyl metal addition, cyclopropanation, metathesis or polymerization. Catalytic C-C bond formation is one of the most actively pursued areas of research in the field of asymmetric catalysis because of its prime importance in the synthesis of fine chemicals, mainly for pharmaceutical and agrochemical products [2]. Most enantioselective catalysts are derived from natural products such as aminoalcohols, alkaloids, amino acids, tartaric acid, terpenes, and enzymes.

Indeed, in his pioneering work on homogeneous enantioselective catalysis, Wynberg prepared optically active Michael adducts with *Cinchona* alkaloids as catalysts (e.e.'s of up to 77%) [3]. In a next phase, the influence of the catalyst structure and the solvent effects were investigated [4]. For instance, in the reaction of the activated indanone **1** and methyl vinyl ketone **2**, either of both enantiomeric Michael adducts **3** may be formed in excess, depending on the choice of catalyst (Scheme 11.1). With quinine **4a**, the levorotatory product was produced, while with **4b** (R=$COCH_3$) or quinidine **5**, an excess of the dextrorotatory Michael adduct was found.

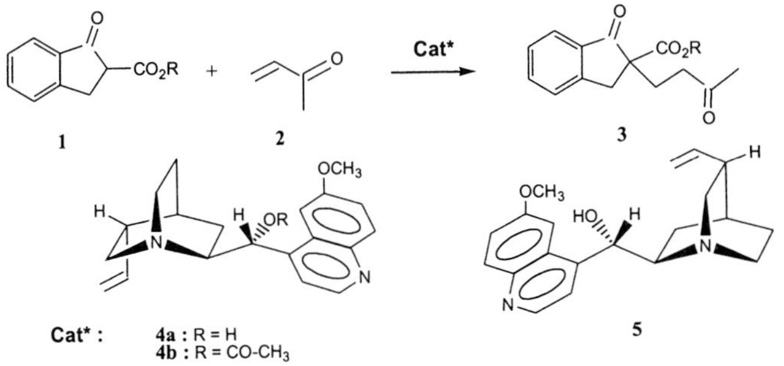

Cat*: **4a** : R = H; **4b** : R = CO-CH_3; **5**

Scheme 11.1. Enantioselective Michael addition catalyzed by Cinchona alkaloids.

The stereochemistry is largely controlled by the configuration at C8 and C9, and by the nature of the R-group. Stereoselection occurs during the irreversible alkylation step following the reversible formation of the enolate (Scheme 11.2).

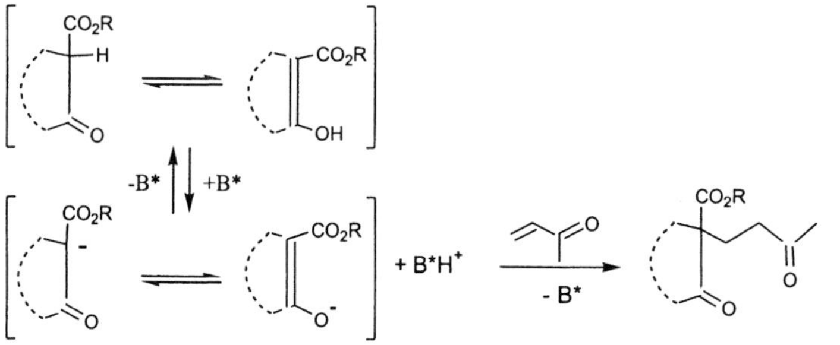

Scheme 11.2. Mechanism of enantioselective Michael addition catalyzed by chiral base.

Cinchona alkaloids are not basic enough to catalyze the addition of the less activated 2-alkylindanone **6** to methyl vinyl ketone, but the use of the quaternary compound **7** and hydroxide bases under phase transfer catalysis conditions resulted in excellent chemical yield and up to 80% e.e. [5]. The same system was also effective in the enantioselective phase-transfer methylation leading to **8** (Scheme 11.3) [6].

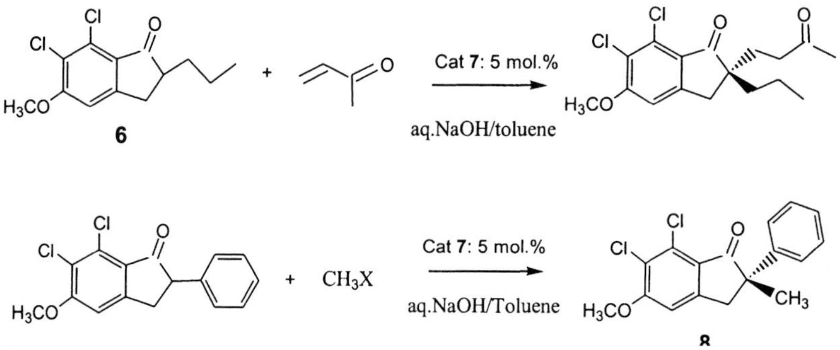

Scheme 11.3. Michael addition of 2-alkylindanone to methyl vinyl ketone catalyzed by a quaternary Cinchona alkaloid.

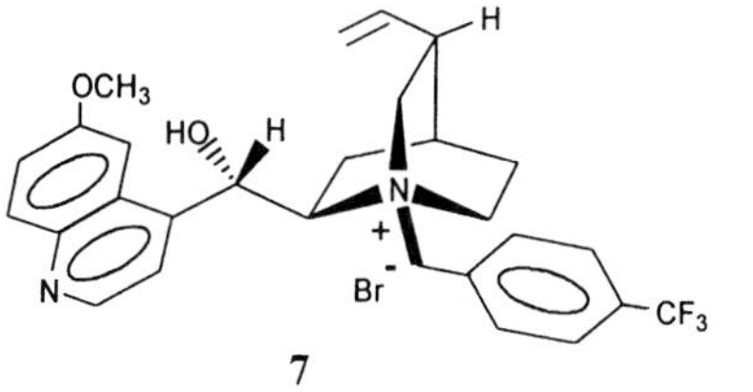

Unfortunately, immobilization of alkaloid derivatives on polymers did not result in worthwhile catalysts for enantioselective Michael addition [7]. The heterogeneous enantioselective Michael and aldol reactions still constitute challenging problems. Nevertheless, there is rapid progress being made in the design of new chiral complexes, e.g. based on binaphthyl, that are active in enantioselective reactions such as Diels-Alder reactions, alkylation of aldehydes (*vide infra*), and Michael reactions [8]. This sustains the interest in the future development of heterogeneous catalysts.

The Diels-Alder reaction has long been recognized as one of the most important methods for constructing cyclohexene derivatives. Highly "asymmetric" synthesis has been achieved with the help of a chiral auxiliary bound to the diene or dienophile, but these reactions must be considered as diastereoselective rather than as enantioselective (see Chapter 12) [9]. In their pioneering work on enantioselective Diels-Alder reactions, Koga et al. used a catalytic amount of **9** for the condensation of cyclopentadiene and methacrolein. The catalyst is easily prepared from (–)-menthol and ethylaluminum dichloride (Scheme 11.4) [10].

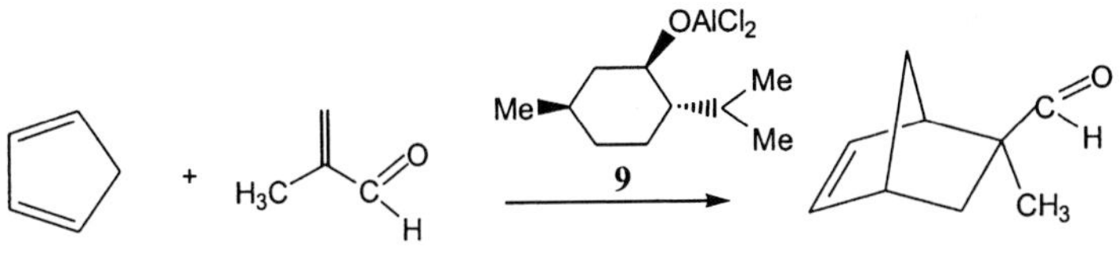

Scheme 11.4. Diels-Alder reaction of cyclopentadiene and methacrolein catalyzed by an Al-ethyl-menthol complex.

Adduct **10** (mainly exo stereoselectivity) was obtained with 70% e.e. Kagan et al. prepared chiral Lewis acid catalysts such as **11** consisting of aluminum alcoholates by reaction of $EtAlCl_2$ with several synthetic chiral diols [11].

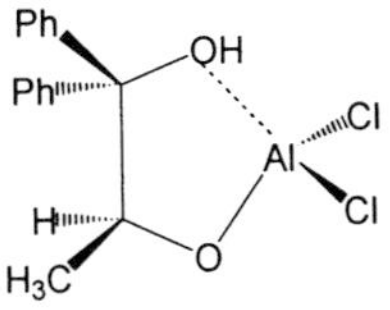

11

The hetero-Diels-Alder addition (Danishefsky reaction) between benzaldehyde and a diene has been investigated by Yamamoto et al. [12]. In this case, the best catalysts were the 3,3′-*bis*(triarylsilyl) derivatives of the methylaluminum complex of a chiral binaphthol (Al-BINOL, **12**). The reaction provides the *cis* dihydropyrone **13** and the *trans*-product **14** in 90% yield with excellent selectivity (*cis*/*trans*=30:1) and with enantioselectivity for the (2*R*,3*R*) configuration (97% e.e.) (Scheme 11.5).

Other chiral Lewis acid catalysts have been proposed for asymmetric Diels-Alder reactions, such as the BINOL-based compound **15** [13].

Narasaka's group reported on the chiral Ti complex **16**, prepared from *a*,*a*,*a*′,*a*′-tetraphenyl-4,5-di(hydroxymethyl)-1,3-dioxolane (TADDOL). The ligand is derived from tartaric acid. The reaction with the Ti-TADDOL complex gave the Diels-Alder adduct

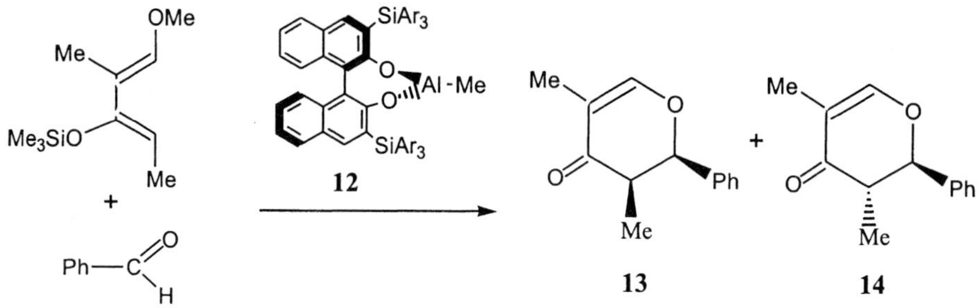

Scheme 11.5. Hetero-Diels-Alder addition between benzaldehyde and diene catalyzed by Al-BINOL derivative.

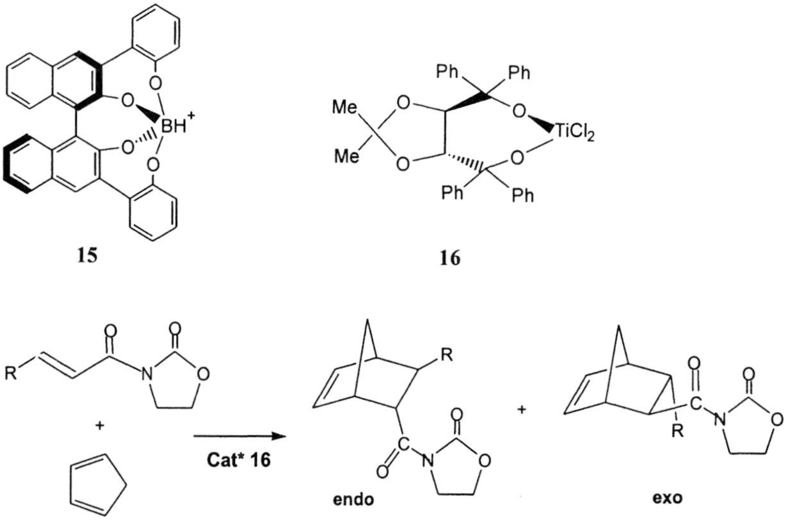

Scheme 11.6. Diels-Alder addition catalyzed by Ti-TADDOL complex.

in 90% yield, with an *exo/endo* ratio of 92/8 and with excellent optical purity (94% e.e.) (Scheme 11.6) [14].

The substituents in position 2 of the dioxolane ring influence the stereoselectivity. Mayoral's group related the experimental results with the relative stabilities of the transition states [15]. Solid catalysts for enantioselective Diels-Alder reactions were developed by immobilizing these catalysts on various supports (polymers, dendrimers and mineral oxides) and will be discussed in Section 11.3.

The enantioselective addition of dialkylzinc to an aldehyde in the presence of a small amount of a chiral ligand (L*) is an important reaction for the formation of chiral alcohols via the zinc alcoholates (Scheme 11.7). These alcohols are of great synthetic use as precursors, for instance, for α-tocopherol [16] or insecticides [17].

The first truly catalytic enantioselective addition was performed by Oguni et al. using chiral β-aminoalcohols [18]. Many other ligands, mostly based on natural prod-

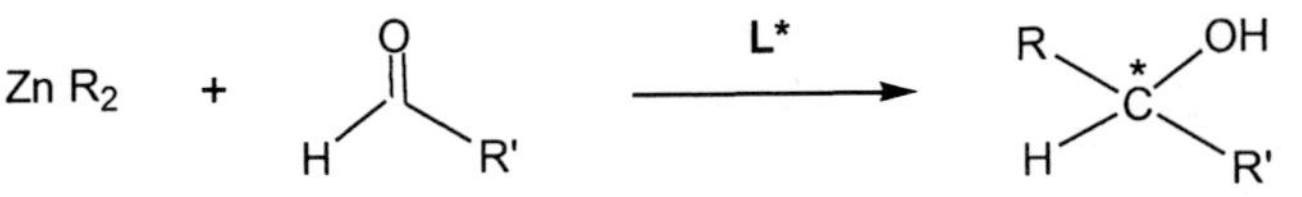

Scheme 11.7. Enantioselective addition of dialkylzinc to aldehydes catalyzed by chiral ligands.

ucts, have been tested in this reaction, e.g. dimethylaminoisoborneol **17** [19], L-proline derivatives such as **18** [20], cinchona alkaloids [21], and (+)-ephedrine **19** [22].

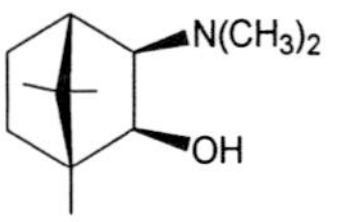

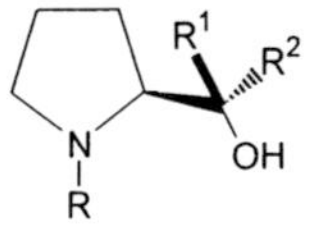

These studies have been reviewed by Noyori and Soai [23, 24]. Furthermore, synthetic catalysts such as (1*R*,2*R*)-1-(dialkylamino)-1-phenyl-3-alkoxy-2-propanols **20** [25], Ti-BINOLs [26–28], Ti-TADDOLs [29] and titanium complexes based on (1*R*,2*R*)-1,2-diaminocyclohexane **21** revealed excellent chemoselectivity and enantioselectivity [30, 31].

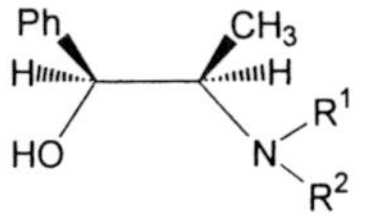

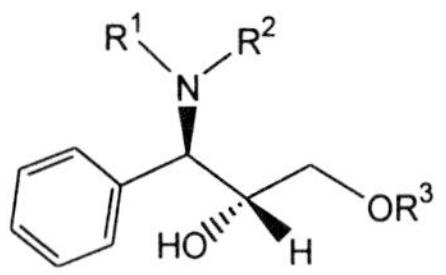

It should be noted that the binaphthol derivative **22** gives a very high turnover frequency even in the absence of $Ti(O^iPr)_4$. Enantiomeric excesses were very high even with aliphatic aldehydes and also with α,β-unsaturated aldehydes.

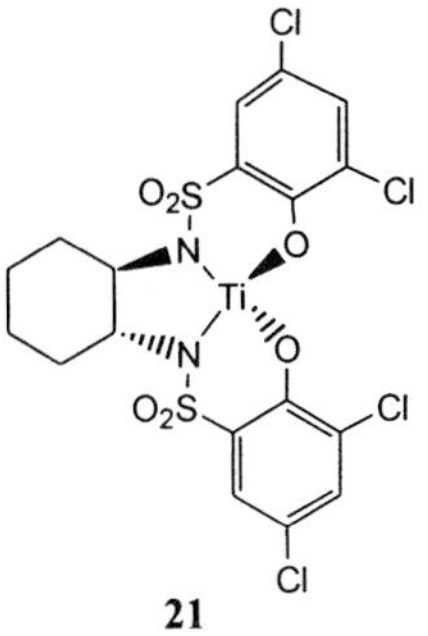

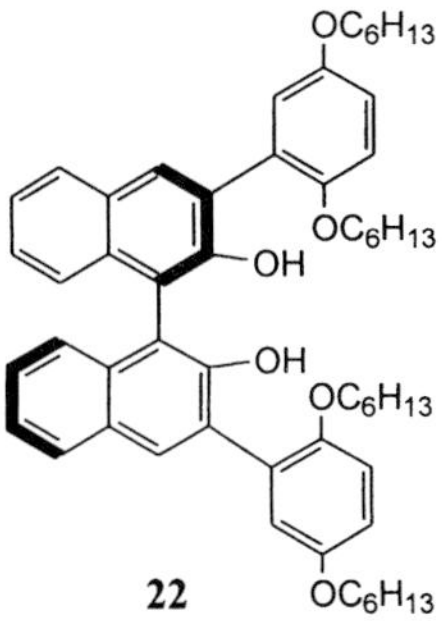

Many homogeneous catalysts for addition on aldehydes have been immobilized on various supports, and the main results are presented in the next section.

11.2 Enantioselective Alkylation of Aldehydes by Organozinc Reagents with Immobilized Catalysts

Attaching a chiral ligand to a solid support offers general advantages such as easy recovery, potential recycling of the generally expensive chiral catalyst, and the possible use of a flow reactor. The first heterogenization attempts showed that catalytic activities and enantioselectivities were lower than for the homogeneous catalysts. Hence, an attempt has been made to improve the catalytic performance by optimally adapting the support to the catalytic site, and by optimizing the nature of the link to the surface as well as the accessibility and dispersion of the sites.

11.2.1 Heterogenization of Chiral Aminoalcohols on Polymeric Supports

Frechet first immobilized chiral aminoalcohols on a polymer support [32]. 1–2% Cross-linked, partly chloromethylated polystyrene was chemically modified with ephedrine, proline, 3-exo-aminoisoborneol and N-methyl-3-exo-aminoisoborneol. These catalysts were used in the reaction between benzaldehyde and diethylzinc. The best result (91% yield in (*S*)-1-phenyl-1-propanol with 92% e.e.) was obtained with catalyst **23**, derived from (–)-N-methyl-3-exo-aminoisoborneol. 80% e.e. was obtained with the ephedrine-derivative **24**.

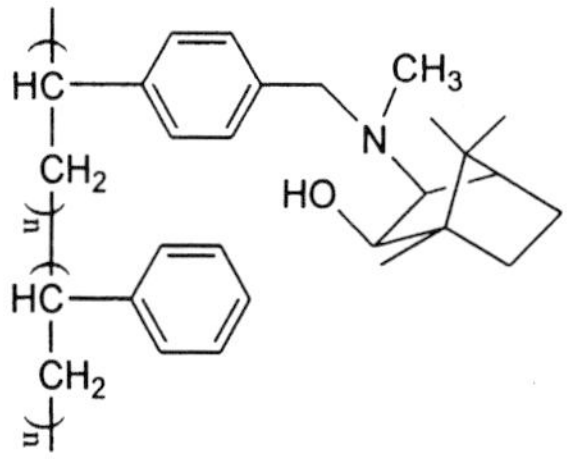

23 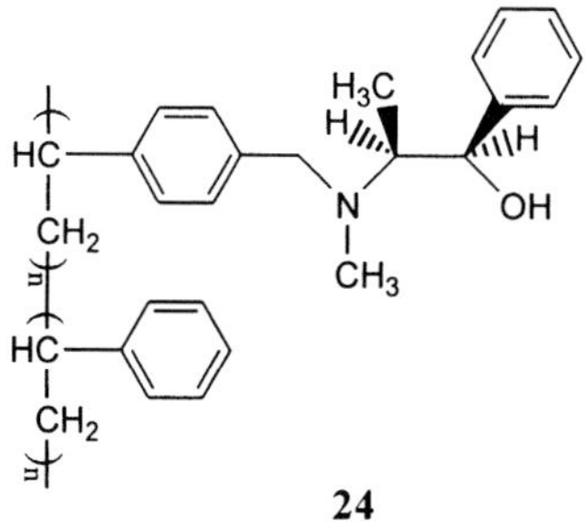 **24**

Removal of the polymer after the first run afforded a filtrate containing soluble chiral alkoxide. The latter was hydrolyzed in good yield to (*S*)-1-phenyl-1-propanol. If recycling is performed in anhydrous conditions, the polymer can often be reused without significant loss of activity or enantioselectivity.

The catalytic cycle is depicted in Scheme 11.8. In a first step, diethylzinc reacts with the immobilized catalyst **23** to form a 1:1 chelate **25**. In the absence of additional diethylzinc, **25** slowly reduces benzaldehyde to benzyl alcohol. Alkylation only proceeds if additional equivalents of dialkylzinc are added. If dibutylzinc is added to a polymer-bound ethylzinc chelate, only butylation and no ethylation of benzaldehyde is observed. Based on such results and on the work of Soai, Noyori and Corey [19, 33–36], it was proposed that benzaldehyde and a second dialkylzinc molecule are bound on the Zn and oxygen atoms of the active site **25**. This enhances the electrophilicity of the aldehyde and the nucleophilicity of the alkyl group. The reaction then proceeds through a six-centered bimetallic transition state such as **26** (Scheme 11.8).

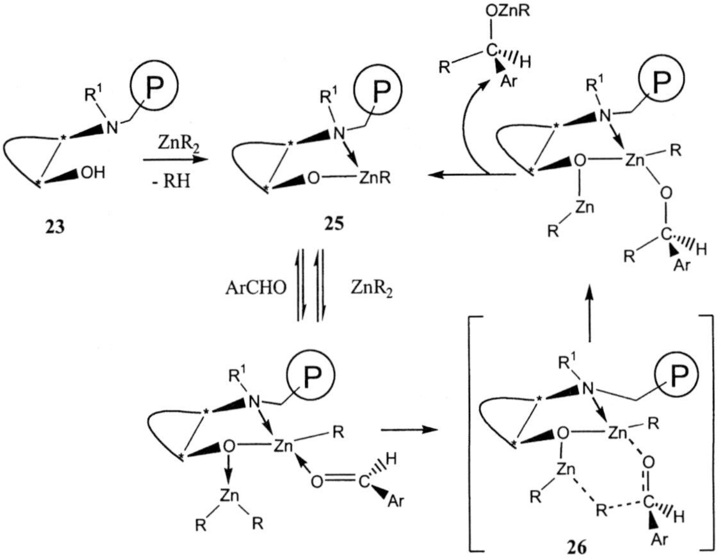

Scheme 11.8. Catalytic cycle of dialkylzinc addition to benzaldehyde.

In the preferred transition state (**28**), the six-membered ring has an anti-conformation with respect to the five-membered chelate ring. Based on calculations, a transition state with the six-membered ring in the *syn*-conformation (**27**) would be less stable [37]. The alkyl transfer from ZnR_2 to the coordinated aldehyde preferentially occurs with the least steric hindrance. Therefore, **28 a** is preferred over **28 b**, in which there is steric interaction between the phenyl group and the alkyl group attached to the chelated zinc. This attack determines the configuration of the resulting zinc alkoxide.

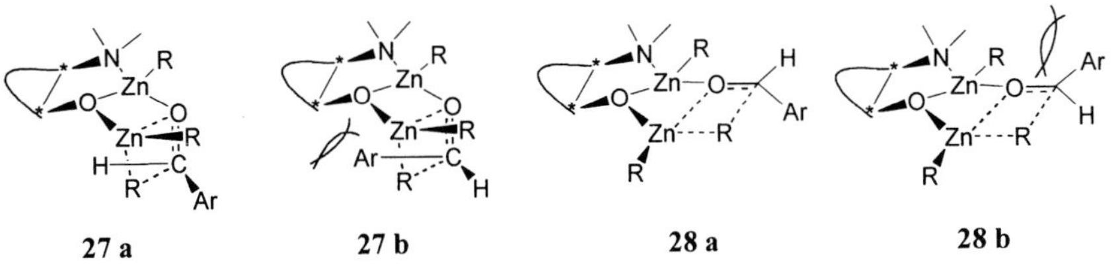

For homogeneous aminoalcohol catalysts, subtle effects determine the eventually observed e.e. For instance, Noyori observed a non-linear relationship between the product e.e. and the e.e. of the chiral auxiliary **17** which forms the active species **29** [19 b–d, 35]. In this case, the catalyst is in equilibrium with a catalytically inactive dimer (Scheme 11.9). When the chiral catalyst is not enantiomerically pure, two diastereoisomeric dimers are possible. The minor enantiomer (+)-**29** gives an anti (+,–)-**30**

dimer which is relatively stable but inactive. The major enantiomer (–)-**29** forms a less stable dimeric species (–,–)-**30**. Hence, the use of **29** with a low e.e.% gives an unexpectedly high enantioselectivity, because the minor enantiomer form is passivated in a catalytically inactive dimer.

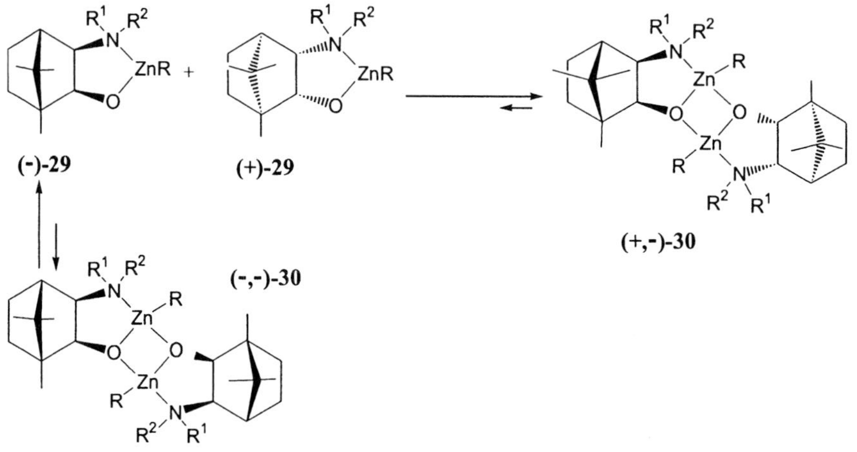

Scheme 11.9. Formation of diastereoisomeric dimers from aminoborneol enantiomers.

For catalysts with heterogenized chiral auxiliaries, many more factors influence not only the e.e. but also the recoverability and stability of the catalyst. Some important elements that will be discussed are:

- The chemical modification of the auxiliary by binding to the polymer,
- The use of a spacer between auxiliary and polymer,
- The swellability and other structural characteristics of the polymer,
- The use of soluble polymers,
- The chemical nature of the polymer backbone,
- The loading of the polymer structure with an aminoalcohol auxiliary.

Effects of chemical modification of the auxiliary by binding to the polymer

If the bond between auxiliary and polymer is in near proximity of the stereogenic centres of the auxiliary, the heterogenization may well influence the e.e. For instance, in the reaction of PhCHO and $ZnEt_2$, (*S*)-1-phenylpropanol was obtained in higher e.e. with a polymer-bound ephedrine than with the corresponding homogeneous auxiliary [38]. Such changes are probably related to an increased steric hindrance around the nitrogen atom induced by the polymer linkage. Nevertheless, it should be noted that for a catalyst such as **31**, the relation between the size of the N-substituents and the e.e. is not straightforward [22a, 34].

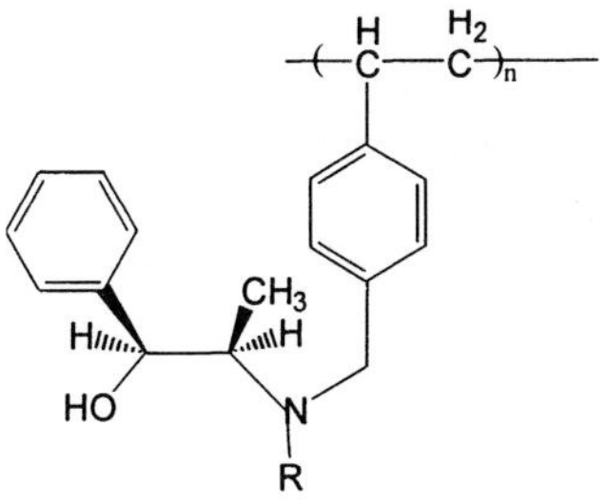

31 R = C_2H_5, C_3H_7, C_4H_9 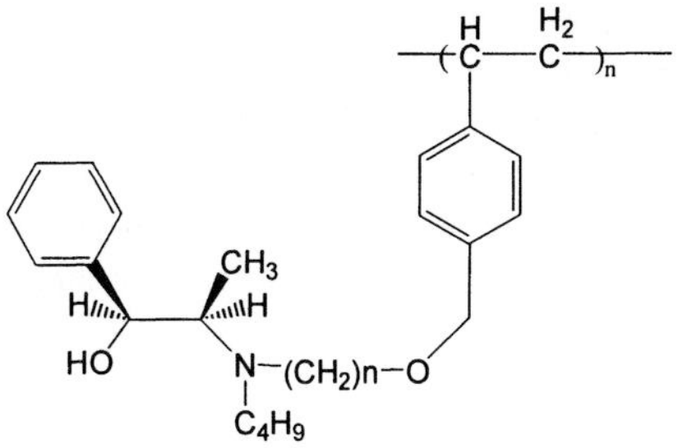 **32**

Use of a spacer between auxiliary and support

Other reports discuss the effects of a spacer on the enantioselectivity [39, 40]. With catalysts of the general formula **32**, the use of a spacer with n=6 increases the e.e. from 17% (without a spacer) to 82%. The spacer may facilitate the approach of the substrate to the chiral auxiliary. Even with aliphatic aldehydes, excellent activities and enantioselectivities are observed (Table 11.1).

Table 11.1. Chiral alcohol formation by enantioselective addition of Et_2Zn to aromatic and aliphatic aldehydes using polymer-supported N-butylnorephedrine **32**.

Chiral catalyst	Aldehyde	(S)-Alcohol	
		Yield (%)	e.e. (%)
(1*S*,2*R*), n=4	PhCHO	91	96
(1*S*,2*R*), n=6	PhCHO	88	99
(1*S*,2*R*), n=6[a]	PhCHO	91	82
(1*S*,2*R*), n=6	$CH_3(CH_2)_7CHO$	79	73
(1*S*,2*R*), n=6[a]	$CH_3(CH_2)_7CHO$	80	71

[a] Recycled catalyst.

Swellability and other structural characteristics of the polymer

Frechet and coworkers studied in detail the influence of the polymer micro-environment on the catalytic activity [41]. In this work, primary aminoalcohols (**33**, **34**) were used as auxiliaries. This leads to high e.e.'s, possibly attributed to the formation of a Schiff base between the primary amine and the aldehyde reactant. These primary aminoalcohols were incorporated into the polymers via different routes. For instance, **33** reacts as a phenoxide anion with a chloromethylated, cross-linked polymer; **34** is copolymerized with styrene and a cross-linker (Fig. 11.1). Particular attention was given to the type of cross-linker. Apart from divinylbenzene (DVB, **35**), a new flexible cross-linking agent was used which contains an oxyethylene chain (**36**).

This flexible cross-linking agent (**36**) proved most valuable to enhance the swellability of the polymer and to ensure easy access to the catalytic sites. Moreover, the reaction of auxiliary **33** with the chloromethyl groups proceeded almost quantitatively in a polymer prepared with the flexible cross-linker **36** (Fig. 11.1, top). Excellent catalytic activities and good enantioselectivities were obtained with the chiral polymers **37** and **38** (Table 11.2). **38** preserves its activity in several consecutive runs and was even used in a continuous flow system. With *p*-chlorobenzaldehyde and $ZnEt_2$ in a 1:1.4 ratio, the corresponding *S*-alcohol was obtained with 94% e.e.

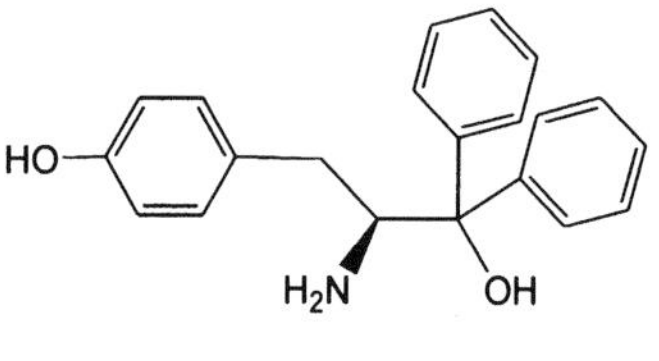

HO H2N OH **33**

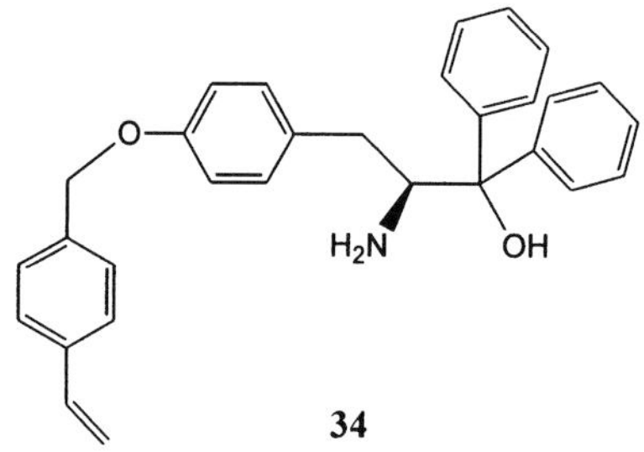

O H2N OH **34**

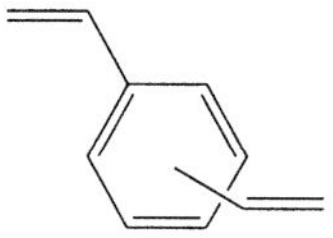

35

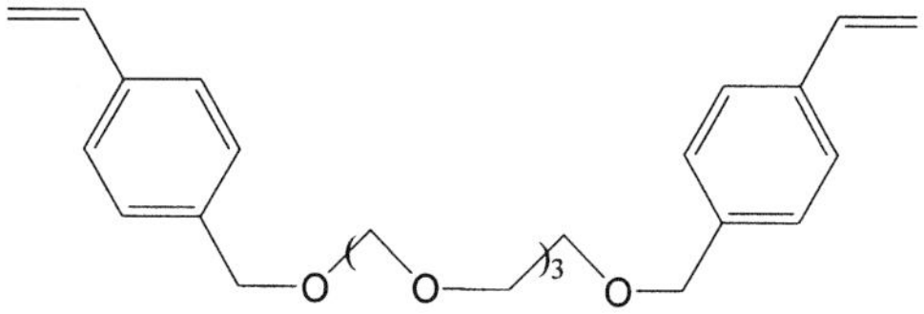

O O O 3 **36**

Figure 11.1. Preparation of polymer-bound primary *β*-aminoalcohols. The polymer contains a flexible polyoxyethylene cross-linker.

Table 11.2. Asymmetric ethylation of benzaldehyde using polymer supported N-butylnorephedrine **38** in a batch system.

Batch	Yield (%)	E.e.% (*S*) 1-phenylpropanol
1	72	86
2	83	87
–	–	–
5	79	85

Experimental conditions: benzaldehyde: 10 mmol; catalyst: 0.5 mmol; solvent: toluene-hexane; temperature: 0 °C.

The polymerization procedure can also be important. In the case of a microemulsion polymerization, improved catalysts were obtained when the polystyrene seed particles were pre-swollen with a polymerizable aminoalcohol monomer [42].

Use of soluble polymers

Soluble polymers may be difficult to handle and to recover quantitatively after the reaction. However, a soluble polymer such as **39** was recently used in a continuously operated reactor equipped with an ultrafiltration membrane. This results in a large increase of the turnover number [43].

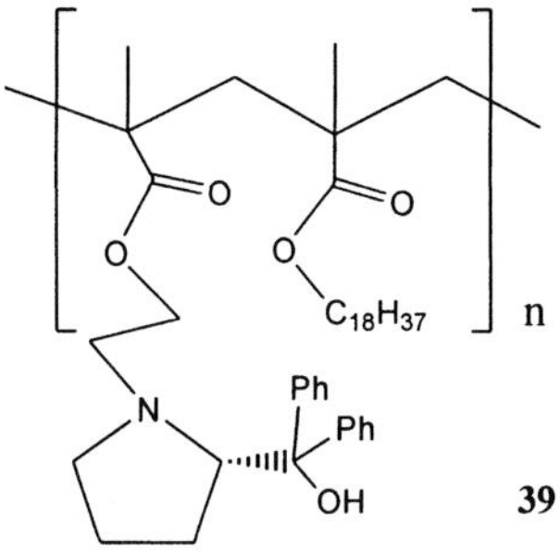

Chemical modification of the polymer backbone

While heterogenized aminoalcohols mostly contain only polystyrene, Hodge's group reported a new type of polymer-supported *N*-methyl ephedrine, based on a polystyrene framework grafted with a methyl siloxane polymer **40** [44]. Residual silanes of the mixed polymer were quenched by treatment with 1-octene (Scheme 11.10).

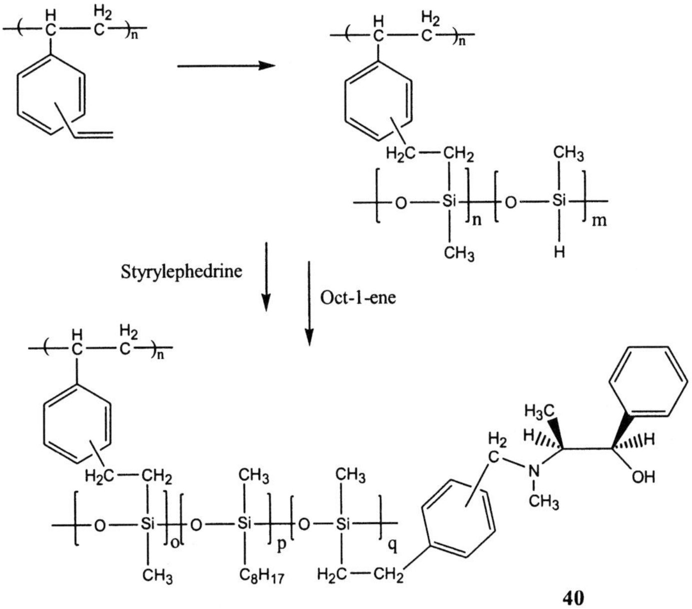

Scheme 11.10. Preparation of polymer-supported N-methyl ephedrine starting from a polystyrene backbone grafted with a methyl siloxane polymer.

Effect of concentration of aminoalcohol on the polymer

In studies with polystyrene-supported ephedrine and aminocamphor derivatives, lower loadings lead to higher e.e.'s [45, 46]. A good dispersion of the Zn sites probably prevents formation of nonselective Zn dimers. Enantiomeric excesses of up to 98% were attained in a simple top-flow system which can be used with even fragile polymers.

Similar trends were observed for the polymeric aminoalcohol catalyst **41** containing *cis*-2,6-dimethylpiperidino groups [47]. The lower the functionalization/reticulation ratio of the ligands, the higher the observed enantioselectivity was. Excellent results have also been obtained with the related polymer **42**.

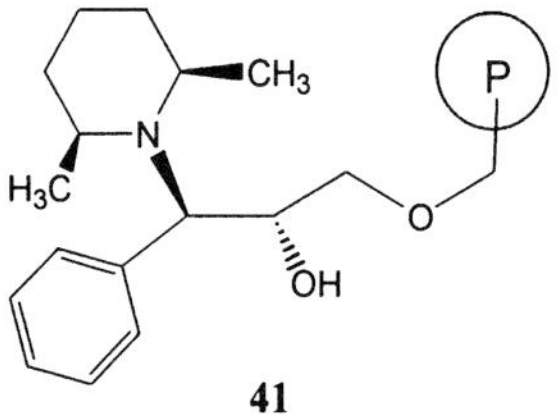

41 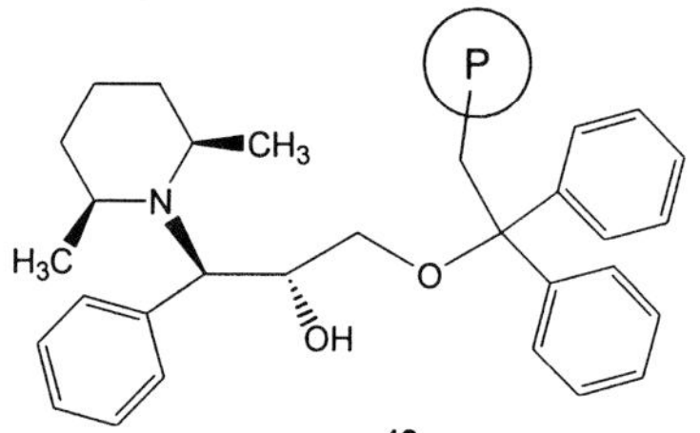 **42**

11.2.2 Heterogenization of TADDOLates and Binaphthols on Polymeric Supports

TADDOL units (cfr. Section 11.1) can be attached to polymer backbones using -CH_2OH or -CH_2Br substituted TADDOLs, leading to **43**. Alternatively, styrene and divinylbenzene can be copolymerized with TADDOLs containing a *para*-styryl group at the C2-position of the dioxolane ring or at the α-position. The resulting catalysts are **44** and **45** [48].

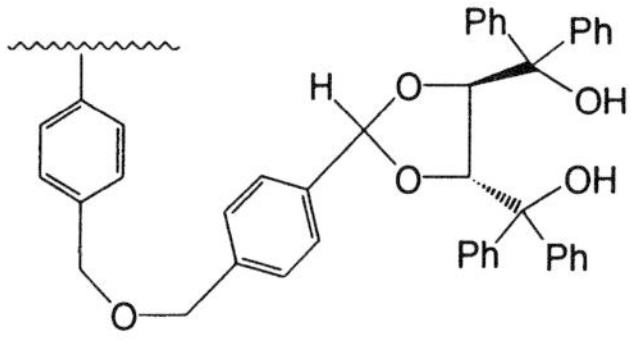

43 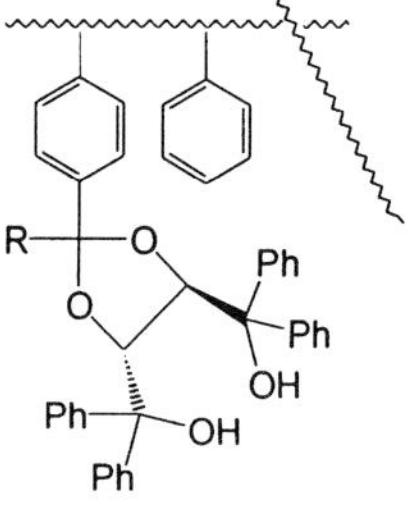 **44** 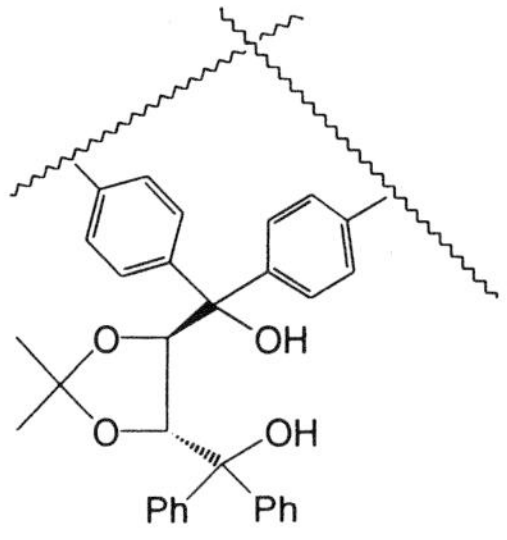 **45**

As in solution, the nucleophilic addition of dialkylzincs to aldehydes requires the presence of excess $Ti(O^iPr)_4$, and the selectivity drastically decreases when less than 5% Ti-TADDOLate is used. Yields of chiral alcohols and enantioselectivities are almost as high as those observed with soluble TADDOLates (95–98% e.e.). Amazingly, the enantioselectivity depends very little on the type of polymer, the degree of polymerization or the degree of cross-linking. Polymer-bound TADDOLates are quite stable and can be recycled several times, e.g. by simple filtration and washing. If the Zn-alkoxides are worked up with acid before removal of the polymeric catalyst, regeneration of the catalyst requires drying and titanation with $Ti(O^iPr)_4$. In either case, the enantioselectivity is preserved.

A new generation of enantioselective polymeric catalysts was prepared by polymerization of chiral binaphthol-type monomers. The polymers have rigid helical structures. Depending on the functionalization (3,3′ or 6,6′), a minor-groove structure **46** and a major-groove polymer **47** are obtained [49, 50]. A remarkable feature of these helical structures is that all catalytic sites have identical chemical and steric environments, which is often not the case for irregular polymer structures.

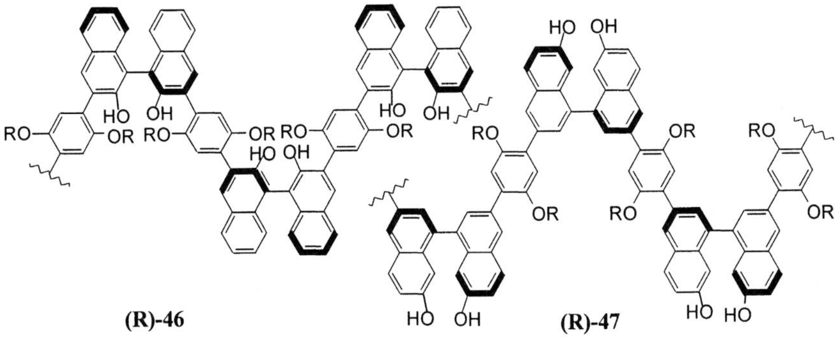

(R)-46 **(R)-47**

In the ethylation of aldehyde with $ZnEt_2$, the minor-groove polymer **46** displays better catalytic properties than **47**. Such differences may be due to an increased steric hindrance in the minor groove structure **46** and a higher degree of coordinative unsaturation of the Zn. In contrast, structure **47,** with its more exposed Zn coordination sites, may be prone to interchain coordination of oxygen atoms, leading to coordinative saturation.

The (*R*)-polybinaphthyl polymer **46** produces the *R*-alcohol (90% yield, 93% e.e.) while the *S*-alcohol is formed in the presence of the corresponding (*S*)-polybinaphthyl (93% yield and 93% e.e.). The polymeric catalyst is easily recovered from the reaction mixture by precipitation with methanol.

11.2.3 Heterogenization of Chiral Ligands on Mineral Supports

Mineral supports display higher mechanical resistance against stirring than organic supports. Moreover, their textural properties are less solvent-dependent than those of organic polymers. However, only three groups have focused on the use of such supports to perform enantioselective dialkylzinc additions.

11.2.3.1 Immobilization on Alumina and Silica Gel

Soai et al. used alumina and silica gel to prepare chiral insoluble catalysts for the enantioselective addition of dialkylzinc to aldehydes [51]. (–)-Ephedrine and N-alkylnorephedrine were immobilized via nucleophilic substitution on a previously grafted 3-chloropropylsilyl chain (Fig. 11.2). However, the catalytic activities and enantioselectivities were only moderate (33–59% e.e.) in comparison with those of homogeneous and polymer-supported catalysts. A change of the alkyl group in the N-alkyl-norephedrine from methyl to propyl resulted in a reversal of the e.e. (from 37% *R* to 29% *S*).

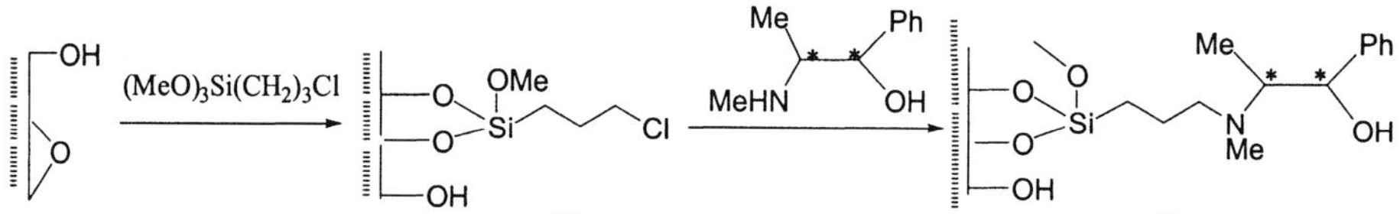

Figure 11.2. Anchorage of a chiral *β*-aminoalcohol on an oxidic surface.

A hybrid organic/inorganic catalyst was obtained by coating silica gel with chloromethyl-polystyrene and subsequent immobilization of (–)-ephedrine. Moderate e.e.'s were obtained in the alkylation of benzaldehyde and *n*-octanal to aromatic and aliphatic secondary alcohols. The solvent effects with this catalyst were significantly different from those observed with the modified polystyrene; instead of pure hexane (36% e.e.), a mixed benzene/hexane solvent gave the best e.e. (56%) [34]. This catalyst was successfully recycled.

11.2.3.2 Immobilization on Zeolites

Starting from *β*-aminoalcohols such as prolinol, Corma et al. anchored chiral transition metal complexes on USY-zeolites [52]. The zeolite-supported Ni complex **50** catalyzes the conjugate addition of $ZnEt_2$ to enones **51**, with similar yields of the *β*-ethylated ketones **52** as in homogeneous conditions (Scheme 11.11) [53].

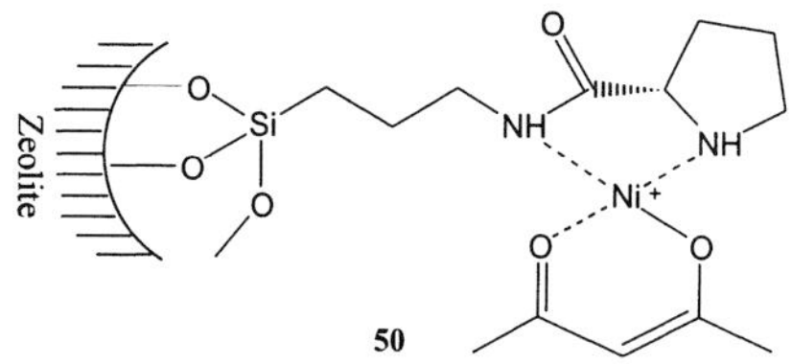

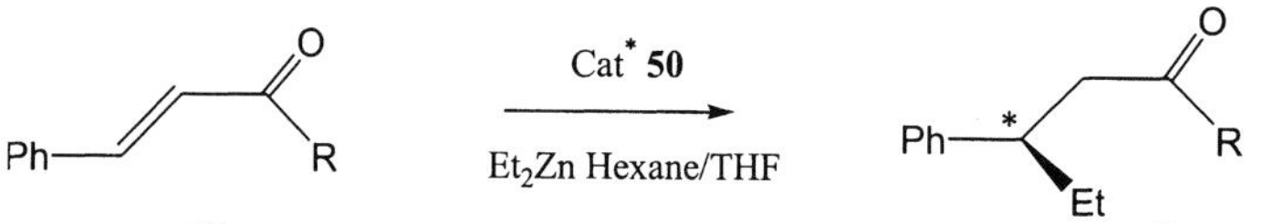

Scheme 11.11. Catalyzed addition of diethylzinc to a benzylidene ketone

The e.e.'s of the saturated ketones **52** are higher with the zeolite-supported Ni complex (91% e.e., R=Ph) than with the homogeneous catalyst in the same conditions (77% e.e., R=Ph). These enhanced enantioselectivities may result from the additional steric constraints imposed by the zeolite pores.

11.2.3.3 Immobilization in Micelle-Templated Silicas (MTS)

In view of our results with functionalized micelle-templated silicas in base catalysis [54, 55], we investigated the anchorage of (–)-ephedrine on MCM-41-type MTS materials. The major advantages of these new silicas – a large surface area, a mesoporous system with narrow pore-size distribution with variable diameter (2–10 nm), and the uniform chemical properties of the surface – allow the preparation of well-defined mesoporous hybrid materials [56].

After calcination of the MTS material, its surface can be functionalized by means of organotrialkoxysilanes (Fig. 11.3) [54]. In anhydrous conditions and using apolar solvents, silylation mainly proceeds on the hydrophobic part of the surface, constituted by the siloxane and isolated silanol groups. Hence, the grafted chains are surrounded by hydrophilic domains, containing adjacent silanol groups. The latter part of the surface can be rendered hydrophobic by 'end-capping' procedures [57, 58].

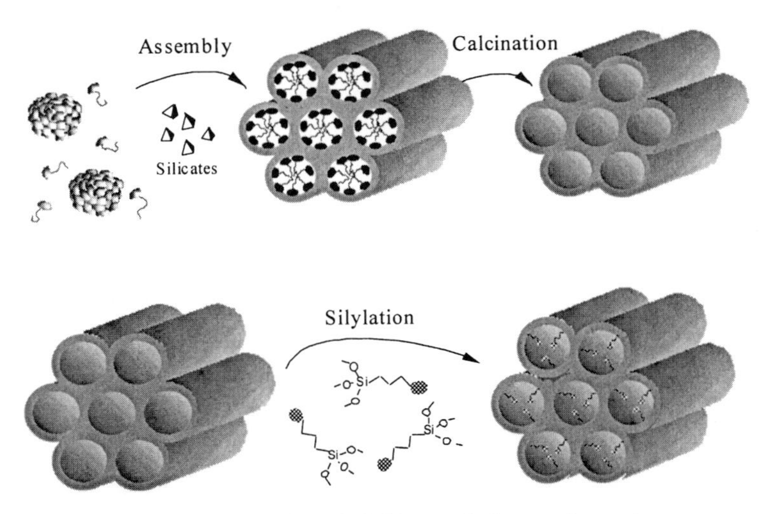

Figure 11.3. Preparation of the hybrid organic-inorganic catalysts.

Initially, (–)-(1*R*,2*S*)-ephedrine and (+)-(1*S*,2*R*)-ephedrine were immobilized in an MCM-41 (3.6 nm pore diameter) following the procedure of Soai [51], and the material was tested in the reaction of benzaldehyde and diethylzinc [59]. Even if kinetic results (first-order in aldehyde, zero-order in $ZnEt_2$ with excess $ZnEt_2$) were similar to those of homogeneous reactions, the e.e.'s were always moderate (e.e. <37%). In order to investigate the role of lateral interaction between the chiral ligands, the loading of the materials was varied and the grafted ephedrine diluted with unfunctionalized ethyl chains [60]. On the other hand, MTSs with varying pore diameter (3.6, 5.2 nm) were functionalized in order to investigate the role of internal diffusion limitation.

For materials prepared with diluting agents, it was observed that sometimes the ephedrine reacted directly with the residual surface silanol groups (Fig. 11.4). Such O-alkylated ephedrine is not a suitable catalytic site, since it cannot form a monoethyl Zn alcoholate. In order to prevent immobilization of ephedrine via O-alkylation, it was attempted to passivate the surface by treatment with hexadimethylsilazane (HMDS) (Fig. 11.5).

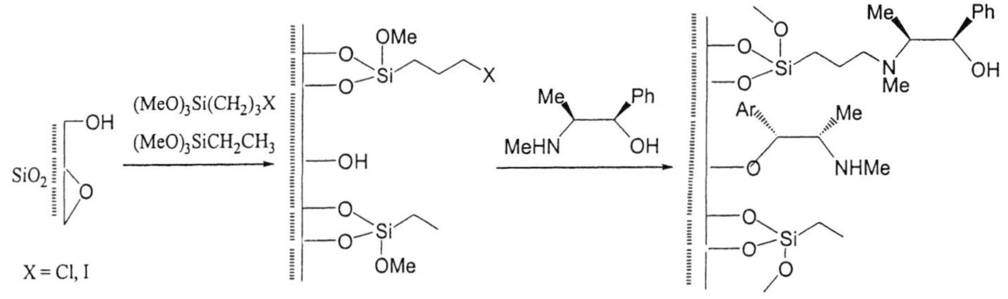

Figure 11.4. Dilution of functional grafted chains and anchorage of ephedrine in excess on the MTS surface.

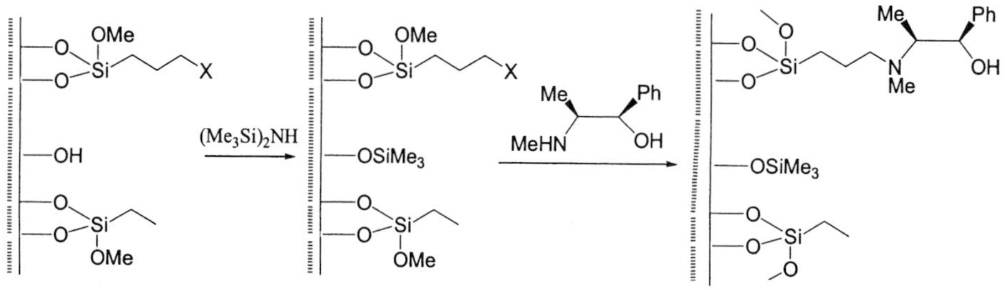

Figure 11.5. Passivation of the uncovered MTS surface.

Even after such a HMDS treatment, e.e. values were still moderate (33%), partly because passivation of residual silanols is never quite complete, as shown by quantitative NMR measurements on Q^3-type silicon species [58]. In general, more addition of diluting agent resulted in less surface coverage and in a higher catalytic activity. This suggests that the uncovered surface is capable of activating the dialkylzinc compound, resulting in formation of racemic alcohols which decrease the e.e. value. Therefore, the role of the inorganic surface was investigated in more detail, using mesoporous materials with different textural characteristics, different chemical compositions (Al-MTS and Si-MTS) and various ephedrine loadings [61]. Al-MTS-bound ephedrine samples were much more active than the corresponding Si-MTS samples.

In recent work, Al-MTS (8 nm pore diameter) was functionalized by a new grafting process, resulting in a high surface coverage with 3-chloropropyl silane chains (3.8 chains/nm^2). After substitution of the halogen by ephedrine, the resulting materials catalyzed alkylation of benzaldehyde by diethylzinc. Both the activity (0.40 h^{-1}, vs. 0.53 h^{-1} under homogeneous conditions) and the enantioselectivity (62% e.e.) compared excellently to those of the dissolved chiral auxiliary. After removal of the soluble product, the solid catalyst featured the same activity and enantioselectivity without further regeneration [62].

11.3 Diels-Alder Reactions with Immobilized Catalysts

Heterogeneous diastereoselective Diels-Alder reactions have been extensively studied in the last decade by Kabalka and Mayoral (Scheme 11.12) [63–71] (see Chapter 12). There is a recent review by Mayoral et al. on heterogeneous catalysis of Diels-Alder reactions [72].

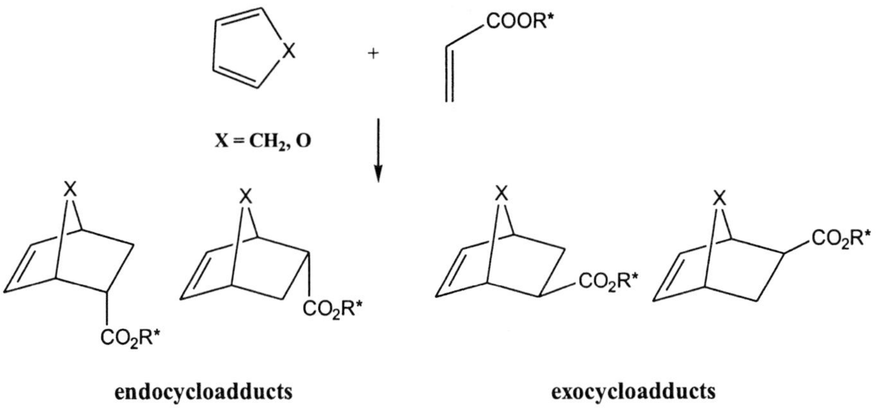

COOR*

$X = CH_2, O$

endocycloadducts

exocycloadducts

Scheme 11.12. Diels-Alder adducts from cyclopentadiene or furan and acrylates.

There are also a few reports on supported chiral catalysts for enantioselective Diels-Alder cycloadditions [48, 73–78]. Polymers and oxides such as silica or alumina have been proposed.

11.3.1 Heterogenization of Chiral Lewis Acids on Polymers

Insoluble polymer-supported chiral Lewis acids were first prepared by Itsuno by treating polymers bearing amino acids with borane [73]. This results in polymer-bound oxazaborolidinones (Scheme 11.13). Catalysts were prepared using the sulfonamides of L-valine, L-isoleucine, D-2-phenylglycine or L-threonine, and BH_2Br or BH_3.

These materials catalyze the addition of methacrolein to cyclopentadiene in CH_2Cl_2/THF, with excellent yield and exo:endo selectivity (Scheme 11.14). For in-

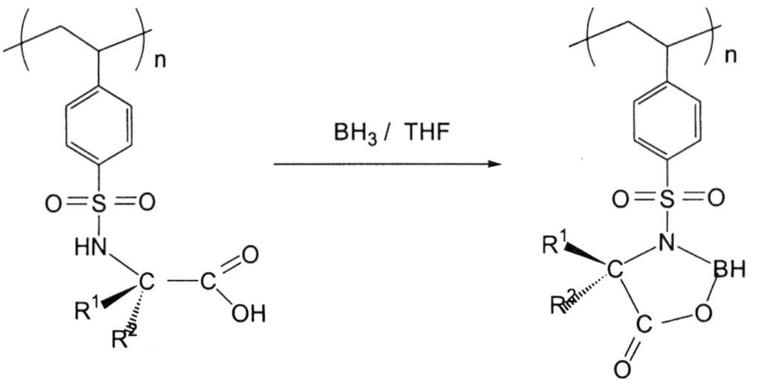

Scheme 11.13. Preparation of polymer-bound chiral oxazaborolidone.

stance, the polymer resulting from anchored L-valine (R_1 = H and R_2 = ^{i}Pr) (15 mol%) gave an *endo/exo* ratio of 1:99 and (3×*R*)-adduct with 64% e.e. at −78 °C. The catalyst was recycled without selectivity loss [74]. An improved catalyst was obtained by the use of the flexible cross-linking agent (36) [75].

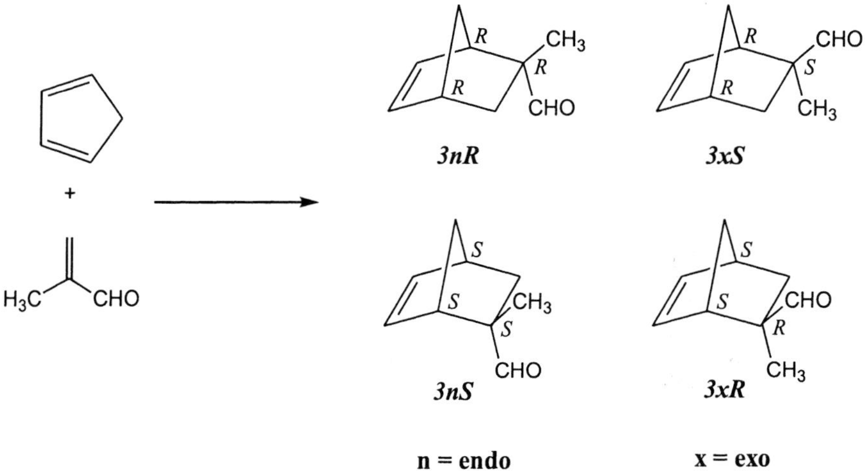

n = endo **x = exo**

Scheme 11.14. Diels-Alder adducts from methacrolein and cyclopentadiene.

Fraile et al. prepared several chiral aminoalcohols supported on chloromethylated polystyrene-divinylbenzene and used the chiral Lewis acid resulting from the complexation with $AlEtCl_2$ in the same Diels-Alder reaction [76]. Among various immobilized chiral catalysts, only polymer-bound chiral Al/proline complex **53** showed high activity, but the enantioselectivities were disappointing (7–14% e.e.). Supported aluminium complexes prepared from (2*R*,3*R*)-tartaric acid moieties anchored on polystyrene gave a low enantiomeric excess but displayed a higher activity than the homogeneous catalyst due to site isolation [76].

Polymer-bound Ti-TADDOLates were developed by Seebach et al. [48] and were successfully applied to the Diels-Alder addition of 3-crotonoyloxazolidinone to cyclo-

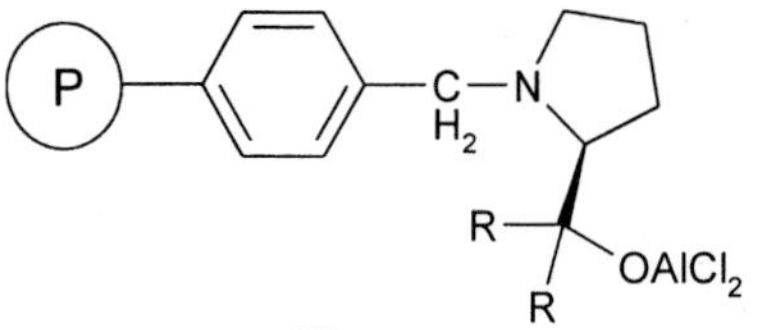

53

pentadiene (Scheme 11.6). TADDOLs immobilized by copolymerization – regardless of the type of copolymerization – demonstrated the same activities as in homogeneous conditions, while polymers resulting from chlorine substitution of a Merrifield resin were inactive. However, the *endo*-enantioselectivities were much lower than those obtained with the soluble TADDOLates. Note that a bimolecular cationic complex is thought to be involved in the homogeneous enantioselective reaction; it is unlikely that such a complex can be formed from immobilized TADDOL complexes.

Mayoral et al. prepared polymer-linked TADDOL according to Scheme 11.15 [77].

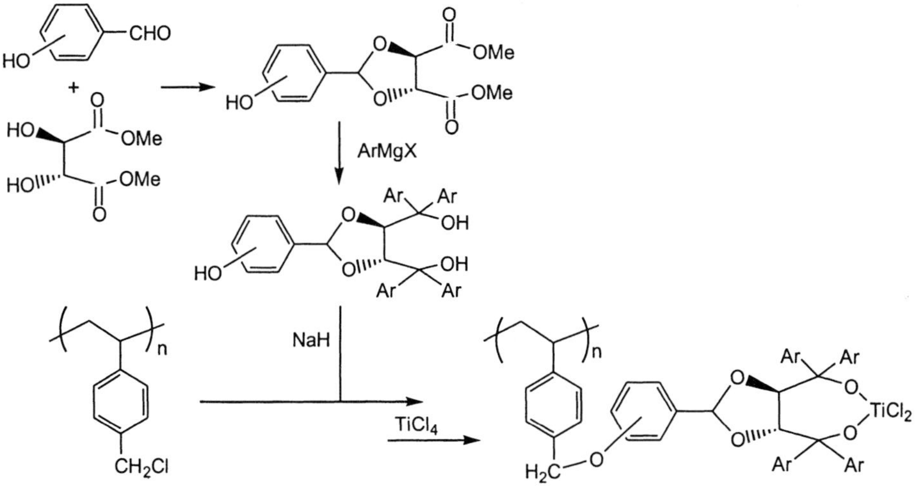

Scheme 11.15. Synthesis of polymer-linked Ti-TADDOL.

With this type of catalyst, both the support and the TADDOL-type influence the catalytic performance. For instance, only the complex with Ar=3,5-dimethylphenyl gave the (2*R*,3*S*)-cycloadduct **54** with an e.e. of 25% (Scheme 11.16).

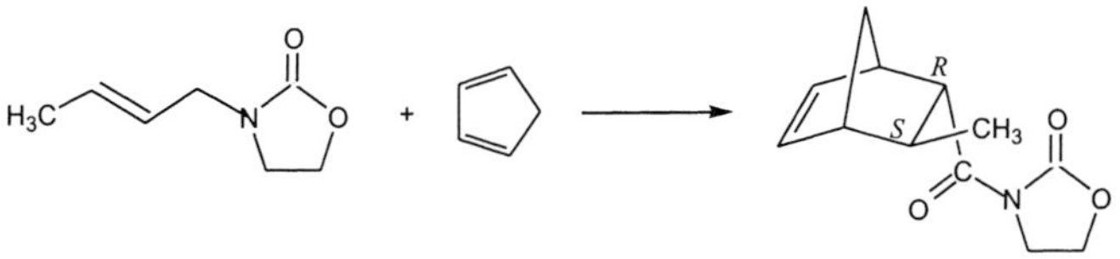

54

Scheme 11.16.

11.3.2 Heterogenization of Chiral Lewis Acids on Mineral Supports

Reports on chiral Lewis acids supported on inorganic materials are scarce [78]. N-tosyl-(*S*)-tyrosine was anchored on functionalized silica and, after reaction with BH_3, the corresponding oxazaborolidinone **55** was obtained (Fig. 11.6). A very low e.e. was observed in the reaction of methacrolein and cyclopentadiene.

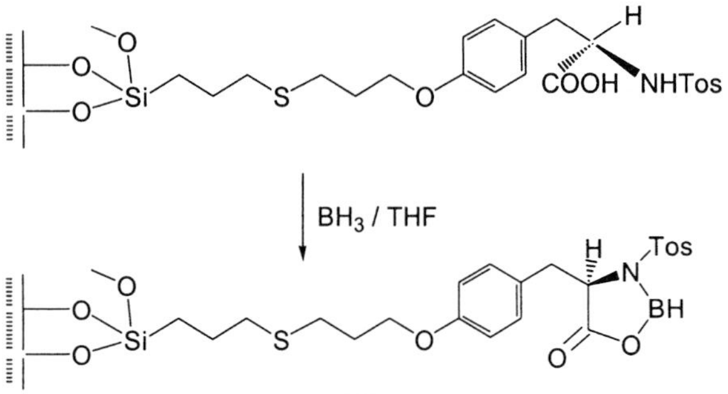

Figure 11.6. Preparation of silica-anchored chiral oxazoborolidinone.

Similar results were found by using anchored proline derivatives such as **56**, but slightly better results were obtained with the anchored chiral (–)-menthoxyaluminium complex **57** (up to 31% e.e. of 3×*S*).

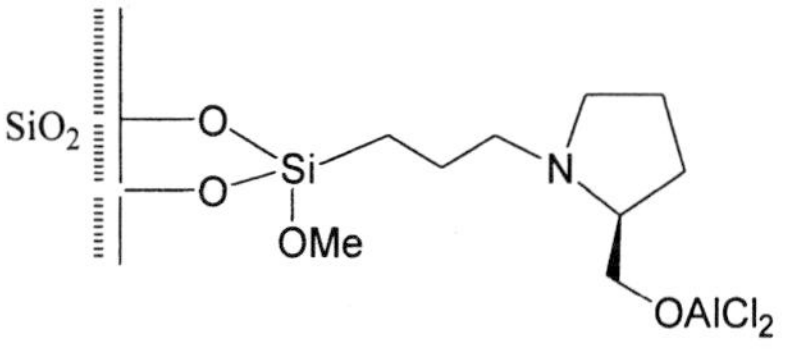

References

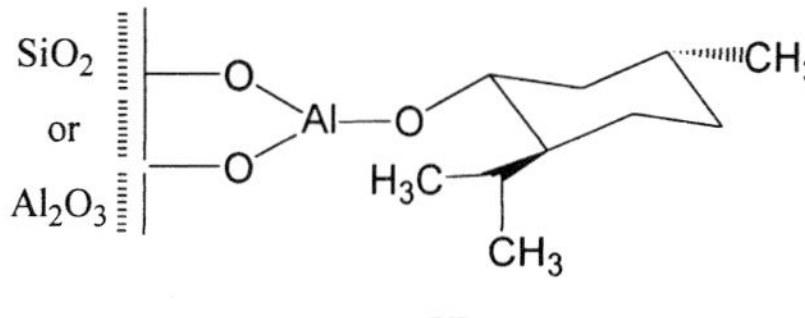

[1] J. Mathieu and J. Weill Reyn, *Formation of C-C bonds*, Georg Thieme Publishers, Stuggart (1973), Vol. I and II.

[2] a) R. Noyori and M. Kitamura in *Modern Synthetic Methods*, Sheffold Ed., Springer, Berlin (1989), pp. 115–198; b) R. Noyori, *Asymmetric Catalysis in Organic Synthesis*, Wiley, New York (1994).

[3] H. Wynberg and R. Helder, *Tetrahedron Lett.*, (1979) 2238.

[4] K. Hermann and H. Wynberg, *J. Org. Chem.*, 44 (1979) 2238.

[5] R.S.E. Conn, A.V. Lowell, S. Karady and L.M. Weinstock, *J. Org. Chem.*, 51 (1986) 4710.

[6] D.L. Hugues, U.-H. Dolling, K.M. Ryan, E.F. Schoenewaldt, and E.J.J. Grabowski, *J. Org. Chem.*, 52 (1987) 4745.

[7] K. Hermann and H. Wynberg, Helv. *Chim. Acta*, 60 (1977) 2208.
[8] a) H. Sasai, T. Suzuki, S. Arai, T. Arai, M. Shibasaki, *J. Am. Chem. Soc.*, 114 (1992) 4418; b) H. Sasai, T. Arai, M. Shibasaki, *J. Am. Chem. Soc.*, 116 (1994) 1571; c) H. Sasai, T. Arai, Y. Satow, K.N. Houk, M. Shibasaki, *J. Am. Chem. Soc.*, 117 (1995) 6194; d) M. Yamaguchi, A. Shiraishi, M. Hirama, *J. Org. Chem.*, 61 (1996) 3520.
[9] J. Jacques, *Bull. Soc. Chim. Fr.* 112 (1995) 353.
[10] S. Hashimoto, N. Komeshima and K. Koga, *J. Chem. Soc., Chem. Comm.* (1979) 437.
[11] F. Rebiere, O. Riant and H.B. Kagan, *Tetrahedron Asymm.*, 1 (1990) 199.
[12] K. Maruoka, T. Itoh, T. Shirasaka and H. Yamamoto, *J. Am. Chem. Soc.*, 110 (1988) 310.
[13] K. Ishihara and H. Yamamoto, *J. Am. Chem. Soc.*, 116 (1994) 1561.
[14] K. Narasaka, N. Iwasawa, M. Inoue, T. Yamada, M. Nakashima and J. Sugimori, *J. Am. Chem. Soc.*, 111 (1989) 5340.
[15] B. Altava, M.I. Burguette, J.M. Fraile, J.M. Fraile, S.V. Luis, J.A. Mayoral, A.J. Royo and M.J. Vicent, *Tetrahedron Asymm.*, 8 (1997) 2561.
[16] J. Hübsher and R. Barner, *Helv. Chim. Acta*, 73 (1990) 1068.
[17] G.M.R. Tombo, E. Didier and B. Loubinoux, *Synlett.* (1990) 547.
[18] N. Oguni and T. Omi, *Tetrahedron Lett.*, 25 (1984) 2823.
[19] a) M. Kitamura, S. Suga, K. Kawai and R. Noyori, *J. Am. Chem. Soc.*, 108 (1986) 6071; b) M. Kitamura, S. Okada, S. Suga and R. Noyori, *J. Am. Chem. Soc.*, 111 (1989) 4028; c) M. Kitamura, S. Suga, M. Niwa, R. Noyori, Z.-X. Zhai and H. Suga, *J. Phys. Chem.*, 98 (1994) 12776.; d) M. Kitamura, S. Suga, M. Niwa and R. Noyori, *J. Am. Chem. Soc.*, 117 (1995) 4832.
[20] a) K. Soai and A. Ookawa, *J. Chem. Soc., Chem. Comm.*, (1986) 412; b) K. Soai, A. Ookawa, T. Kaba and K. Ogawa, *J. Am. Chem. Soc.*, 109 (1987) 7111; c) S. Conti, M. Falorni, G. Giacomelli and F. Soccoli, *Tetrahedron Lett.*, 48 (1992) 8993.
[21] A.A. Smaardijk and H. Winberg, *J. Org. Chem.*, 52 (1987) 135.
[22] a) K. Soai, S. Yokoyama and T. Hayasaka, *J. Org. Chem.*, 56 (1991) 4264; b) M. Watanabe and K. Soai, *J. Chem. Soc.Perkins Trans.1*, (1994) 3125.
[23] R. Noyori and M. Kitamura, *Angew. Chem. Int. Ed. Engl.*, 30 (1991) 49–69.
[24] K. Soai and S. Niwa, *Chem. Rev.*, 92 (1992) 833–856.
[25] A. Vidal-Ferran, A. Moyano, M.A. Pericas and A. Riera, *J. Org. Chem.*, 62 (1997) 4970.
[26] H. Kitajima, K. Ito and T. Katsuki, *Chem. Lett.*, (1996) 343.
[27] F.-Y. Zhang, C.-W. Yip, R. Cao and A.S.C. Chen, *Tetrahedron Asymm.*, 8 (1997) 585.
[28] W.-S. Huang, Q.-S. Hu and L. Pu, *J. Org. Chem.*, 63 (1998) 1364.
[29] B. Schmidt and D. Seebach, *Angew. Chem. Int. Ed. Engl.*, 30 (1991) 99.
[30] H. Takahashi, T. Kawakita, M. Yoshioka, S. Kobayashi and M. Ohmo, *Tetrahedron Lett.*, 30 (1989) 7095.
[31] X. Zhang and C. Guo, *Tetrahedron Lett.*, 36 (1995) 4947.
[32] S. Itsuno and J.MJ. Frechet, *J. Org. Chem.*, 52 (1987) 4140.
[33] S. Soai, S. Niwa and M. Watanabe, *J. Org. Chem.*, 53 (1988) 927.
[34] S. Soai, S. Niwa and M. Watanabe, *J. Chem. Soc. Perkin Trans 1*, (1989) 109.
[35] M. Yamakawa and R. Noyori, *J. Am. Chem. Soc.*, 117 (1995) 6327.
[36] E.J. Corey and F.J. Hannon, *Tetrahedron Lett.*, 28 (1987) 5233 and 5237.
[37] B. Goldfuss and K.N. Houk, *J. Org. Chem.*, 63 (1998) 8998.
[38] P.A. Chaloner, S.A. Renuka Perera, *Tetrahedron Lett.*, 28 (1987) 3013.
[39] S. Soai and M. Watanabe, *Tetrahedron Asymm.*, 2 (1991) 97.
[40] M. Watanabe and S. Soai, *J. Chem. Soc., Perkin Trans I*, (1984) 837.
[41] S. Itsuno, Y. Sakurai, K. Ito, T. Maruyama, S. Nakahama and J.M.J. Frechet, *J. Org. Chem.*, 55 (1990) 304.
[42] K. Hosoya, S. Tsuji, K. Yoshisato, K. Kimata, T. Araki and N. Tanaka, *React. Funct. Polymer*, 29 (1996) 159.
[43] C. Dreisbach, G. Wischnewski, U. Kragl and C. Wandrey, *J. Chem. Soc. Perkin Trans. 1* (1995) 875.
[44] Z. Zhengpu, P. Hodge and P.W. Stratford, *React. Polymers*, 15 (1991) 71.
[45] D.W.L. Sung, P. Hodge and P.W. Stratford, *J. Chem. Soc. Perkin Trans. 1*, (1999) 1463.
[46] P. Hodge, D.W.L. Sung, and P.W. Stratford, *J. Chem. Soc. Perkin Trans. 1*, (1999) 2335.

[47] A. Vidal-Ferran, N. Bampos, A. Moyano, M.A. Pericás, A. Riera and J.K.M. Sanders, *J. Org. Chem.*, 63 (1998) 6309.
[48] D. Seebach, R.E. Marti and T. Hintermann, *Helvetica Chim. Acta*, 79 (1996) 1710.
[49] W.-S. Huang, W.-S. Hu, X.-F. Zheng, J. Anderson and L. Pu, *J. Am. Chem. Soc.*, 119 (1997) 4313.
[50] W.-S. Hu, W.-S. Huang, D. Vitharana, X.-F. Zheng and L. Pu, *J. Am. Chem. Soc.*, 119 (1997) 12454.
[51] K. Soai, M. Watanabe and A. Yamamoto, *J. Org. Chem.*, 55 (1990) 4832.
[52] a) A. Corma, M. Iglesias, C. del Pino and F. Sanchez, *J. Chem. Soc., Chem. Comm.*, (1991) 1253; b) A. Corma, M. Iglesias, C. del Pino and F. Sanchez, *J. Organometal. Chem.*, 431 (1992) 233.
[53] A. Corma, M. Iglesias, M.V. Martin, J. Rubio and F. Sanchez, *Tetrahedron Asymm.*, 3 (1992) 845.
[54] a) D. Brunel, A. Cauvel, F. Fajula and F. Di Renzo, *Stud. Surf. Sci. Catal.*, 97 (1995) 173; b) A. Cauvel, D. Brunel, F. Di Renzo and F. Fajula, *Am. Inst. Phys.*, 354 (1996) 477.
[55] a) A. Cauvel, G. Renard and D. Brunel, *J. Org. Chem.*, 62 (1997) 749; b) M. Laspéras, T. Lorett, L. Chaves, I. Rodriguez, A. Cauvel, D. Brunel, *Stud. Surf. Sci. Catal.*, 108 (1997) 75.
[56] A. Cauvel, D. Brunel, F. Di Renzo, E. Garrone and B. Fubini, *Langmuir*, 13 (1997) 2773.
[57] D. Brunel, A. Cauvel, F. Di Renzo, F. Fajula, B. Fubini, B. Onida and E. Garrone, *New J. Chem.*, submitted.
[58] P. Sutra, F. Fajula, D. Brunel, P. Lentz, G. Daelen and J. B. Nagy, *Colloids Surf.*, 158 (1999) 21.
[59] M. Laspéras, N. Bellocq, D. Brunel, P. Moreau, *Tetrahedron Asymm.*, 9 (1998) 3053.
[60] N. Bellocq, S. Abramson, M. Laspéras, D. Brunel and P. Moreau, *Tetradedron Asymm.*, 10 (1999) 3229.
[61] S. Abramson, N. Bellocq, M. Laspéras, *Topics in Catalysis*, in press.
[62] S. Abramson, M. Laspéras and D. Brunel, in preparation.
[63] G.W. Kabalka, R.M. Pagni, S. Bains, G. Hondrogiannis, M. Plesco, R. Kurt, D. Cox and J. Green, *Tetrahedron Asymm.*, 2 (1991) 1283.
[64] C. Cativiela, F. Figuéras, J.M. Fraile, J.I. Garcia and J.A. Mayoral, *Tetrahedron Asymm.*, 2 (1991) 953.
[65] C. Cativiela, P. Lopez, J.A. Mayoral, *Tetrahedron Asymm.*, 2 (1991) 1295.
[66] C. Cativiela, J.M. Fraile, J.I. Garcia, J.A. Mayoral, E. Pires, F. Figuéras and L.C. de Ménorval, *Tetrahedron*, 48 (1992) 6467.
[67] C. Cativiela, F. Figuéras, J.M. Fraile, J.I. Garcia and J.A. Mayoral, *Tetrahedron Asymm*, 4 (1993) 223.
[68] C. Cativiela, J.M. Fraile, J.I. Garcia and J.A. Mayoral, J.M. Campelo, D. Luna, J.M. Marinas, *Tetrahedron Asymm*, 4 (1993) 2507.
[69] J.I. Garcia, J.A. Mayoral, E. Pires, D.R. Brown and J. Massam, *Catal. Lett.*, 37 (1996) 261.
[70] J.M. Fraile, J.I. Garcia, D. Gracia, J.A. Mayoral and E. Pires, *J. Org. Chem.*, 61 (1996) 9479.
[71] F. M. Bautista, J.M. Campelo, A. Garcia, D. Luna, J.M. Marinas, J.I. Garcia, J.A. Mayoral and E. Pires, *Catal. Lett.*, 36 (1996) 215.
[72] J.M. Fraile, J.I. Garcia and J.A. Mayoral, *Recent Res. Devl. in Synth. Organic Chem.*, 1 (1998) 77–92.
[73] S. Itsuno, K. Kamahori, K. Watanabe, T. Koizumi and K. Ito, *Tetrahedron Asymm.*, 5 (1994) 523.
[74] K. Kamahori, S. Tada, K. Ito and S. Itsuno, *Tetrahedron Asymm.*, 6 (1995) 2547.
[75] K. Kamori, K. Ito and S. Itsuno, *J. Org. Chem.*, 61 (1996) 8321.
[76] J.M. Fraile, J.A. Mayoral, A.J. Royo, R.V. Salvador, B. Altava, S.V. Luis and M.I. Burguete, *Tetrahedron*, 52 (1996) 9853.
[77] B. Altava, M.I. Burguete, B. Escuder, S.V. Luis, R.V. Salvador, J.M. Fraile, J.A. Mayoral and A.J. Royo, *J. Org. Chem.*, 62 (1997) 3126.
[78] J.M. Fraile, J.I. Garcia, J.A. Mayoral and A.J. Royo, *Tetrahedron Asymm.*, 7 (1996) 2263.

12 Heterogeneous Diastereoselective Catalysis

Dirk E. De Vos, Mario De bruyn, Vasile I. Parvulescu, Florian G. Cocu, and Pierre A. Jacobs

12.1 Introduction

The previous chapters in this book deal with enantioselective catalysis: starting from a prochiral substrate molecule, a product is formed that contains one or more new stereogenic centers. Since the catalyst itself is chiral, one of the configurations at the new stereocenter(s) is formed in excess.

Alternatively, one might perform reactions with a non-enantioselective catalyst, and with a substrate that already contains at least one stereogenic center besides the prochiral group. Transformation of this prochiral group (C=C, C=O, C=N, ...) may then result in preferential formation of one of the diastereomers. In that case, one of the configurations (*R* or *S*) prevails at the new stereocenter(s). This concept is known as diastereoselective synthesis.

For instance, imagine that an *α,β*-substituted product is formed, and that the reactant already contains a stereogenic carbon at *α*. If the reaction of (*αS*) leads e.g. largely to (*αS, βR*) and hardly to the (*αS, βS*) diastereomer, the reaction is diastereoselective (Fig. 12.1). Starting from a racemic mixture of (*αS*) and (*αR*), the same diastereoselective reaction will mainly lead to the racemic mixture of the enantiomers (*αS, βR*) and (*αR, βS*). The minor products are then (*αS, βS*) and (*αR, βR*).

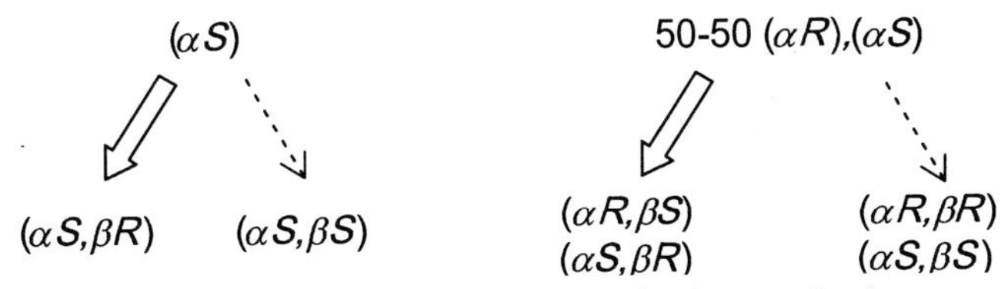

Figure 12.1. The concept of diastereoselective synthesis.

Evidently, diastereoselectivity only leads to enantiomerically pure products if the starting product is enantiomerically pure. With a non-enantioselective catalyst, the absolute value of the d.e. (diastereomeric excess) is the same whether starting from (*αS*), (*aR*) or a 50–50 mixture of (*aS*) and (*aR*). In contrast, if the catalyst is itself enantioselective, the d.e. values may well be different depending on the configurations in the starting product.

Molecules with a prochiral group and one or more stereocenters are frequently encountered in complex organic syntheses, and these molecules can be used as such in diastereoselective transformations. Alternatively, relatively simple molecules with a prochiral group can be derivatized with a 'chiral auxiliary'. Well-known chiral auxiliaries are proline and its derivatives, menthol, or any other enantiomerically pure compound from the 'chiral pool'. Availability at a low price is essential for a chiral auxiliary, even if it is often possible to regain the auxiliary after the diastereoselective reaction.

The present paper gives a wide overview of classical and recent examples of diastereoselective reactions with heterogeneous catalysts. Not only the large group of reductive transformations, but also other reaction types will be discussed. Many basic concepts of the stereochemistry of heterogeneously catalyzed reactions can be found in the book by Bartok [1]. Diastereoselective hydrogenations have been covered rather recently in an excellent review by Besson and Pinel [2].

12.2 Diastereoselective Heterogeneously Catalyzed Reactions

12.2.1 Hydrogenations

12.2.1.1 Hydrogenation of C=C Bonds

Unsaturated hydrocarbons, alcohols and amines

Diastereoselectivity is often associated with complex, multifunctional molecules. However, even in a simple carbon skeleton such as in *α*-pinene, the configuration of the stereogenic carbon atoms can induce the selective formation of a new stereogenic center. For instance, in the hydrogenation of (1*S*,5*S*)-2-pinene (**1**), (1*S*,2*R*,5*S*)-pinane (**2**, (–)-*cis*-pinane) is formed with high diastereoselectivity [3, 4]. For example, d.e. values are 96% with a Raney Ni catalyst, or 99% with Rh/C.

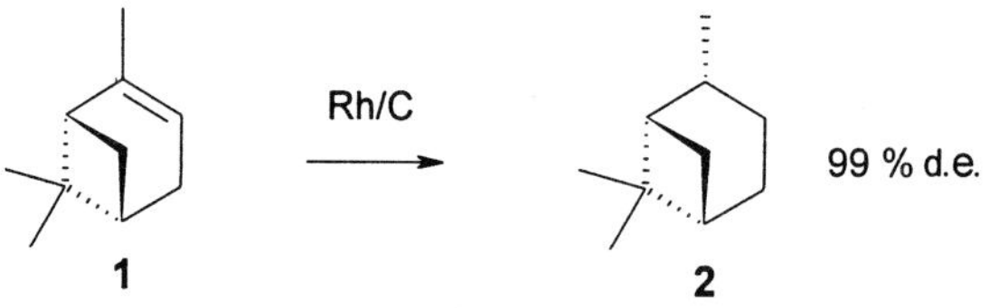

Rh/C

1

2

99 % d.e.

Scheme 12.1.

More complex examples are frequently encountered in the literature. In the following hydrogenation, leading to (+)-royleanone **3**, only Pd/C exhibited some diastereoselectivity [5]:

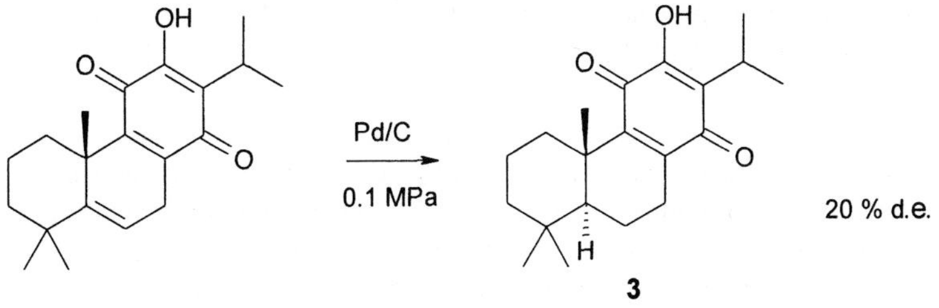

Pd/C

0.1 MPa

3

20 % d.e.

Scheme 12.2.

When the molecule contains a second functional group besides the double bond, this group might have a decisive effect on the stereochemistry of the C=C bond hydrogenation. Generally, there is a complex interplay between the nature of the functional group, the catalyst type and the solvent, and one has to look for the optimum combination for each case.

Amine and alcohol groups strongly interact with the surface of many catalysts, and thus are useful in obtaining diastereoselectivity [6–8]. For instance, the following alkylidene cyclopentanol **4** is hydrogenated with excellent diastereoselectivity over Raney Ni in ethanol [6]:

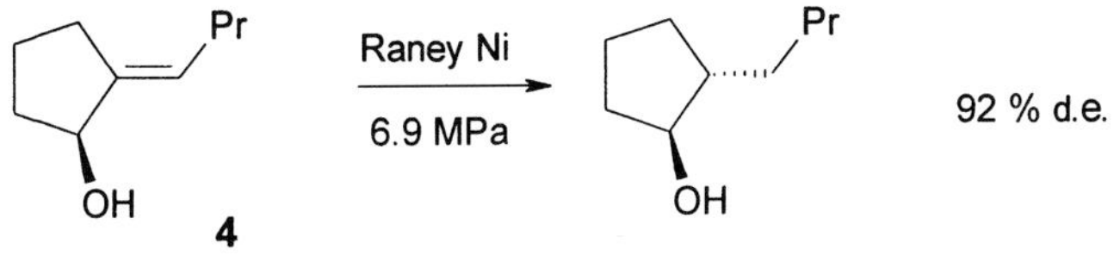

Raney Ni

6.9 MPa

4

92 % d.e.

Scheme 12.3.

Slightly lower d.e. values are obtained in the reaction of the following unsaturated six-ring [6]:

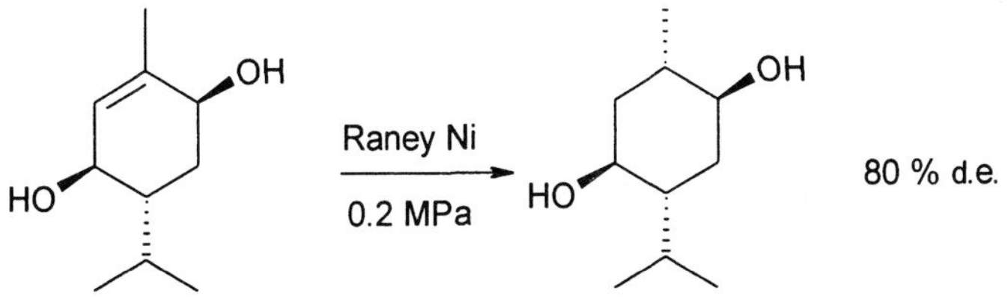

Raney Ni

0.2 MPa

80 % d.e.

Scheme 12.4.

In the hydrogenation of indenols like **5**, the effect of the size of the R-group was investigated [9]:

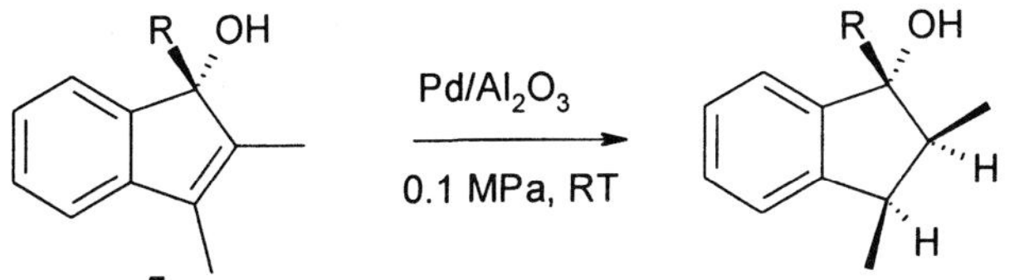

Pd/Al_2O_3

0.1 MPa, RT

5

Scheme 12.5.

With R=i-C_4H_9, the reaction proceeds with 100% diastereoselectivity over a Pd/Al_2O_3 catalyst in toluene, while lower d.e. values are found for smaller groups.

The polarity of the solvent may determine whether or not a -CH_2OH functional group can be adsorbed on a catalytic surface. In the hydrogenation of 7-methoxy-10a-hydroxymethyl-1,2,3,9,10,10a-hexahydrophenantrene (**6**) with Pd/C, a *cis*- or a *trans*-hydrogenation product may be obtained [10]. In a polar solvent such as dimethylformamide, the hydroxymethyl group is not adsorbed on the surface, due to competitive solvent adsorption, and the *trans* product is obtained. For a reaction in hexane, the hydrogen is delivered to the molecule from the same side as the adsorbed -CH_2OH group, leading to the *cis* product:

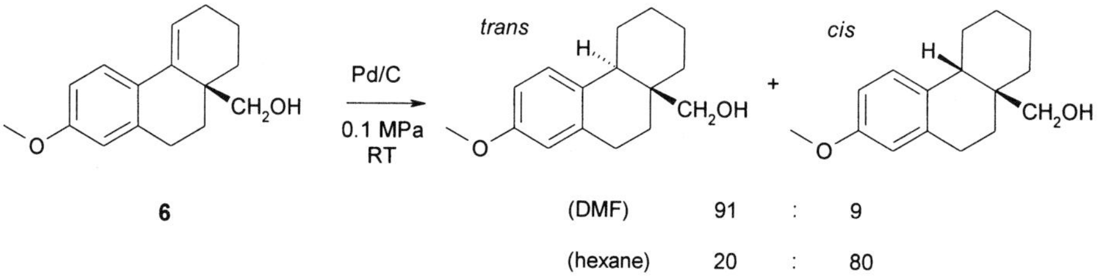

Pd/C, 0.1 MPa, RT: **6** → *trans* + *cis*

(DMF) 91 : 9

(hexane) 20 : 80

Scheme 12.6.

Unsaturated amides and esters; synthesis of amino acids

Diastereoselectivity can be induced in the C=C hydrogenation of α,β-unsaturated amides or esters by enantiomerically pure alcohols or amines, which are easily linked to the carbon skeleton via the amide or ester groups. For instance, β-methylcinnamic acid can be esterified with (–)-(1*R*,2*S*)-*N,N*-dimethylephedrinium chloride. Subsequent hydrogenation of the ester **7** on Raney Ni leads to substantial diastereoselectivity, with an excess of the *R*-hydrogenated product [11]:

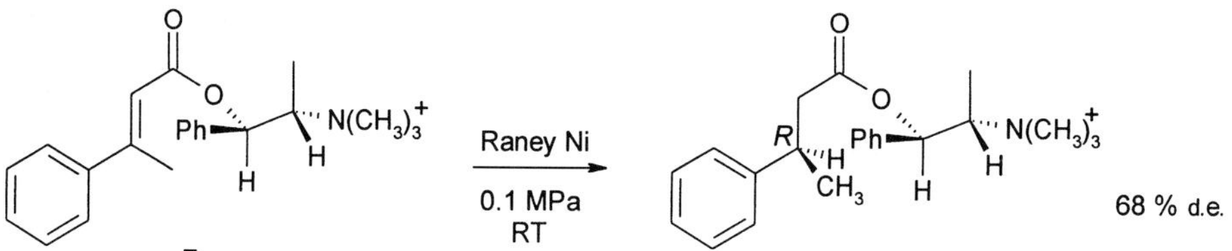

Raney Ni, 0.1 MPa, RT: **7** → 68 % d.e.

Scheme 12.7.

Similar reactions with α-acylamino-α,β-unsaturated amides are straightforward routes to enantiomerically pure amino acids. Note that the *S*-configuration is the natural L-form of the amino acid, while *R* corresponds to the D-form. In a first approach, one starts from the amide **8** of (*R*)-α-methylbenzylamine and α-benzamido-β,β-dimethylacrylic acid. After hydrogenation and hydrolysis, L-valine was obtained with an e.e. of 18% [12]:

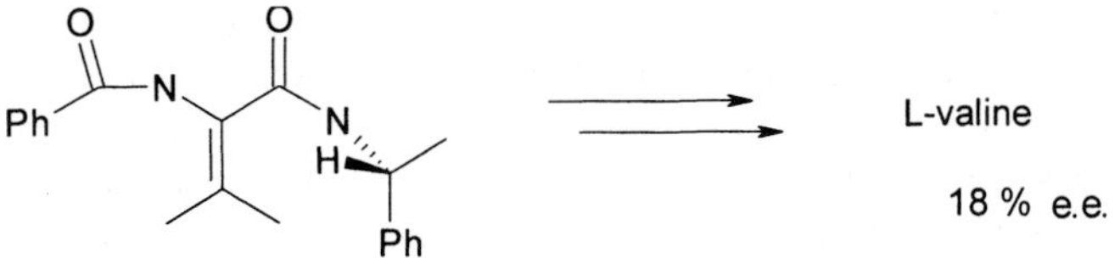
Ph
N
N
H
Ph
L-valine
18 % e.e.

8

Scheme 12.8.

Much higher d.e. values are obtained when proline or its amides are used to derivatize the *α*-acylamino-*α*,*β*-unsaturated acids. The starting compound is then actually a dehydrodipeptide such as **9**. In an early report, the amide of L-proline and *α*-acetylamidocinnamic acid was hydrogenated in EtOH, using a polymer-supported Pd catalyst. After hydrolysis, this leads to an e.e. of 88% for the unnatural D-phenylalanine [13]. In an improved version of this technique, proline methylamide is used as a derivatizing agent, and an *α*-acylamino group is used in cinnamic acid or *α*,*β*-unsaturated aliphatic acids. Starting from dehydrodipeptides like **9,** the diastereomeric excesses are systematically over 90%, e.g. [14, 15]:

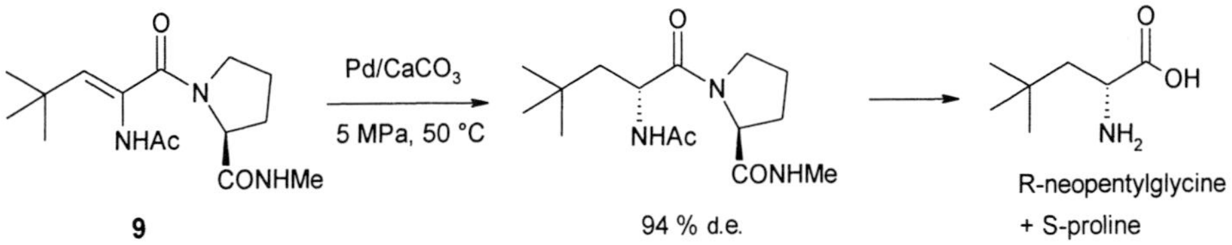
NHAc
CONHMe
Pd/CaCO3
5 MPa, 50 °C
NHAc
CONHMe
94 % d.e.
OH
NH2
R-neopentylglycine
+ S-proline

9

Scheme 12.9.

Alternatively, cyclic dehydrodipeptides may be used as amino acid precursors. For instance, **10** is a cyclic dipeptide consisting of L-proline and dehydroalanine. The rigidity of the cyclic structure allows a better conformational control and, hence, optimal diastereoselectivity. A general entry to this route is formation of an amide between L-proline methyl ester and an *α*-keto acid such as pyruvic acid. After reaction with an anhydrous amine and dehydration, an alkylidene diketopiperazine **10** is obtained. Hydrogenation over PtO_2 in ethanol and subsequent hydrolysis results in natural amino acids of high purity [16]:

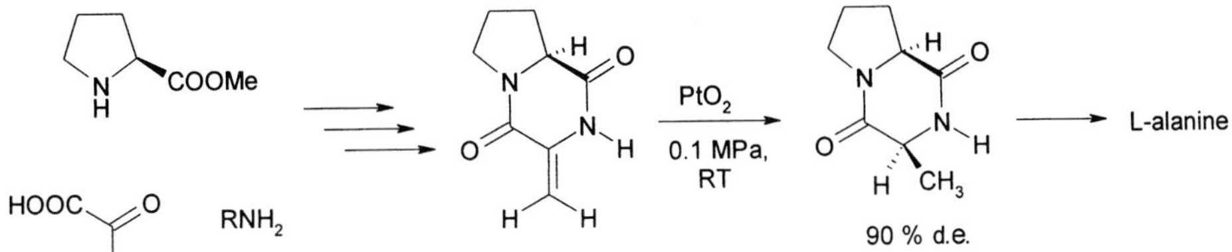
COOMe
HOOC
RNH2
PtO2
0.1 MPa,
RT
CH3
90 % d.e.
L-alanine

10

Scheme 12.10.

A similar approach was used even a few years earlier by Schmidt and Poisel [17]. Starting from a benzylidene-substituted, proline-containing dioxopiperazine, they prepared aromatic amino acids of over 90% enantiomeric purity, such as L-phenylalanine and L-DOPA (3-(3,4-dihydroxyphenyl)-(*S*)-alanine). The preferred hydrogenation catalyst was PdO for a hydrogenation in glacial acetic acid.

Unsaturated diketopiperazines (or cyclic dehydrodipeptides) were also used by Japanese workers. However, in their procedure, it is not necessary to use proline derivatives. Even alanine can successfully be used to effect a diastereoselective hydrogenation [18]. The stereoselectivity in the hydrogenation of the cyclic dipeptide **11** containing dehydroalanine and (*S*)-valine is illustrated in Scheme 12.11.

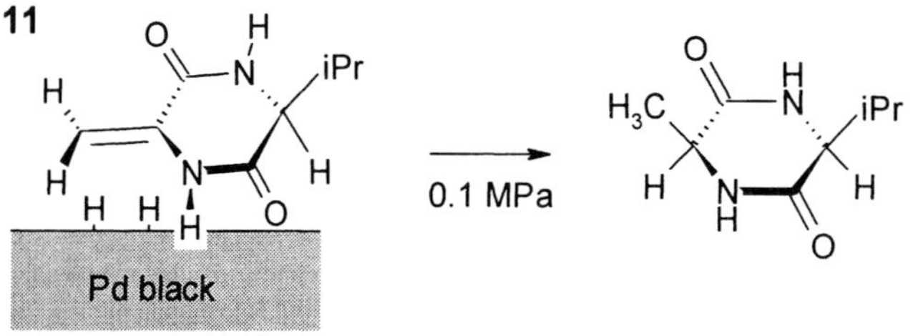

Scheme 12.11.

In a reaction over Pd black in methanol, a diastereomeric excess of 98% was thus obtained for L-alanine. Similar routes have been described for the synthesis of aromatic amino acids such as O-methyl-L-tyrosine [19].

The adsorption of a diketopiperazine, like **11,** might be disturbed by the presence of functional groups in the side chain of the amino acid. Nevertheless, even L-ornithine was synthesized in high purity via this route, starting from the diketopiperazine **12** [20]:

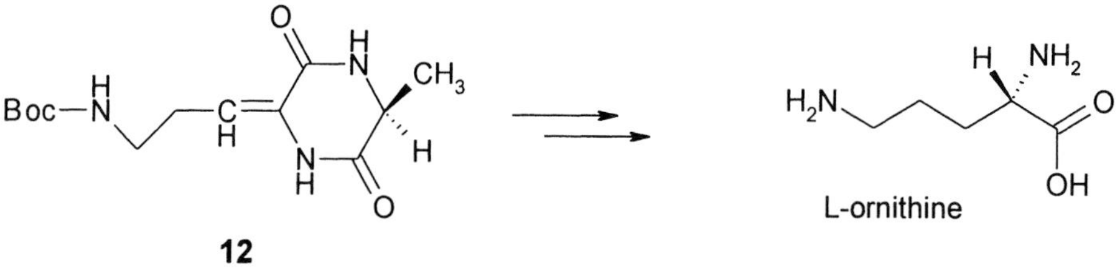

Scheme 12.12.

During the synthesis, the δ-amino group is protected. The excellent d.e. proves that the stereochemical outcome of the reaction is not affected by adsorption of the urethane group.

Another route towards optically pure amino acids involves diastereoselective hydrogenation of enantiomerically pure alkylidene imidazolidinones (e.g. **14**). The synthesis starts from commercially available Boc-BMI (**13**, 1-Boc-2-*tert.*-butyl-3-methyl-4-imidazolidinone). The diastereoselective hydrogenation of **14** proceeds best over Pd/C in ethyl acetate [21, 22]:

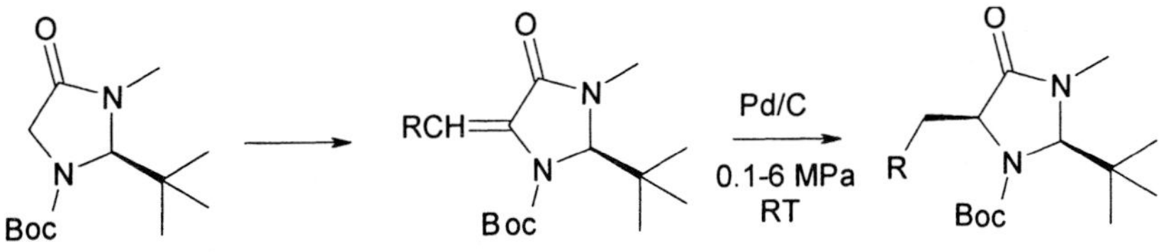

Scheme 12.13.

However, the subsequent hydrolysis of **15** to obtain the desired amino acid is difficult, and the optically active auxiliary is lost in the overall process.

Hydrogenation of steroids

Many steroids such as 4-cholesten-3-one, testosterone or 19-nortestosterone contain a 3-keto-4-ene group. Hydrogenation can lead to the 5*α*- or the 5*β*-isomers. The outcome of the reaction is strongly dependent on a series of factors, such as the support, the metal, the acidity and the other functional groups in the molecule [23]. In the case of 4-cholesten-3-one, the 5*β*/5*α* isomer ratio varies between 8, for a Pd/C catalyst, and 0.8 for Pd supported on a Dowex resin [24]. The same reaction was also performed with a heterogeneous Ni catalyst, prepared by NaH reduction of $Ni(OAc)_2$, and resulted in a 5*β*/5*α* ratio of 7 [25]. Addition of acid to a Pd-catalyzed reaction tends to increase the 5*β*/5*α* ratio. This has been ascribed to protonation of the ketone group [26].

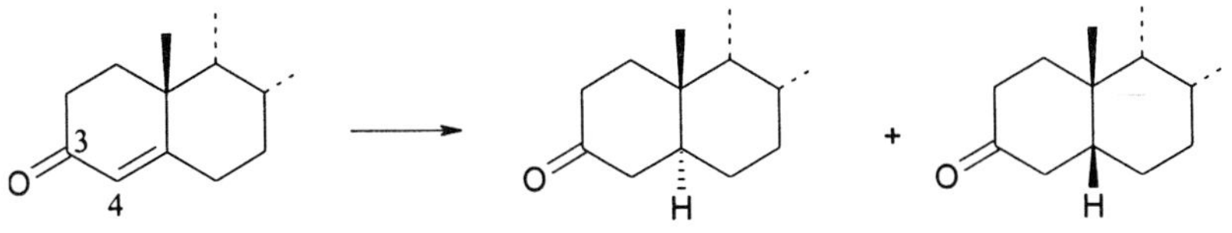

Scheme 12.14.

Finally, the 5*β*/5*α* ratio is considerably lower (<1) in the reaction of testosterone. This might well be due to an interaction between the catalyst and the distant 17-OH group in the testosterone molecule. This hypothesis was proved by acetylation of the 17-OH group; testosterone acetate gives again a relatively high 5*β*/5*α* ratio of 2.6 [26].

Even heterogeneous hydrogen transfer reactions have been applied to reduction of steroids. Thus, reaction of testosterone over Pd/C in benzyl alcohol resulted again in a preferential formation of the 5*β*-isomer 17*β*-OH-5*β*-androstan-3-one, with a 60% d.e. [27]. A practically complete stereoselectivity was observed in the reaction of 7-methyl-6-dehydrotestosterone acetate **16**, again with benzyl alcohol and Pd/C [28]:

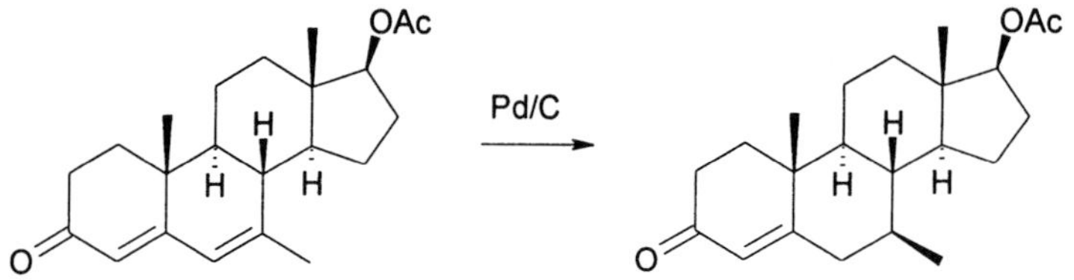

Scheme 12.15.

Hydrogenation of unsaturated lactones

There are several examples of diastereoselective hydrogenation of a C=C double bond in a lactone structure. In the reaction of **17**, an 80% d.e. for the *cis* product was obtained:

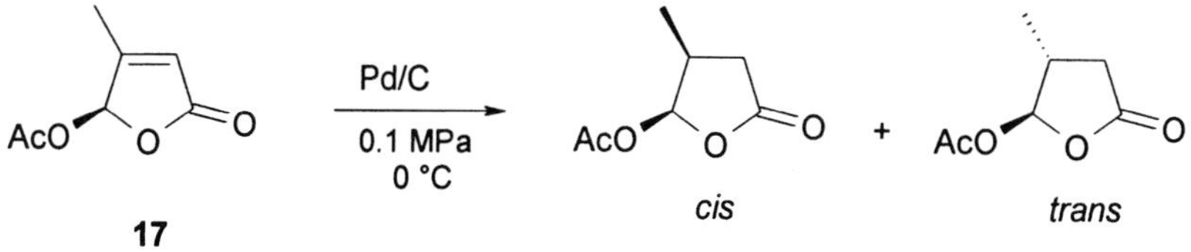

Scheme 12.16.

This reaction was used in the synthesis of the antitumor drug (–)-acetomycin [29]. Another unsaturated lactone substrate is L-ascorbic acid **18**. Hydrogenation with Pd/C in water resulted in the exclusive formation of L-gulono-1,4-lactone **19**, with an almost complete diastereoselectivity [30]. With a Pt catalyst, a mixture of products is obtained; Raney Ni does not give any products.

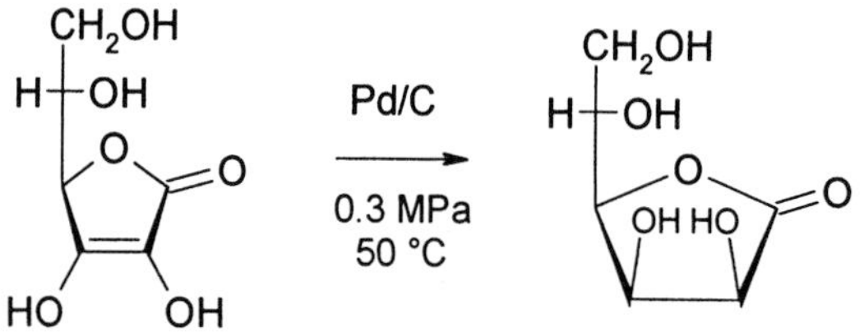

Scheme 12.17.

12.2.1.2 Hydrogenation of C=N Bonds

Synthesis of amino acids

Hydrogenation of C=N double bonds is another obvious track to enantiomerically pure amino acids [31–35]. An early approach makes use of the formation of imines between α-keto acids and optically pure amines, such as L-phenylglycine. Hydrogenation of these imines (e.g., **20**) leads to variable degrees of diastereoselective induction.

The reaction is directly followed by hydrogenolysis of the product **21**, so that the enantiomerically pure amino acid is immediately available:

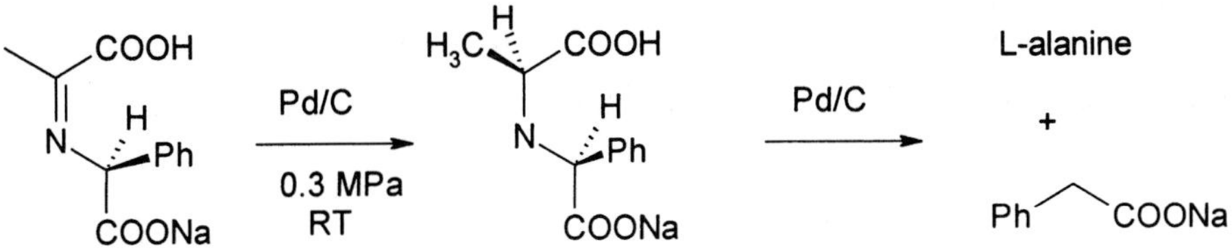

Scheme 12.18.

An important disadvantage of such a reductive amination with an enantiomerically pure amine is that the chiral auxiliary cannot be recovered after reaction [31].

On the other hand, the -COOH residue of for instance pyruvic acid may be derivatized with a chiral auxiliary. In initial reports, menthyl pyruvates were converted into their oximes or imines. Hydrogenation of these C=N containing esters over Pd/C resulted in moderate diastereoselectivities (maximally 45%) [36]. Far better results were obtained with pyruvamides. For instance, the amide of pyruvic acid and (*R*)-*α*-ethylbenzylamine was converted into an oxime **22** by reaction with hydroxylamine. Subsequent hydrogenation with Pd/C in methanol proceeded with 70% d.e., leading to fairly pure L-alanine [37]:

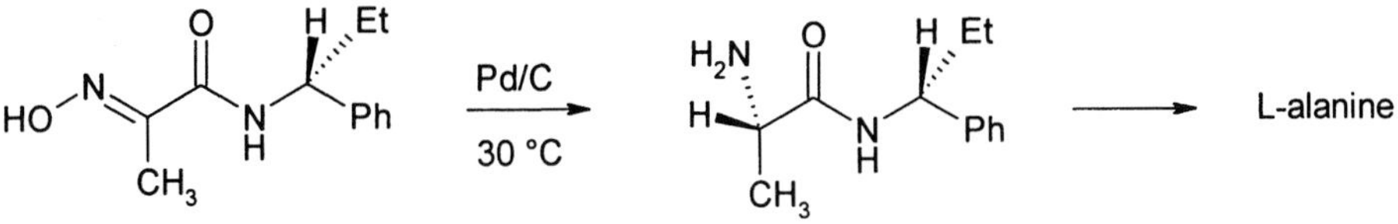

Scheme 12.19.

A somewhat similar approach is the hydrogenation of oximes of chiral *α*-keto alcohols (e.g. the (*E*)-oxime of (*R*)-benzoin) [38].

Synthesis of secondary amines

The synthesis of amines with two stereogenic centers can be based on the same principles of reductive diastereoselective amination. Diastereomeric excesses of over 90% are easily achieved, for instance in the preparation of (*S*)-N-[1-(o-methoxyphenyl)ethyl]-(*S*)-*α*-methylbenzylamine **23** [39]:

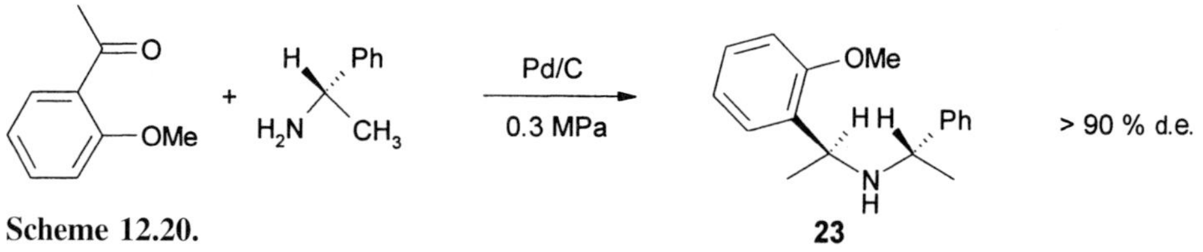

Scheme 12.20.

Similar results were obtained by Blackmond and coworkers, using catalysts prepared from inorganic Grignard reagents. In the reaction of ethyl pyruvate with various chiral α-substituted ethyl amines, the maximum d.e. was 87% with at most 2% selectivity for the hydrogenolysis products [40]. Note that in these cases the hydrogenation is not followed by hydrogenolysis. However, if desired, the latter reaction can be conducted with high regioselectivity [41].

A recent application of these methods is the synthesis of Enalapril (**26**) [42]. The last step in the preparation is the reductive amination between ethyl 4-phenyl-2-keto-butyrate (**24**) and the dipeptide alanyl proline (**25**). With the Raney Ni catalyst, the reaction can be stopped after the hydrogenation, without subsequent hydrogenolysis:

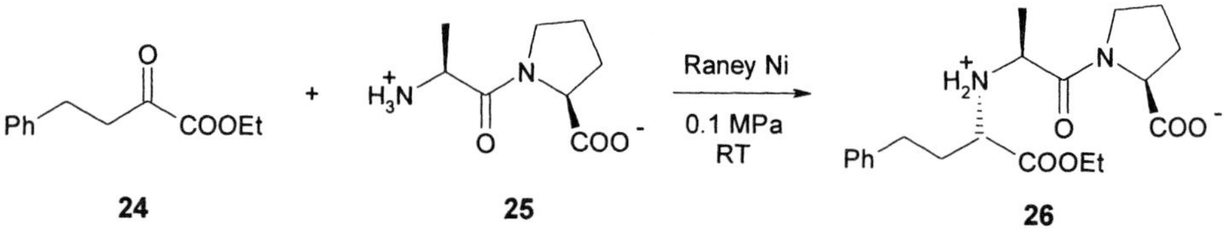

Scheme 12.21.

Careful tuning of the reaction conditions via combinatorial techniques allowed the increasing of the diastereoselectivity. Simultaneous addition of acetic acid and KF increased the d.e. to 94% [42].

A beautiful example comprising C=N hydrogenation can be found in the synthesis of an indolizidine (**27**). The hydrogenation of the alkene, the removal of the CBz (benzyloxycarbonyl) protecting group, the cyclization and finally, the diastereoselective reduction of the iminium intermediate are all accomplished with H_2 and a PtO_2 catalyst [43]:

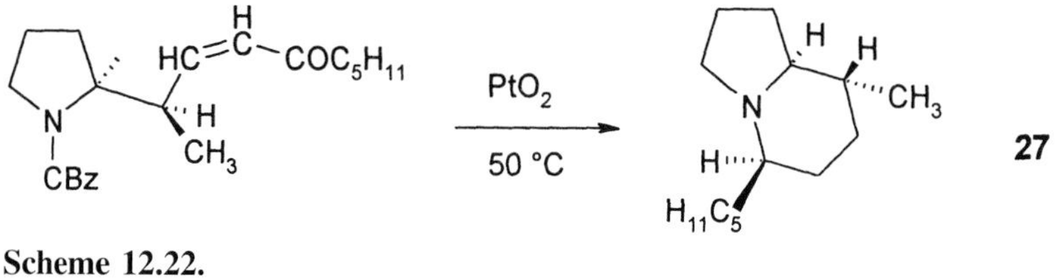

Scheme 12.22.

12.2.1.3 Hydrogenation of C=O Bonds

Enantiomerically pure lactic acid can be obtained by hydrogenation of chirally modified pyruvates. As in the synthesis of amino acids, proline derivatives such as proline esters and proline amides have been used. However, the d.e. values did not exceed 59 and 77%, respectively [44, 45]. A much better diastereoselectivity was mentioned in an early report by Harada [46]. He used the amide **28** of pyruvic acid and (*S*)-*a*-methylbenzylamine:

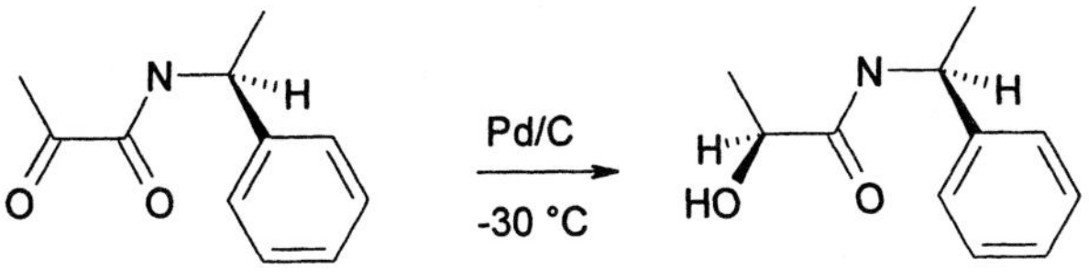

Pd/C, -30 °C; **28** → **29**

Scheme 12.23.

In a hydrogenation in methanol at –30 °C, a d.e. value of 98% was obtained, with preferential formation of the *S*-configuration in the lactic amide **29**. Even at 30 °C, the d.e. is still 86%.

Apart from the pyruvic acid derivatives, there are several reports on diastereoselective ketone reduction. (*R,R*)-Tartrate/NaBr-modified Raney Ni was used in the reduction of methyl 2-oxocyclopentanecarboxylate **30** [47]:

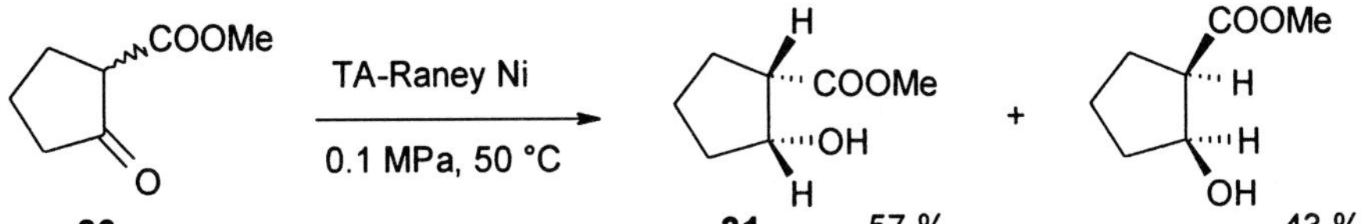

TA-Raney Ni, 0.1 MPa, 50 °C; **30** → **31** 57 % + 43 %

Scheme 12.24.

The major product was the *cis* hydrogenation product, but the d.e. for the (1*S*,2*R*) isomer (**31**) was rather disappointing (14%). In contrast, complete diastereoselectivity at 100% conversion was obtained in the hydrogenation of a steroidic *a*-diketone **32** [48]:

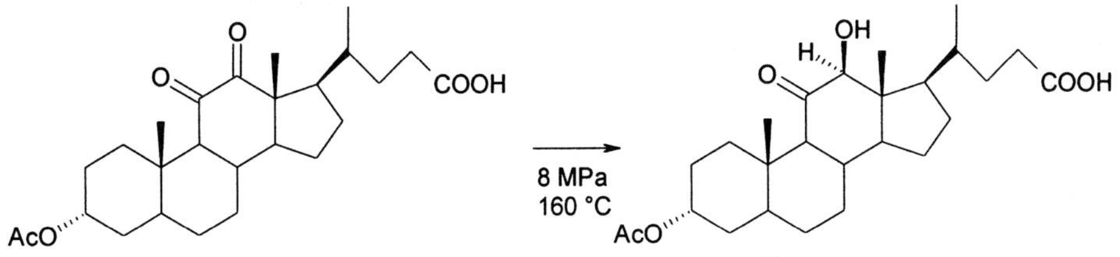

8 MPa, 160 °C; **32** → **33**

Scheme 12.25.

The product, 3*a*-acetyl-11-oxo-12*β*-hydroxo-5*β*-cholan-24-oic acid (**33**) is obtained with complete regioselectivity. The reaction is relatively independent of the type of the metal, since Rh/SiO_2, Ru/SiO_2, Pd/SiO_2, Pt/SiO_2 and Pd/C all give the same result.

In the same context of biologically active compounds, hydrogenation of prostaglandin intermediates such as **34** has attracted much attention [49–51]. The reduction of the enone in the side chain of **34** is performed up to now with stoichiometric Al-reduction agents. Use of H_2 is highly attractive, provided that a satisfactory chemoselectivity for the allylic alcohol **35** and diastereoselectivity for the biologically relevant isomer can be obtained. Encouraging diastereoselective effects have been observed, for instance, with supported Ru catalysts, with or without modifying agents:

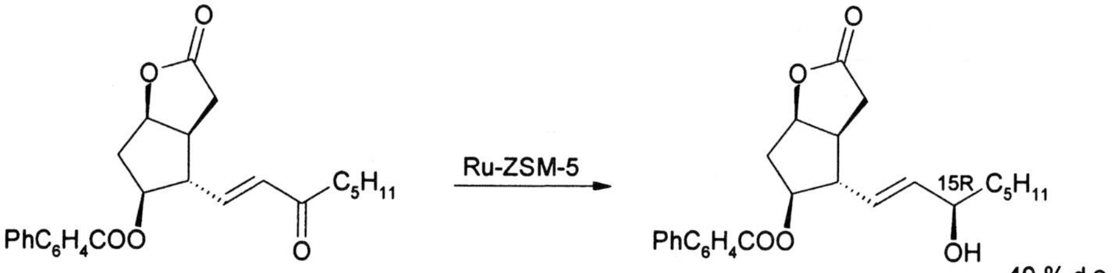

Scheme 12.26.

Finally, hydrogen transfer reactions can also result in diastereoselective ketone hydrogenations. Of course, the structure of the hydrogen donor is an essential factor. Ethanol, cyclohexanol and even chiral alcohols such as (–)-menthol were used as hydrogen donors in a study using liquid In as the catalytic surface at 470 °C [52]. Ketone substrates were camphone and fenchone. The hydrogen transfer follows the Cram rule, which states that attack at the less sterically crowded side is favored. In the case of camphone (**36**) reduction, this leads to attack at the *endo*-side, resulting in isoborneol (**37**), particularly if a bulky reductant such as tetralol is used:

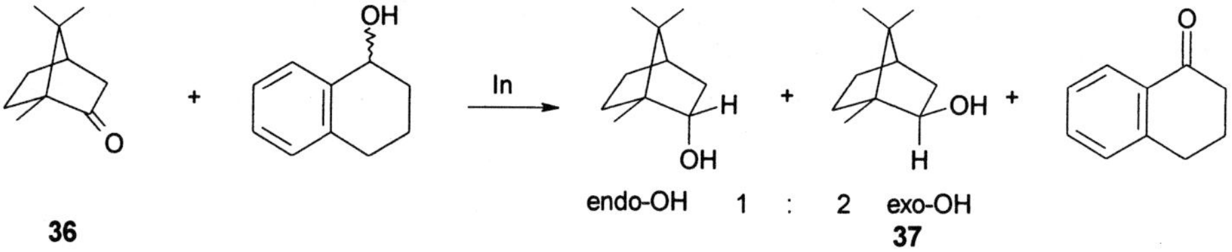

Scheme 12.27.

12.2.1.4 Hydrogenation of Aromatics

Substituted aromatic compounds seldomly contain stereogenic centers. Hence, if one desires to obtain enantiomerically pure products from hydrogenation of an aromatic compound, one of the substituents must be functionalized with a chiral auxiliary. On the other hand, it is known that reduction of aromatics can be a multistep process. Stereogenic centers formed in the first step might then influence the stereochemistry at stereogenic centers formed in a subsequent step. Such influences are essentially diastereoselective effects.

The latter case is nicely illustrated by the hydrogenation of thymol. With a Pt/C catalyst in cyclohexane, the reaction proceeds mainly, though not exclusively, *via* the isomenthones **38a** and the menthones **38b**, which are formed in a ratio of roughly 1:2 [53]. In a second step, the isomeric neo- and isomenthols are formed (only the (1*R*)-3-menthanone isomers are shown):

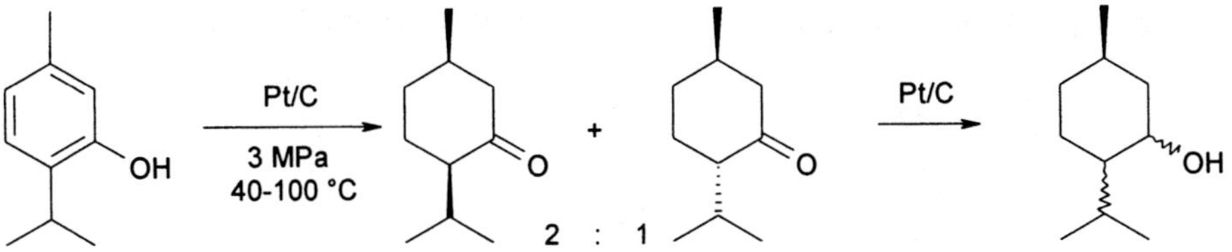

Pt/C, 3 MPa, 40-100 °C; 2 : 1; **38a**, **38b**; Pt/C

Scheme 12.28.

A high diastereoselectivity (93%) is observed in the second step, for the hydrogenation of isomenthone **38a** into mainly neoisomenthol **39**:

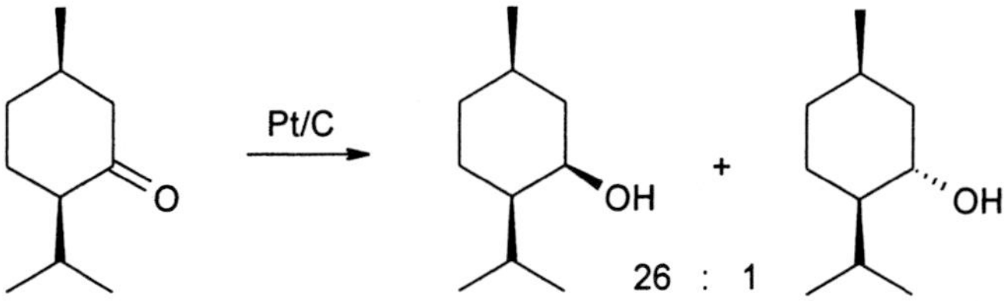

38a; Pt/C; 26 : 1; **39**

Scheme 12.29.

Unfortunately, the desired natural product (±)-menthol is formed with a fairly low overall selectivity (~15%). Nevertheless, the example of thymol clearly demonstrates that in a heterogeneously catalyzed hydrogenation, diastereoselective effects can favor the formation of one couple of enantiomers over their diastereomers.

Enantiomerically pure 2-methylcyclohexanecarboxylic acids **41** can be obtained by hydrogenation of toluic acid derivatives on a non-enantioselective catalyst, if the toluic acid is derivatized with a suitable auxiliary [54–56]. While Pd and Pt give relatively large amounts of the *trans*-2-methyl-1-COOH-cyclohexanes, high *cis*-selectivities (>97%) are observed with Rh and Ru. In many experiments the chiral auxiliary was a proline ester, but the very best results have been obtained with the toluamide **40** of related ketone, namely methyl (*S*)-pyroglutamate [54]:

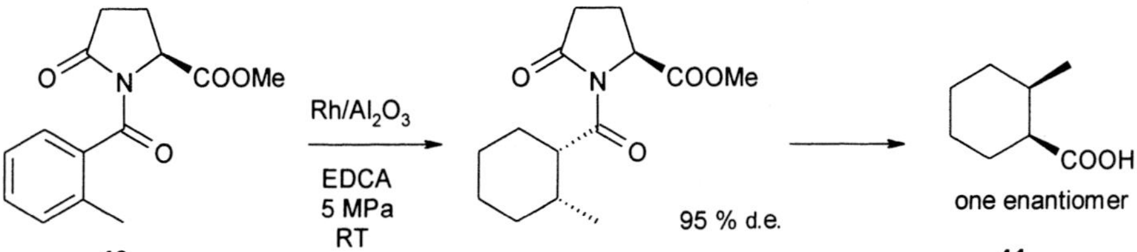

40; Rh/Al_2O_3, EDCA, 5 MPa, RT; 95 % d.e.; one enantiomer **41**

Scheme 12.30.

These reactions are frequently performed in the presence of a bulky amine, such as ethyldicyclohexylamine (EDCA). The role of the amine is to partially cover the surface, so that the *ortho*-toluamide reactant is forced to adsorb in a conformation that occupies the least space on the surface. Thus amine addition is able to increase the d.e., even if rates are smaller with amine-modified catalysts than with the catalysts as such [55].

Another important parameter is the nature of the support, and results may be quite different for Rh/C or Rh/Al_2O_3. For instance, in the absence of EDCA, the reaction with N-(2-methylbenzoyl)-(*S*)-proline methyl ester on Rh/C in ethanol is unselective. In the same conditions, the reaction with Rh/Al_2O_3 gives a d.e. of 40% [56]. Pretreatment of the catalyst, for instance under flowing H_2, is another important variable.

While the best results were obtained with Rh/C or Rh/Al_2O_3 catalysts, it should be noted that Ru catalysts generally give high levels of diastereoselective induction, even if the trends are quite different from those observed with Rh. With the latter metal, the toluamide adsorbs on the catalyst via the aromatic rings. At contrast, on the more oxophilic Ru surface, the adsorption likely takes place via the oxygen atoms of the amide and ester groups.

In analogy to the reactions with *o*-toluic acid, the hydrogenation of proline-derivatized aniline and anthranilic acid has been studied [57, 58]. For instance, starting from the anilide **42**, the *cis*-2-CH_3-cyclohexylamide **43** was obtained in 55% d.e. over a Rh/Al_2O_3 catalyst:

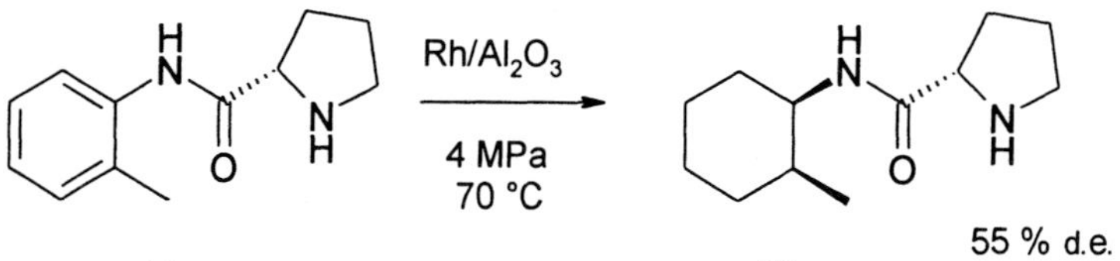

Rh/Al_2O_3, 4 MPa, 70 °C; **42** → **43** (55 % d.e.)

Scheme 12.31.

Results for other chirally substituted aromatic compounds are generally less exciting. For instance, d.e. values in the hydrogenation of chiral esters of *ortho*-cresol did not exceed 10%. In a similar approach, esters were prepared from vanillic acid and chiral alcohols such as (–)-menthol. However, the d.e.'s were rather low [2].

12.2.1.5 Hydrogenation of Heterocyclic Compounds

Furans

2,5-Disubstituted furans are of interest in the synthesis of natural products. Hydrogenation of such a furan normally leads to the *cis*-products. If one of the side chains of the furan contains a stereogenic center, the formation of one of the diastereomeric hydrogenation products may be favored. A beautiful example can be found in the hydrogenation of furylcarbinols such as **44** over Raney Ni [59]:

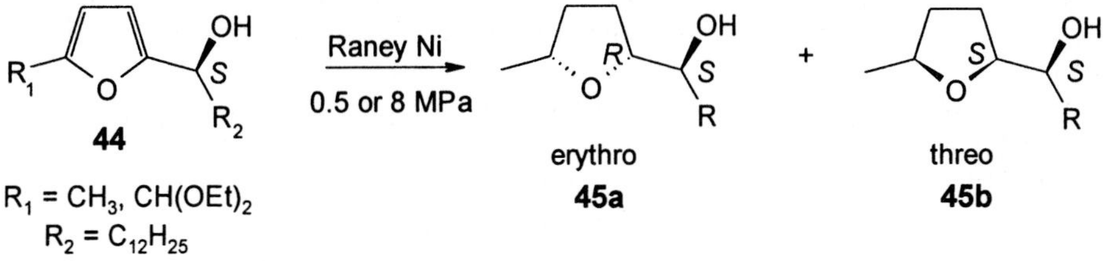

Raney Ni
0.5 or 8 MPa

44
R_1 = CH_3, $CH(OEt)_2$
R_2 = $C_{12}H_{25}$

erythro
45a

threo
45b

Scheme 12.32.

There are remarkable effects of the solvent on the diastereoselectivity. In isopropanol and with $R_1=CH_3$, the *erythro*-isomer **45a** predominates with a d.e. of 69%. In contrast, there is a large selectivity for the *threo*-isomer (**45b**, d.e. of 70%) if the reaction is run in methanol, with $R_1=CH(OEt)_2$. Thus, the size of the substituents on the furan ring also seems important. Both the substituents and the solvent can have an effect on the preferred conformations of the molecule.

Pyrroles and pyrrolines

The synthesis of (±) lycorane nicely illustrates the power of diastereoselective synthesis in creating new stereogenic carbon atoms [60]. Before hydrogenation, the substrate molecule **46** contains only a single stereogenic carbon atom, and both enantiomers are present in the solution. Reduction of the pyrrole ring creates two new stereocenters **47**:

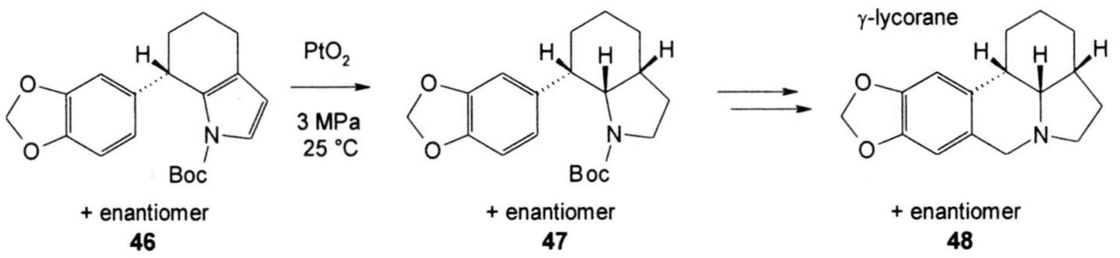

PtO_2
3 MPa
25 °C

γ-lycorane

+ enantiomer
46

+ enantiomer
47

+ enantiomer
48

Scheme 12.33.

The reaction proceeds with 97% diastereoselectivity on PtO_2 in $CHCl_3$, so that eventually the two enantiomers of lycorane **48** are isolated, without contamination by diastereomers.

In the hydrogenation of the dihydropyrrole **49**, the choice between the formation of *R,R*- or *S,S*-diastereomers is determined by the chiral substituent on the nitrogen atom [61]. The relatively inexpensive chiral auxiliary (*S*)-*a*-methylbenzylamine can eventually be removed by hydrogenolysis, and a pyrrolidine **50** of high optical purity is obtained. Such homochiral pyrrolidines are important in insecticide preparation:

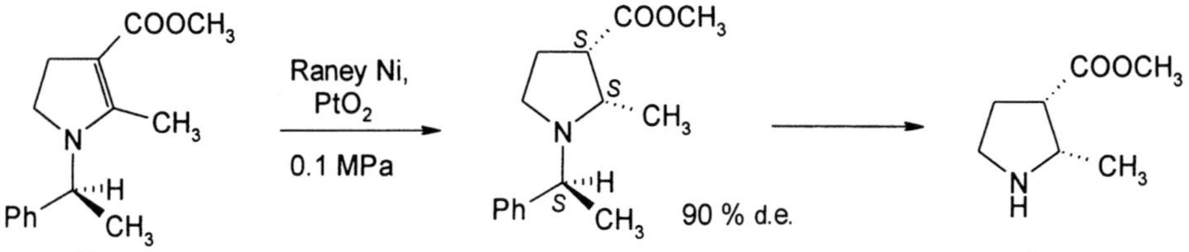

Scheme 12.34.

Pyridines

Hydrogenation of (*R*)-2-pyridylethylcarbinol **51** results in the formation of conhydrine, with the new stereocenter in the *S*-configuration [62]. Similarly, reduction of (*R*)-picolylmethylcarbinol **52** gives the natural alkaloid sedridine as the sole product [63]. In the latter case, formation of a hydrogen bonded 6-ring between the nitrogen and oxygen atoms may lock the desired conformation during the hydrogenation. Adsorption on the surface of the Pt catalyst and subsequent hydrogenation then lead to an *S*-configuration for the ring asymmetric carbon atom:

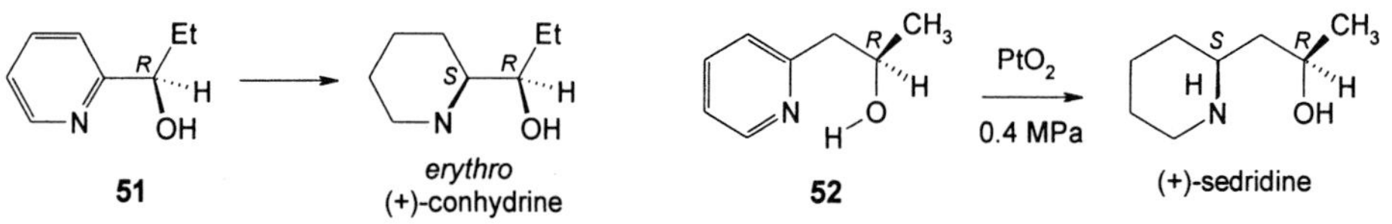

Scheme 12.35.

Pyrazines

A final example of the hydrogenation of a heterocyclic ring is the partial reduction of the pterine moiety in folic acid **53** [64]. In human cells, stepwise reduction by the folic acid reductases exclusively yields the (6*S*,*S*) isomer **54**. Reduction with PtO_2 in glacial acetic acid or with Rh/SiO_2 in water produces both diastereomers in more or less equal amounts. Thus, it seems that the stereogenic carbon atom in the (*S*)-glutamate residue is too far away to obtain substantial diastereoselectivity.

Derivatization at the N-5 position with (–)-menthyl chloroformate is one option to separate the diastereomers. Alternatively, the reaction may be performed with various chiral Rh-diphosphine complexes immobilized on SiO_2. Ligands of the BDPP, DIOP and NORPHOS types gave moderate d.e. values of up to 24%. Significantly, the absolute value of the d.e. obtained with (+)-NORPHOS (22% for the (6*S*,*S*) isomer **54**) is clearly larger than the d.e. in the reaction with (–)-NORPHOS (12% for the (6*R*,*S*) isomer **55**).

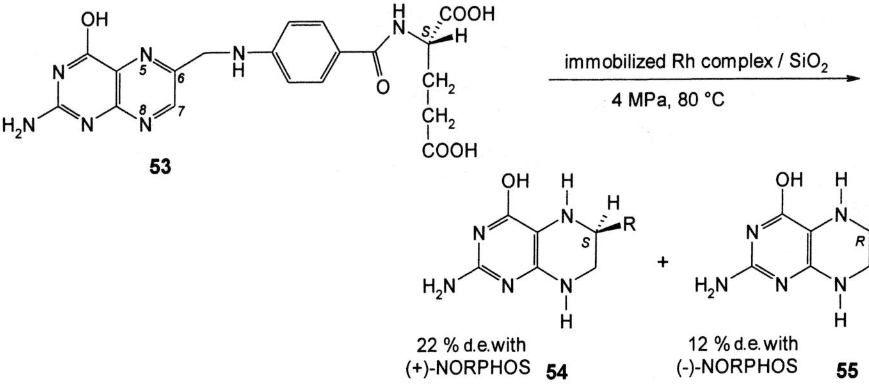

Scheme 12.36.

12.2.2 Miscellaneous Reactions

12.2.2.1 Hydrogenolysis

Hydrogenolysis of an amine, for example, may result in the loss of a stereocenter and is therefore often an undesired reaction. However, one may imagine cases in which selective hydrogenolysis of one of two C-X bonds results in a new stereocenter, for instance in the hydrogenolysis of 6,6-dihalopenicillanates [65]. The following reaction was performed with Pd or Rh catalysts in ethyl acetate/methanol, starting from (pivaloyloxy)methyl 6,6-dibromopenicillanate **56**:

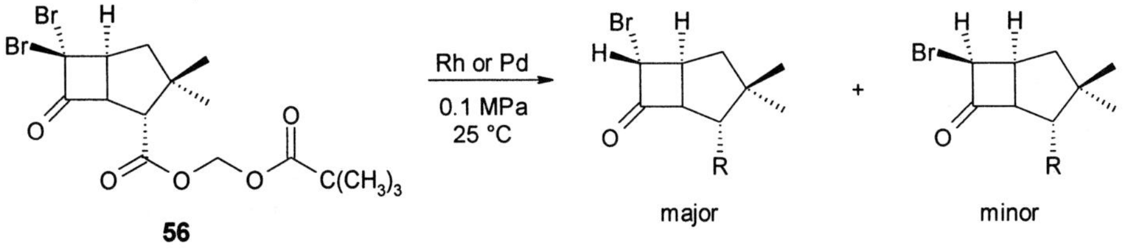

Scheme 12.37.

Diastereomeric excesses were 80% with 5% Pd/$CaCO_3$, and 75% with 5% Rh/Al_2O_3/$CaCO_3$. Methanol seems to enhance both rate and site selectivity of the reaction.

12.2.2.2 Pd-Catalyzed Cyclizations

Pd-catalyzed reaction of *cis*-1,2-divinylcyclohexane (**57**) with an acid results in 9-methylenebicyclo[4.3.0]nonanes with an esterified alcohol group in position 7. If the acid contains a stereogenic center, such as in derivatives of (*R*)-lactic or (*R*)-mandelic acid (e.g. **58**), one of the diastereomeric products may be formed in excess [66, 67]. The reaction is performed in acetone, and requires addition of MnO_2 and benzoquinone as reoxidants for the Pd. In the absence of molecular sieves, diastereomeric excesses are limited. Starting from (*R*)-lactic acid, either 7*R*- or 7*S*-isomers may be formed. However, addition of some molecular sieves spectacularly improves the diastereomeric induction. First, the prevailing configuration at C7 is now always *S* (**59**), and secondly, the d.e. values are much higher, reaching 76%:

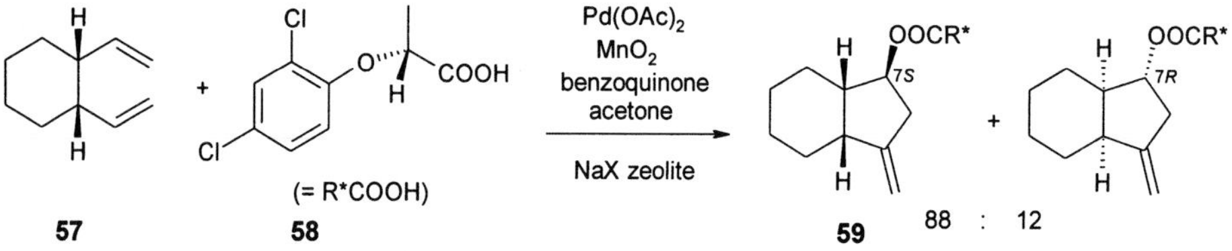

Scheme 12.38.

The reasons for this unexpected zeolite effect on the diastereoselectivity were investigated in detail via physicochemical means. Water scavenging by the zeolite and adsorption of the diene are not relevant to the stereochemistry. Only zeolites with a high Na^+ content, such as 4A and 13X are effective. Solid-state ^{13}C-NMR spectroscopy showed that the acid nucleophile is adsorbed as a Na carboxylate on the surface of the zeolite, even if, for 4A, the carboxylate has only access to the outer crystal surface. The ether oxygen atom (in **58**) probably provides a second interaction spot between nucleophile and surface. Thus, adsorption of the nucleophile in well-defined conformations seems to determine the stereochemistry. This hypothesis was confirmed by the observation that the diastereomeric product esters were adsorbed on the surface of the molecular sieves to a different extent.

12.2.2.3 Diels-Alder Cycloadditions

Several heterogeneous Lewis acid catalysts have been used in the Diels-Alder condensation of cyclopentadiene and enantiomerically pure dienophiles, such as (–)-menthyl acrylate [68–73]. A properly catalyzed reaction leads to a high *endo:exo* ratio of 10 or more. Moreover, the diastereomer with the 1*R*-configuration is usually formed in clear excess [68]:

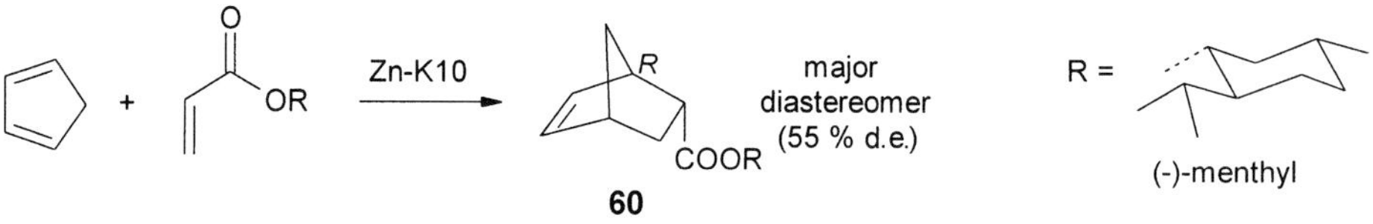

Scheme 12.39.

Several parameters such as solvent, catalyst type and catalyst pretreatment determine the outcome of the reaction. For instance, it is known that activation of alumina at 800 °C creates very strong Lewis acid sites that can catalyze the retro-Diels-Alder reaction. Hence, an excessive calcination temperature strongly diminishes the d.e. values [69]. Among the metal ion-exchanged K10 montmorillonites, Zn^{2+}- and Ti^{4+}-containing clays are superior to Fe^{3+} clays; the latter contain not only Lewis but also Brönsted acid sites which cause side reactions [70].

Reactions with fumarates such as (–)-menthyl methyl fumarate, or di-(–)-menthyl fumarate were generally less successful [71]. The use of very large dienophiles, such as neopentoxyisobornyl acrylate, leads to poor reaction rates. However, a remarkable result was obtained with the acrylate ester **61** of (*R*)-pantolactone. With this auxiliary, the 1*S*-*endo*-isomer **62** is formed with appreciable diastereoselectivity:

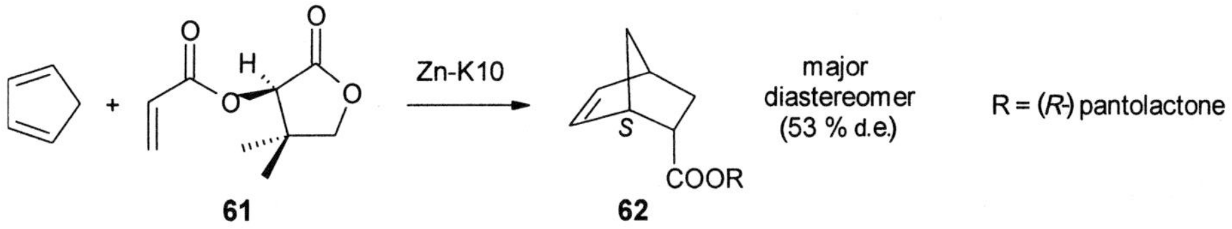

Scheme 12.40.

This demonstrates that (–)-menthol and (*R*)-pantolactone are complementary auxiliaries, leading to respectively the *endo*-(1*R*)- and *endo*-(1*S*)-5-norbornene-2-carboxylates [72].

12.2.2.4 Stereoselective Protonation of Enolates

Protonation of chiral enolates can lead to different diastereomers, as exemplified in the following reaction sequence [74]. Reaction of (*R*)-carvone (**63**) with lithium dimethylcuprate leads to a chiral, endocyclic keto-enolate (**64**). In a next step, a polymeric proton donor is added, and the two diastereomers **65** and **66** are formed:

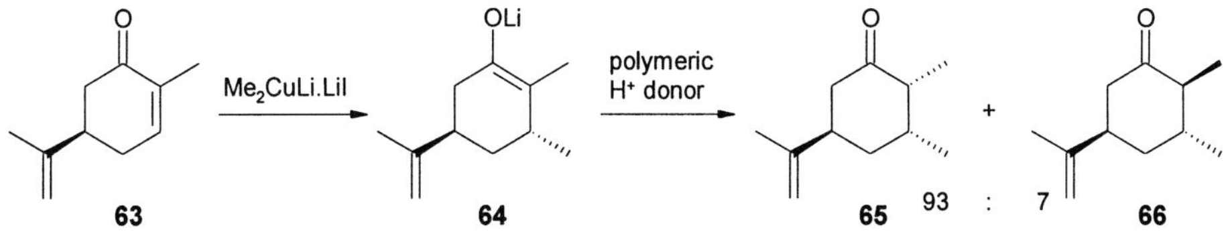

Scheme 12.41.

As proton donor, a polymer of methyl 5-vinylsalicylate was used. Due to the *ortho*-disposition of the -OH and -COOMe groups in the polymer, such a material is actually a chelating proton donor.

12.2.2.5 Thio-Claisen Rearrangement

The thio-Claisen rearrangement in Scheme 12.42 is the conversion of a γ-hydroxy ketene dithioacetal (**67**) into an α-allyl β-hydroxy dithioester (**68** and **69**). Essentially, the reaction is the migration of an allyl group. The configuration of the C(OH) stereogenic center in the starting product may determine the configuration at the α-carbon, to which the allyl substituent migrates. If the reaction is run over acid zeolites such as H-Y, H-ZSM-5 or H-β, only one diastereomer (**68**, *threo*) is formed [75]:

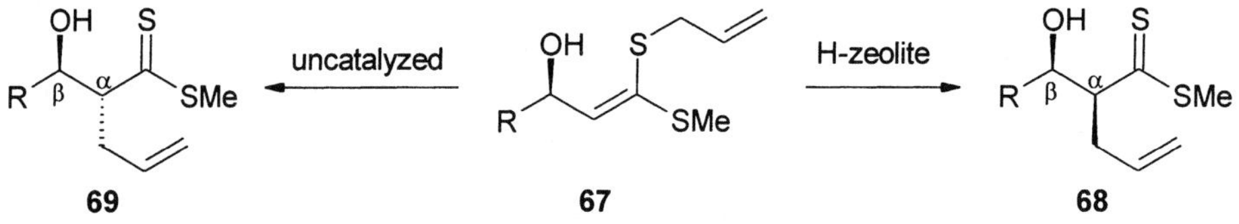

Scheme 12.42.

In contrast, the *erythro*-diastereomer (**69**) predominates in the uncatalyzed reaction. The particular selectivity of the zeolite-catalyzed reactions may be attributed to adsorption of the reactant molecule on the surface.

12.2.2.6 Epoxidation and Subsequent Epoxide Rearrangement

Epoxidation with amorphous or crystalline Ti-Si oxides is a very well researched field, and the catalysts have mostly been employed for selective oxygenation of small molecules such as allyl alcohol or propylene, either with H_2O_2 or with an organic peroxide as the oxidant. On the other hand, the stereochemistry of the epoxidation of more complicated olefins has been documented, for instance in the case of variously substituted allylic alcohols [76–79]. Such studies can be helpful in order to elucidate whether the transition states of the heterogeneous Ti reactions are similar to those encountered with e.g. peracids, or $Ti(iOPr)_4$/*t*BuOOH, or $VO(acac)_2$/*t*BuOOH.

Within the group of amorphous Ti-Si oxides, the mixed titania-silica aerogels have been characterized in much detail [76]. Such catalysts are obtained by a sol-gel process, followed by solvent extraction with supercritical CO_2. These materials can catalyze epoxidation reactions with *t*BuOOH. In the epoxidation of 2-cyclohexen-1-ol, slightly more of the *cis*-epoxide is formed than of the *trans*-epoxide:

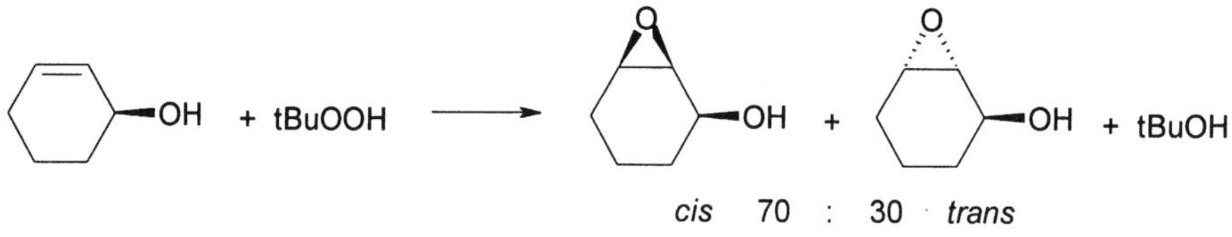

Scheme 12.43.

Much higher *cis*/*trans* ratios are obtained if the allylic alcohol group coordinates to the metal center in the transition state, as is the case for Sharpless' $VO(acac)_2$/*t*BuOOH system. This proves that in the TiO_2/SiO_2 aerogel, the alcohol group interacts only weakly with the peroxide-activated metal center.

Detailed stereochemical studies have been performed for the epoxidation of aliphatic allylic alcohols with H_2O_2 and the zeolites TS-1 and Ti-β [77–79]. Several transition states can be envisaged for such molecular sieve-promoted epoxidations, and some are shown in Fig. 12.2 for the simplest case, i.e. allyl alcohol. In model A (Fig. 12.2), the transition state is quite similar to that of a stoichiometric peracid epoxidation. Note that the allylic alcohol group is hydrogen-bonded to one of the oxygen atoms of the Ti-bound hydroperoxy group. On the other hand, model B shows a substrate molecule that is coordinated to the Ti center *via* an alkoxy group. The latter situation is known to occur with the homogeneous enantioselective Sharpless epoxidation catalyst, which contains $Ti(iOPr)_4$ in combination with diethyltartrate.

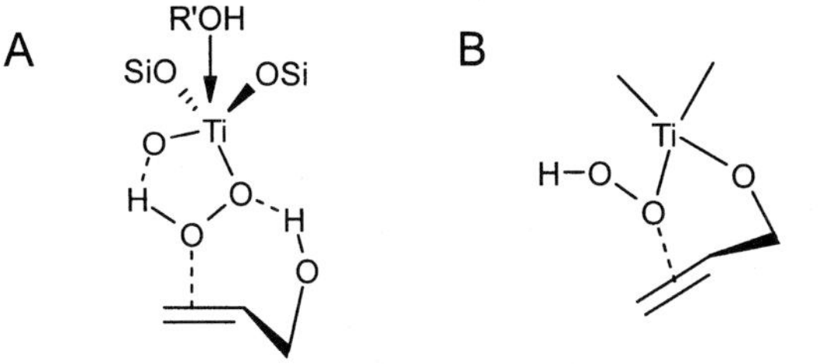

Figure 12.2. Transition states in the epoxidation of an allylic alcohol with (A) a solid TS-1 catalyst, and (B) a soluble Ti catalyst.

If one looks at the distances between the oxygen atom in the alcohol group and the oxygen atom to be transferred to the olefin, it is obvious that these distances are quite different in models A and B. Consequently, the dihedral angles (O-C-C=C) for the optimal conformations in the allylic alcohols are different too. Peracid type mechanisms (model A) require an angle of about 120°, while oxygen transfer in model B is optimal with a dihedral angle of 45°.

Practical implications become obvious if one looks at the epoxidation of (*R*)-2-methyl-1-buten-3-ol **70**:

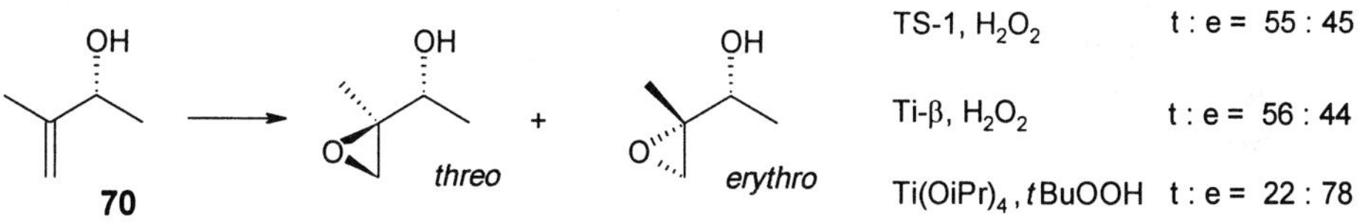

TS-1, H_2O_2	t : e = 55 : 45
Ti-β, H_2O_2	t : e = 56 : 44
$Ti(OiPr)_4$, *t*BuOOH	t : e = 22 : 78

Scheme 12.44.

If an intermediate with a coordinated alcohol is involved (model B), the dihedral angle must be ~45°. The conformer leading to the *threo*-epoxyalcohol then has the two methyl groups in an almost eclipsed conformation (Scheme 12.45 right part):

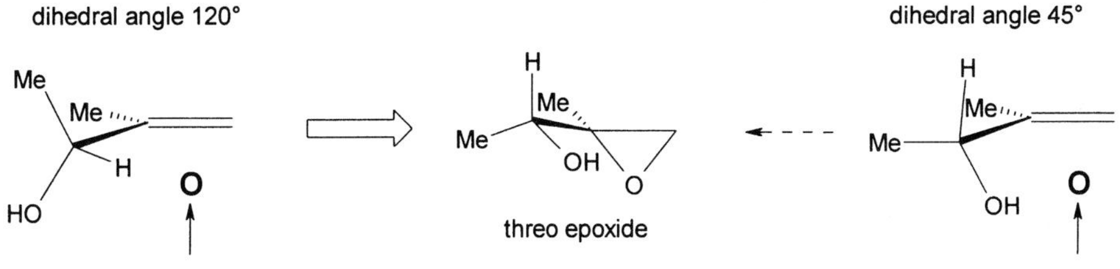

Scheme 12.45.

Consequently, this conformer is not easily formed, and formation of the *threo*-epoxide is disfavored. In contrast, the conformer leading to *threo*-epoxide is easily formed with a 120° dihedral angle (model A), since the vicinal methyl groups are staggered. Thus, the t:e ratios with the Ti zeolites favor a 120° angle over the 45° situation. Based on such arguments, Adam and Corma proposed that the transition state in TS-1 and Ti-β most resembles that of a peracid (model A) [77, 78].

A quite different mechanism must be operative in the zeolite A-catalyzed epoxidation of allylic alcohols with *t*BuOOH [80, 81]. In this reaction, both the allylic alcohol and the organic peroxide are probably coordinated on an unsaturated Al^{3+} atom at the surface of the zeolite. Hence, a Sharpless-type mechanism (model B in Fig. 12.2) applies. This is confirmed by the diastereoselectivity in the epoxidation of 3-CH_3-2-nonen-4-ol (**71**):

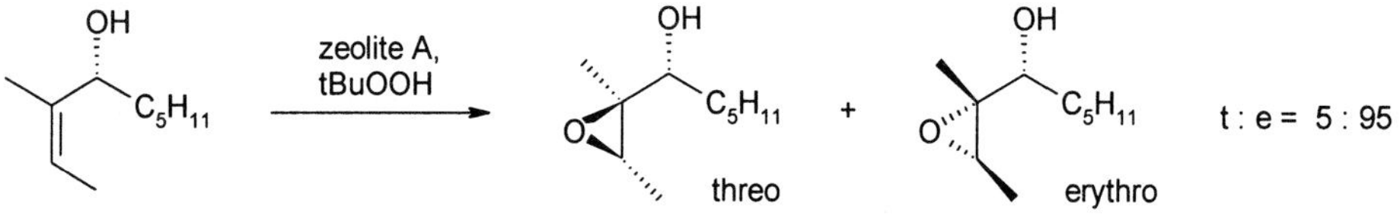

OH C_5H_{11} **71** —zeolite A, tBuOOH→ OH C_5H_{11} threo + OH C_5H_{11} erythro t : e = 5 : 95

Scheme 12.46.

The strong *erythro*-preference is due to a 1,2 strain between the vicinal pentyl and methyl groups. As shown previously, the low *threo*-selectivity is typical for coordination of the allylic alcohol with a small dihedral angle.

Finally, epoxidation with TS-1 or Ti-β may be instantly followed by intramolecular rearrangement, due to the acidity of the Ti-OOH groups. For instance, from (*R*)-4-penten-2-ol, two diastereomers of 2-methyl-4-OH-tetrahydrofuran are obtained in a 72:28 ratio [82].

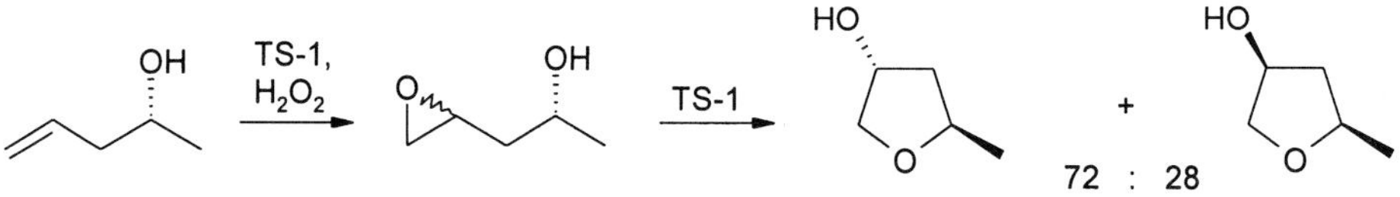

OH —TS-1, H_2O_2→ OH —TS-1→ HO + HO 72 : 28

Scheme 12.47.

However, it is unclear from the available data whether the zeolite framework decisively affects the *cis-trans* ratios.

References

[1] M. Bartok, J. Czombos, K. Felföldi, L. Gera, Gy. Göndös, A. Molnar, F. Notheisz, I. Palinko, Gy. Wittmann, A.G. Zsigmond, *Stereochemistry of Heterogeneous Metal Catalysis*, J.Wiley&Sons, Chichester, **1985.**
[2] M. Besson and C. Pinel, *Top. Catal.*, **1998**, *5*, 25.
[3] M.S. Pavlin, US Patent 4310714 (1982).
[4] L.A. Canova, US Patent 4018842 (1977).
[5] M. Carmen Carreno, J. Garcia Ruano and M. Toledo, *Chem. Eur. J.*, **2000**, *6*, 288.
[6] J. M. Brown, *Angew. Chem., Int. Ed. Engl.*, **1987**, *26*, 190.
[7] H. Thompson and J. Wong, *J. Org. Chem.*, **1985**, *50*, 4270.
[8] T. Howard, *Rec. Trav. Chim. Pays Bas*, **1964**, *83*, 992.
[9] K. Borszeky, T. Mallat and A. Baiker, *J. Catal.*, **1999**, *188*, 413.
[10] H. Thompson, E. McPherson and B. Lences, *J. Org. Chem.*, **1976**, *41*, 2903.
[11] L. Horner, H. Ziegler and H.-D. Ruprecht, *Liebigs Ann. Chem.*, **1979**, 341.
[12] J. Sheehan and R. Chandler, *J. Am. Chem. Soc.*, **1961**, *83*, 4795.
[13] I. N. Lisichkina, A. I. Vinogradova, B. O. Tserevitinov, M. B. Saporovskaya, V. K. Latov and V. M. Belikov, *Tetrahedron: Asymmetry*, **1990**, *2*, 567.
[14] U. Schmidt, S. Kumpf and K. Neumann, *J. Chem. Soc. Chem. Commun.* **1994**, 1915.
[15] A. Tungler, A. Fürcht, Z. Karancsi, G. Toth, T. Mathé, L. Hegedus and A. Sandi, *J. Mol. Catal. A*, **1999**, *139*, 239.
[16] B. Bycroft and G. Lee, *J. Chem. Soc. Chem. Commun.*, **1975**, 988.
[17] H. Poisel and U. Schmidt, *Chem. Ber.*, **1973**, *106*, 3408.
[18] N. Izumiya, S. Lee, T. Kanmera and H. Aoyagi, *J. Am. Chem. Soc.*, **1977**, *99*, 8346.
[19] H. Aoyagi, F. Horike, A. Nakagawa, S. Yokote, N. Park, Y. Hashimoto, T. Kato and N. Izumiya, *Bull. Chem. Soc. Jpn.*, **1986**, *59*, 323.
[20] N. Park, S. Lee, H. Maeda H. Aoyagi and T. Kato, *Bull. Chem. Soc. Jpn.*, **1989**, *62*, 2315.
[21] D. Seebach, M. Bürger and C. Schickli, *Liebigs Ann. Chem.,* **1991**, 669.
[22] C. Schickli and D. Seebach, *Liebigs Ann. Chem.,* **1991**, 655.
[23] R.L. Augustine, *Adv. Catal.*, **1976**, *25*, 56.
[24] J. McQuillin, W. Ord and P. Simpson, *J. Chem. Soc.,* **1963**, 5996.
[25] P. Gallois, J. Brunet and P. Caubere, *J. Org. Chem.*, **1980**, *45*, 1946.
[26] (a) S. Nishimura, M. Shimahara and M. Shiota, *J. Org. Chem.*, **1966**, *31*, 2394. (b) K. Mori, K. Abe, M. Washida, S. Nishimura and M. Shiota, *J. Org. Chem.*, **1971**, *36*, 231.
[27] R. Vitali, G. Caccia and R. Gardi, *J. Org. Chem.*, **1972**, *37*, 3745.
[28] R. Johnstone, A. Wilby and I. Entwistle, *Chem. Rev.*, **1985**, *85*, 129.
[29] S. Kinderman and B. Feringa, *Tetrahedron: Asymmetry*, **1998**, *9*, 1215.
[30] G. Andrews, T. Crawford and B. Bacon, *J. Org. Chem.*, **1981**, *46*, 2976.
[31] K. Harada, *Nature*, **1966**, *212*, 1571.
[32] R. Hiskey and R. Northrop, *J. Am. Chem. Soc.*, **1965**, *87*, 1753.
[33] R. Hiskey and R. Northrop, *J. Am. Chem. Soc.*, **1961**, *83*, 4798.
[34] M. Tamura, S. Shiono and K. Harada, *Bull. Chem. Soc. Jpn.*, **1989**, *62*, 3838.
[35] S. Yamada and S. Hashimoto, *Tetrahedron Lett.*, **1976**, *17*, 997.
[36] K. Matsumoto and K. Harada, *J. Org. Chem.*, **1966**, *31*, 1956.
[37] T. Munegumi and K. Harada, *Bull. Chem. Soc. Jpn.*, **1988**, *61*, 1425.
[38] K. Harada and S. Shiono, *Bull. Chem. Soc. Jpn.*, **1984**, *57*, 1040.
[39] M. Eleveld, H. Hogeveen and E. Schudde, *J. Org. Chem.*, **1986**, *51*, 3635.
[40] G. Siedlaczek, M. Schwickardi, U. Kolb, B. Bogdanovic and D.G. Blackmond, *Catal. Lett.*, **1998**, *55*, 67.
[41] G. Bringmann and J. Geisler, *Tetrahedron Lett.*, **1989**, *30*, 317.
[42] M.Huffman, P. Reider, *Tetrahedron Lett.*, **1999**, *40*, 831.
[43] A. Bardou, J. Célérier and G. Lhommet, *Tetrahedron Lett.*, **1998**, *39*, 5189.
[44] T. Munegumi, M. Fujita, T. Maruyama, S. Shiono, M. Takasaki and K. Harada, *Bull. Chem. Soc. Jpn.*, **1987**, *60*, 249.

[45] T. Munegumi, T. Maruyama, M. Takasaki and K. Harada, *Bull. Chem. Soc. Jpn.*, **1990**, *63*, 1832.
[46] K. Harada, T. Munegumi and S. Nomoto, *Tetrahedron Lett.*, **1981**, *22*, 111.
[47] G. Wittmann, G. Göndös and M. Bartók, *Helv. Chim. Acta*, **1990**, *73*, 635.
[48] V. Gertosio, C. Santini and M. Vivat, *J. Mol. Catal.*, **1999**, *142*, 141.
[49] F. Cocu, S. Coman, C. Tanase, D. Macovei and V. I. Parvulescu, *Stud. Surf. Sci. Catal.*, **1997**, *108*, 207.
[50] S. Coman, F. Cocu, J. F. Roux, V. I. Parvulescu and S. Kaliaguine, *Stud. Surf. Sci. Catal.*, **1998**, *117*, 501.
[51] S. Coman, M. Florea, F. Cocu, V. I. Parvulescu, P.A. Jacobs, C. Danumah and S. Kaliaguine, Chem. Commun., **1999**, 2175.
[52] M. Komiyama, H. Sugawara, Y. Iwamoto and Y. Ogino, *J. Catal.*, **1987**, *104*, 237.
[53] M. Besson, L. Bullivant, N. Nicolaus and P. Gallezot, *J. Catal.*, **1993**, *140*, 30.
[54] M. Besson, P. Gallezot, S. Neto and C. Pinel, *Chem. Commun.*, **1998**, 1431.
[55] M. Besson, B. Blanc, M. Champelet, P. Gallezot, K. Nasar and C. Pinel, *J. Catal.*, **1997**, *170*, 254.
[56] M. Besson, P. Gallezot, C. Pinel and S. Neto, *Stud. Surf. Sci. Catal.*, **1997**, *108*, 215.
[57] V. Ranade and R. Prins, *J. Catal.*, **1999**, *185*, 479.
[58] V. Ranade, G. Consiglio and R. Prins, *Catal. Lett.*, **1999**, *58*, 71.
[59] A. Gypser and H.-D. Scharf, *Synthesis*, **1996**, 349.
[60] S. Angle and J. Boyce, *Tetrahedron Lett.*, **1995**, *36*, 6185.
[61] G. Haviari, J. Célérier, H. Petit, G. Lhommet, D. Gardette and J. Gramain, *Tetrahedron Lett.*, **1992**, *33*, 4311.
[62] G. Fodor, E. Bauerschmidt and J. Craig, *Can. J. Chem.*, **1969**, *47*, 4393.
[63] G. Cooke and G. Fodor, *Can. J. Chem.*, **1968**, *46*, 1105.
[64] H. Brunner and C. Huber, *Chem. Ber.*, **1992**, *125*, 2085.
[65] E. Setti, D. Belinzoni and O. Mascaretti, *J. Org. Chem.*, **1989**, *54*, 2236.
[66] L. Tottie, P. Baeckström, C. Moberg, J. Tegenfeldt and A. Heumann, *J. Org. Chem.*, **1992**, *57*, 6579.
[67] A. Heumann, L. Tottie and C. Moberg, *J. Chem. Soc. Chem. Commun.*, **1991**, 218.
[68] C. Cativiela, J. Fraile, J. Garcia, J. Mayoral, E. Pires and F. Figueras, *Tetrahedron*, **1992**, *48*, 6467.
[69] G. Hondrogiannis, R. Pagni, G. Kabalka, R. Kurt and D. Cox, *Tetrahedron Lett.*, **1991**, *32*, 2303.
[70] C. Cativiela, F. Figueras, J. Fraile, J. Garcia and J. Mayoral, *Tetrahedron: Asymmetry*, **1991**, *2*, 953.
[71] S. Bains, R. Pagni and G. Kabalka, *Tetrahedron Asymmetry*, **1994**, *5*, 821.
[72] C. Cativiela, F. Figueras, J. Fraile, J. Garcia and J. Mayoral, *Tetrahedron: Asymmetry*, **1993**, *4*, 223.
[73] G. Kabalka, R. Pagni, S. Bains, G. Hondrogiannis, M. Plesco, R. Kurt, D. Cox and J. Green, *Tetrahedron: Asymmetry*, **1991**, *2*, 1283.
[74] N. Krause and M. Mackenstedt, *Tetrahedron Lett.*, **1998**, *39*, 9649.
[75] R. Sreekumar and R. Padmakumar, *Tetrahedron Lett.*, **1997**, *38*, 2413.
[76] M. Dusi, T. Mallat and A. Baiker, *J. Mol. Catal.*, **1999**, *138*, 15.
[77] W. Adam, A. Corma, T. Reddy and M. Renz, *J. Org. Chem.*, **1997**, *62*, 3631.
[78] W. Adam, A. Corma, A. Martinez, C. Mitchell, T. Indrasena Reddy, M. Renz and A. Smerz, *J. Mol. Catal. A*, **1997**, *117*, 357.
[79] M. Clerici, *La Chimica e l'Industria*, **1998**, *80*, 1137.
[80] L. Palombi, F. Bonadies and A. Scettri, *Tetrahedron*, **1997**, 11369.
[81] L. Palombi, F. Bonadies and A. Scettri, *Tetrahedron*, **1997**, 15867.
[82] A. Bhaumik and T. Tatsumi, *J. Catal.*, **1999**, *182*, 349.

Index